SCIENCE
for today and tomorrow

M.A. Atherton
T. Duncan
D.G. Mackean

John Murray

© M. A. Atherton, T. Duncan, D. G. Mackean 1983

First published 1983
by John Murray (Publishers) Ltd
50 Albemarle Street, London W1X 4BD

Reprinted 1984, 1987 (twice), 1988, 1990, 1991

All rights reserved. Unauthorized
duplication contravenes applicable laws

Printed in England by Clays Ltd, St Ives plc

British Library Cataloguing in Publication Data

Atherton, M. A.
Science for today and tomorrow.
1. Science
I. Title   II. Duncan, T.   III. Mackean, D. G.
500   Q161.2

ISBN 0 7195 4008 9

PREFACE

This book is designed for students following a course of study leading to a first examination in General or Combined Sciences, in both the U.K. and overseas.

Subject matter—biology, chemistry, physics and earth sciences—is presented in broad sections, comprising 92 chapters each dealing with a particular topic. Each chapter contains essential facts, ideas, details of experiments, everyday applications and questions for study and revision.

The order of presentation suggests a possible two-year course of study, but the organization in no way assumes that this sequence need be followed. Indeed, we have assumed that teachers will develop the subjects in a wide variety of ways, so our primary concern has been to ensure that students have simple access to information under easy-to-recognize headings. Cross references, where applicable, stress the essential interrelationships between the parts of science, but by exposing students to the 'flavours' of the separate disciplines, we hope to prepare the way for later specialist studies in separate sciences.

M. A., T. D., D. M.

ACKNOWLEDGEMENTS

Thanks are due to the following for permission to reproduce copyright photographs:
Figs. 1.9, 6.2, 6.4, 7.7, 46.2, 62.6, 64.3 Crown; 6.1, 6.3 British Museum (Natural History); 7.2, 7.4, 7.8 Geological Museum; 7.5, 7.6, 7.8, 62.7 Institute of Geological Sciences; 12.1 London Electricity Board; 12.2, 37.4 Central Electricity Generating Board; 12.3 British Aircraft Corporation; 30.4, 79.10 Brian Bracegirdle; 31.1a and b British Airways; 32.1, 34.3, 81.1 Popperfoto; 34.6a National Institute of Agricultural Engineering; 34.6b Ford Motor Company Ltd; 35.15 Griffin and George Ltd; 36.10 Italian State Tourist Office; 38.4 Derek Pratt; 39.6 Ministry of Defence; 39.7 The English Electric Company Ltd; 41.3a British Railways; 41.3b G. Maungtill and Partners; 43.7, 47.5 Central Press Photos; 46.3 John Topham; 47.13 Howard Jay; 48.8b Ronan Picture Library; 53.4 Barnes Engineering Company; 53.5 by courtesy of The Post Office; 62.3 Amalgamated Roadstone Corporation; 63.9 The Italian Government Tourist Office; 66.1 British Gas; 67.5 Russel Adams; 68.7, 68.8 Shell Photographic Service; 72.2 Gene Cox; 76.7 Professor W. J. Hamilton; 76.10 Reproduced from the Birth Atlas—published by The Maternity Centre Association, New York; 80.1 Mullard Ltd; 80.5 Hurst Electrics; 85.1 Chloride Automotive Batteries; 88.10 P. R. Dicks; 90.5 LMA Ltd; 91.10 United Kingdom Atomic Energy Authority.

Section Illustrations

Page 17 Barnaby's Picture Library; page 37 Aluminium Laboratories Ltd; page 73 Gene Cox; page 85 Howard Jay; pages 123, 223 Tony Langham; page 171 French Government Tourist Office; page 275 Space Frontiers; pages 291, 375 Pace.

Cover Photographs

Peacock butterfly (Barbara Jay); Launching of Space-Shuttle 'Columbia' (J. Allan Cash); England Cheshire Chemical Refinery (Tony Stone Associates).

CONTENTS

Preface and Acknowledgements ... iii

INVESTIGATING MATTER

1 Separating Substances ... 2
2 Breaking Down Substances ... 7
3 Mixing and Joining ... 11
4 The Particle Nature of Matter ... 14

AIR, EARTH, FIRE AND WATER

5 Investigating Air ... 18
6 Minerals ... 23
7 Rocks ... 27
8 Soil ... 31
9 Two Chemicals from Water: Oxygen and Hydrogen ... 34

CLASSIFYING ELEMENTS AND COMPOUNDS

10 Metals and Non-Metals ... 38
11 Reacting Metals ... 41
12 Using Metals ... 44
13 Families of Elements ... 47
14 Acids and Alkalis ... 52
15 Oxides ... 55
16 Effects of Electricity on Compounds ... 58
17 More about Acids ... 61
18 Making Salts ... 64
19 Soluble and Insoluble Salts ... 68
20 Breaking down Salts ... 71

LIVING ORGANISMS

21 Characteristics of Living Organisms ... 74
22 Cells ... 75
23 Activities in Cells ... 79

STUDYING PLANTS

24 Photosynthesis and Nutrition in Plants ... 86
25 Plant Structure ... 91
26 The Interdependence of Living Organisms ... 96
27 Diffusion and Osmosis ... 100
28 Transport of Materials in Plants ... 106
29 Flowers, Fertilization and Fruits ... 111
30 Germination of Seeds ... 117

FORCES AND ENERGY

31 Measurement ... 124
32 Introduction to Forces ... 132
33 Forces and Turning Effects ... 136
34 Centre of Gravity ... 139
35 Forces and Pressure ... 142
36 Forces and Motion ... 150
37 Energy ... 159
38 Simple Machines ... 163
39 Heat Engines ... 167

HEAT, LIGHT AND SOUND

40 Thermometers ... 172
41 Expansion of Solids and Liquids ... 174
42 Gas Laws ... 177
43 Conduction, Convection, Radiation ... 180
44 Changes of State ... 184
45 Measuring Heat ... 188
46 Light Rays ... 192
47 Reflection of Light ... 195
48 Refraction of Light ... 202
49 Lenses ... 207
50 Optical Instruments ... 210
51 Dispersion and Colour ... 212
52 Waves ... 214
53 The Electromagnetic Spectrum ... 217
54 Sound Waves ... 220

FORMULAS, EQUATIONS AND CHEMICAL COMPOUNDS

55	More about Atoms	224
56	Charged Particles	228
57	Chemical Formulas	232
58	Equations	238
59	Calculations Using Equations	244
60	Chemical Bonding	246

CHEMICALS FROM NATURE

61	Metals from Rocks	252
62	Chemicals from Limestone	256
63	Chemicals from Salt	261
64	Chemicals from Sulphur	266
65	Chemicals and Foods	270

CHEMICAL ENERGY

66	Energy in Chemistry	276
67	Investigating Fuels	280
68	Fossil Fuels	283
69	Rates of Chemical Reactions	287

THE BIOLOGY OF MAN

70	Food and Diet	292
71	Digestion, Absorption and use of Food	295
72	The Blood Circulatory System	301
73	Breathing	306
74	Excretion	310
75	Temperature Regulation	313
76	Sexual Reproduction	315
77	The Skeleton and Locomotion	321
78	The Senses	326
79	The Nervous System	332

MAGNETISM AND ELECTRICITY

80	Magnets and Magnetic Fields	340
81	Static Electricity	346
82	Electric Current	348
83	Potential Difference	351
84	Resistance	353
85	Electric Cells: E.M.F.	356
86	Electric Power	359
87	Electromagnets	362
88	Electric Motors and Meters	365
89	Generators and Transformers	370

ATOMIC PHYSICS

90	Electrons	376
91	Radioactivity	379
92	Nuclear Energy	384

Answers	387
Index	390
Table of Relative Atomic Mass } Periodic Table	inside back cover

Investigating Matter

1 SEPARATING SUBSTANCES

Pure substances and mixtures

Some substances you see around you are *mixtures* of more than one substance, while others consist of one substance only and are said to be *pure*. Soil, air, grass, beer and milk are all mixtures; sugar is a pure substance.

Sometimes different substances in a mixture can be easily seen, e.g. a mixture of maize and beans, or a sample of soil. But many mixtures, such as beer, look as if they contain only one substance. You can show these are mixtures by separating the substances in them.

To find out whether a substance is pure or impure, we examine its *properties*. Each pure substance has a set of properties that help make it look and behave differently from any other pure substance. So pure water is different from other liquids because it is the only one that freezes at 0 °C, boils at 100 °C (at sea-level) and has a density of 1 g/cm³.

If a sample of water boils at 102 °C (at sea-level) and has a density greater than 1 g/cm³, then you know that it must be a mixture of water and something else. Mixtures may sometimes look as if they are pure substances, but their properties are sure to be different in some ways from the pure substances that make them up.

Some separation methods

Figure 1.1 shows how pure sand, salt and water can be separated from a mixture of all three. There are several important words in the diagram and you will find out what they mean as you work through this chapter.

Filtration and evaporation

When salt is shaken with water, it 'disappears'. The salt is said to *dissolve* in the water to make a *solution*. But when sand is added to water, you can still see the grains either 'hanging' in the water or forming a layer at the bottom of the container. Salt is *soluble* in water whereas sand is *insoluble*.

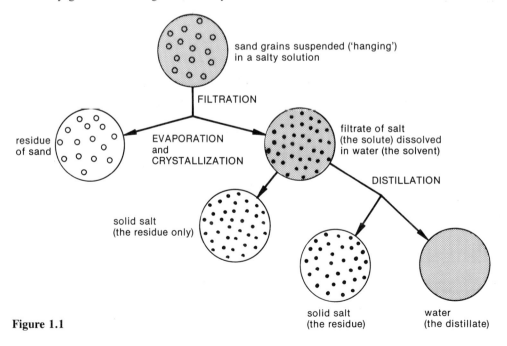

Figure 1.1

Filtration is a method of separating an insoluble solid and a liquid. In some countries, muddy river water is made cleaner by pouring it through a cloth tied to the mouth of a pot (Figure 1.2). The mud and sand grains cannot pass through the very small holes in the cloth but the water can.

Figure 1.2

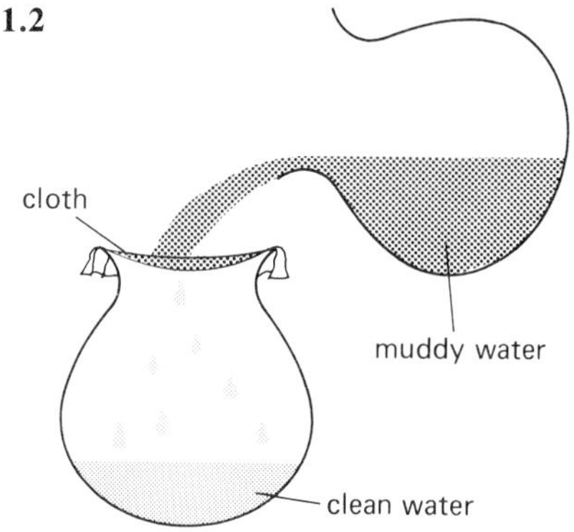

Figure 1.4

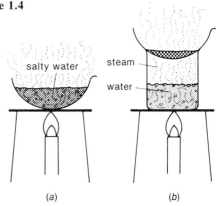

(a)   (b)

You now have to get solid salt from the salty solution (the *filtrate*). Pour the filtrate into an evaporating dish (Figure 1.4a) and heat it until very little solution is left. Then heat the rest of the filtrate on a steam-bath (Figure 1.4b). Look for crystals of salt on the sides of the dish.

The water you get from a tap has often been made cleaner by filtration at a waterworks. As it passes slowly down through the beds of sand and stones, the grains get trapped and the water coming out of the bottom of the beds is clear.

Evaporation is a method of separating a soluble solid from its solution. Heat is used to boil off the liquid (the *solvent*), leaving the solid (the *solute*) behind.

If only part of a solvent is evaporated, a more *concentrated* solution is formed. When this is cooled, crystals of the solute often separate out in a process called *crystallization*. This happens because less solute can dissolve in the same volume of solvent at a lower temperature, and so the extra solute appears as solid crystals.

Distillation

From Figure 1.4 you will see that there are two ways of getting solid salt from a filtrate. Both ways separate a solvent and a solute from a mixture of the two. But, unlike evaporation, *distillation* allows you to get both substances and not just one of them. This is because the vapour coming off the liquid is 'trapped' and made to condense back to the liquid in another container.

Experiment 2 Getting pure water from salty water

In this experiment you are allowed to taste the liquids. Usually you should never try to taste any substances in the laboratory because many of them could be very harmful.

Dissolve a spatula measure of salt in about 25 cm³ of water and taste the solution. Pour the solution into a boiling tube and add a few pieces of pumice. Heat the solution using the apparatus shown in Figure 1.5. When boiling begins, turn down the heat so that the liquid boils gently.

Experiment 1 Separating sand and salt from a mixture

Put a mixture of sand and salt into a beaker of water and stir the water. Filter this mixture using the apparatus shown in Figure 1.3. The filter paper acts like the cloth in Figure 1.2.

- What can you see in the tubing?

Carry on heating until you have got about 2 cm³ of the liquid in the test-tube.

- Does this liquid taste salty?
- Where is the salt now?

In all distillations, the more *volatile* substances (the ones with lower boiling points) form vapours first,

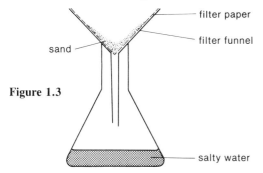

Figure 1.3

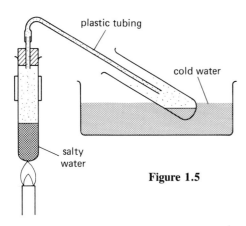

Figure 1.5

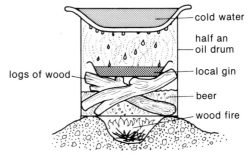

Figure 1.6

so leaving the less volatile substances behind. As the temperature is raised, these less volatile substances may also form vapours. But *involatile* substances like salt only turn into a vapour at very high temperatures.

It is very easy to separate completely a volatile substance like water and an involatile substance like salt by ordinary distillation. It is much more difficult to do this for a mixture of two fairly volatile substances. What usually happens is that the distillate is still a mixture of the two, but it is richer in the more volatile substance than the original mixture.

Figure 1.6 shows the home-made apparatus used in some African villages for the distillation of beer, a mixture of alcohol and water. The 'local gin' that collects as the distillate is richer in alcohol than the beer, but it still contains a lot of water.

The complete separation of alcohol and water cannot be carried out using this apparatus. Nearly pure alcohol can be obtained from a mixture of alcohol and water by a special kind of distillation called *fractional distillation* (Figure 1.7). As the mixture rises up the column packed with glass beads, it gets richer and richer in alcohol. At the top, the liquid contains up to 95 per cent alcohol.

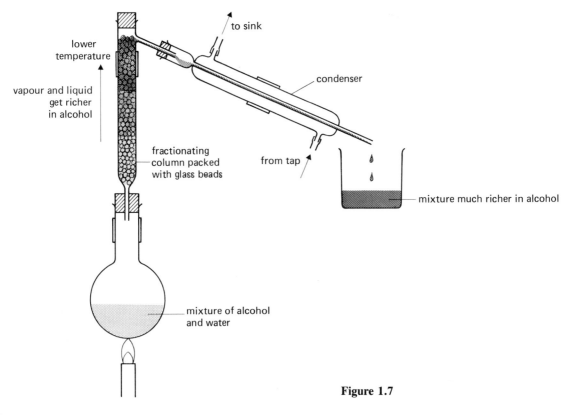

Figure 1.7

Paper chromatography

A solid is left when some ink is evaporated to dryness. This is the substance that gives ink its colour. The next experiment shows that this dye is not just one substance but a mixture.

Experiment 3 Separating the dyes in ink

Add one drop of ink to the centre of a piece of filter paper which is supported on an evaporating dish (Figure 1.8). Then slowly add drops of water from

Figure 1.8

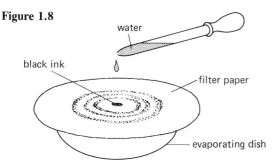

a dropper on to the ink blot. You get the best results when the water does not flow over the surface of the paper but soaks through the pores of the paper.

As the water spreads out through the filter paper, some of the dyes go with it. But they do not spread out to the same extent so that coloured bands form at different distances from the central blot.

This method can be used to separate small amounts of substances in a mixture. It is called *paper chromatography*. After the experiment the filter paper can be cut up so that each dye is separated from the others.

Figure 1.9

Questions

1. Complete the following passage about rock salt by filling in the gaps with the words listed below.

crystals, dissolves, evaporated, filtered, filtrate, impure, pure, solution

In many parts of the world, salt can be found in a rock-like form called rock salt. .... of rock salt can be seen in Figure 1.9. When it is dug out of the ground, rock salt often contains sand and so it is ....

To get .... salt from rock salt, you have to crush the rock to a powder and then add it to water. The salt .... but the sand does not. This mixture is .... and the salt .... collects as the .... in the conical flask. The contents of the conical flask are then .... to drive off the water, so leaving the salt behind.

2. Describe how you would obtain pure water from salty water, using materials from your home.

3. Sugar-cane is a mixture of substances. Some, like the sugar itself, are soluble in water, while others are insoluble.

Describe how you would get a solution of the soluble substances from the sugar-cane using materials from your home. How could the solids be obtained from this solution?

4. Figure 1.10 shows some white flowers that grow in the highlands of Kenya. An insecticide called *pyrethrum* can be extracted from these flowers by using hexane as a solvent.

Describe how you would get (*a*) a solution of pyrethrum in hexane from them, and (*b*) pyrethrum from the hexane solution. (HINT: hexane has a much lower boiling point than pyrethrum.)

Figure 1.10

5. Local gin (page 4) is a colourless liquid but the beer from which it is made is brown. Explain how this difference in colour comes about.

6. Crude oil is a mixture of many liquids, all with different boiling points. In an oil refinery crude oil is separated into different parts (fractions) by fractional distillation. There is a diagram of the process in Figure 1.11.

The fractionating column is a more complicated version of the column in Figure 1.7 (page 4). The more volatile liquids form their gases which then move up the column. At different parts of the column, some of the gases condense to liquids on a 'tray' and are tapped off. A few gases reach the top of the column without condensing.

(*a*) What happens to the temperature inside the column going from the bottom to the top?
(*b*) Why do different gases condense to liquids on different trays in the column?
(*c*) Which must be more volatile, the petrol fraction or the fuel oil fraction?
(*d*) What can you say about the boiling points of the substances that reach the top of the column without condensing to liquids?
(*e*) Why do some liquids collect at the bottom of the column as a residue?

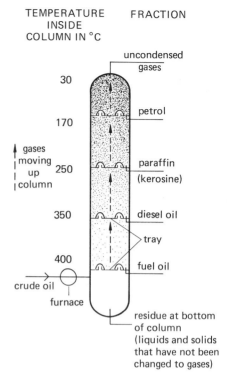

SIMPLIFIED FRACTIONATING COLUMN

Figure 1.11

2 BREAKING DOWN SUBSTANCES

Making charcoal from wood

In many parts of the world, charcoal is a useful fuel. It is made by heating wood without much air being present so that the wood cannot burn (page 18). Because charcoal is completely different from wood in its properties, a *chemical reaction* must take place when wood is heated.

All substances, whether solids, liquids or gases, are made of particles (page 14). But the properties of wood and charcoal are very different. It follows that the particles of which they are made must also be different.

When there is a change in the kinds of particles so that a new substance is formed, the change is called a chemical reaction.

The results of the experiment in Figure 2.1 show that wood must contain the liquids and the gases but that charcoal does not. Because of this, wood and charcoal cannot be the same substance. Wood is made of very large and complicated particles. When it is heated, these are broken down into the much smaller and simpler particles of charcoal, the liquids and the gases.

Changing copper sulphate crystals

In Figure 2.2 blue copper sulphate crystals are being heated in a test-tube that is sealed so that nothing can get in or out. Soon the blue colour fades leaving a white solid, and a colourless liquid condenses on the sides of the test-tube away from the flame of the burner.

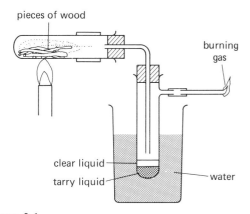

Figure 2.1

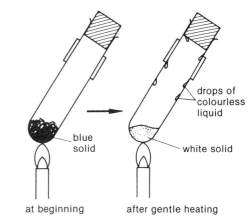

Figure 2.2

When charcoal is made, the wood always loses some mass: 100 kg of wood makes about 25 kg of charcoal. The other 75 kg just seems to 'disappear'. Figure 2.1 shows what happens to this extra mass. It is given off the wood as liquids and some gases that burn. A simple *word equation* for this chemical reaction is:

$$\text{wood} \xrightarrow{\text{heat}} \text{charcoal} + \text{liquids} + \text{gases}$$

This experiment shows that the blue solid is made of the white solid and the colourless liquid, and so the blue solid and the white solid must be different substances (Figure 2.3). This change is another chemical reaction.

It can be shown that the total mass of the test-tube and its contents is the same after heating as before heating. The masses of the white solid and the colourless liquid add up to the mass of the blue solid.

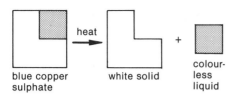

Figure 2.3

In all chemical reactions, the total mass of the substances that react together equals the total mass of the substances formed.

This important idea is known as the *Law of Conservation of Mass*.

All solids that lose mass when they are heated must be giving off a gas. This law shows that the gas must always have a mass equal to the loss in mass of the solid.

Heating some more solids

Before you carry out the next experiment, you need to know how to detect gases. Like all substances, each gas has some properties that help you to decide what it is. Detecting coloured gases is quite easy, but many gases cannot be seen because they are colourless. *Tests* must be used on these; tests that depend on certain properties of the gases.

Figure 2.4 shows the tests you can use to detect two colourless gases called oxygen and carbon dioxide, and also how you should carry them out.

In the experiment that follows, you can try to find out whether heating some solids breaks them down into new substances. You should look for:

(*i*) a change in the colour of the solid
(*ii*) any gases given off
(*iii*) a change in the mass of the solid.

If a solid loses mass by giving off a gas, then it must form a new substance made of different particles.

Experiment 1 What happens when some solids are heated?

Each group should use at least two of these solids:

copper carbonate, potassium chloride, potassium permanganate, sodium hydrogencarbonate

Find the mass of a test-tube with about two spatula measures of the solid in it. Then heat the solid and watch for any change that might take place. If you think a gas is being given off, try to detect it using the tests given above. Lastly, let the test-tube cool and then find its mass again.

• Copy and complete a table with these headings.

Mass before heating in g	Mass after heating in g	Colour before heating	Colour after heating	Gas given off (if any)

Figure 2.4

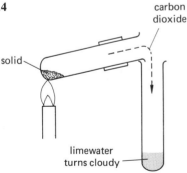

(a) Testing for carbon dioxide

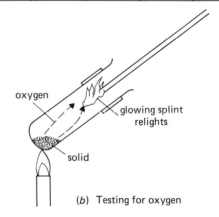

(b) Testing for oxygen

Name of gas	Colour	Test	Property on which test depends
Carbon dioxide	Colourless	Pass the gas into lime water	Turns lime water cloudy
Oxygen	Colourless	Hold a glowing splint in the gas	Relights a glowing splint

- Write a word equation for each of the chemical reactions that take place. For the heating of potassium permanganate, for example, this would be:

$$\text{potassium permanganate} \xrightarrow{\text{heat}} \text{black solid} + \text{oxygen}$$

When solids are broken down into new substances on heating, the chemical reaction is called *thermal decomposition*. This kind of reaction takes place with three of the solids in the experiment. With the other one, no change at all takes place on heating.

A CHEMICAL REACTION THAT GOES BOTH WAYS

Some changes are *reversible*, that is they can go both ways. Suppose that the test-tube shown in Figure 2.2 (page 7) is cooled and the colourless liquid runs down the sides on to the white solid. The white solid then turns back to the original blue solid. The word equation for this reversible chemical reaction is:

$$\text{blue copper sulphate} \underset{\text{cool}}{\overset{\text{heat}}{\rightleftharpoons}} \text{white copper sulphate} + \text{colourless liquid}$$

If some of the colourless liquid is collected (Figure 2.5), some of its properties can be investigated. These properties show that the liquid is pure water.

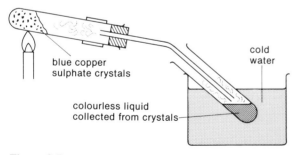

Figure 2.5

Many other crystals also give off water when they are heated. Such crystals are said to be *hydrated*. The new solids formed after the water has been driven off do not have any water in them and these are said to be *anhydrous*.

The water that is 'locked up' in the hydrated crystals is called *water of crystallization*. It is always driven off when the crystals are heated. In one or two cases, as with hydrated sodium carbonate, it may also be given off just by leaving the crystals exposed to the air for a few days.

From compounds to elements

Two ways of breaking down substances are shown in Figures 2.6 and 2.7. When red mercury oxide is heated, it changes to shiny grey drops of mercury and a colourless gas (oxygen) that relights a glowing splint. (WARNING: toxic mercury vapour is given off in this experiment.)

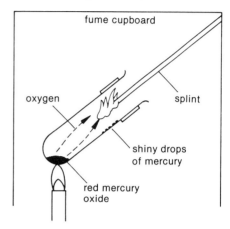

Figure 2.6

Another way of breaking down a substance is to pass electricity through it after it has been melted. Figure 2.7 shows green copper chloride that has been melted to form a liquid. When electricity is passed through it it changes to a green-yellow gas

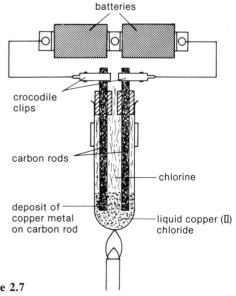

Figure 2.7

(chlorine) with a nasty smell and a red solid (copper metal) that forms around one of the carbon rods.

No matter what is done to the mercury, oxygen, chlorine or copper, none of them can be broken down any further. Even though electricity can pass through both mercury and copper, neither is changed by it. All four substances are called *elements* and the particles they are made of are called *atoms* (page 224).

Elements are the simplest substances that can be made through chemical reactions: they are made of only one kind of atom.

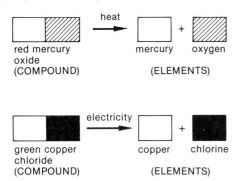

Figure 2.8

Red mercury oxide and green copper chloride are *compounds* made of two elements joined together (Figure 2.8).

Compounds can always be broken down into simpler substances: they are made of the atoms of two or more elements.

Not all compounds can be broken down straight away into their elements by heat or electricity. Many are broken down into simpler compounds or a mixture of elements and simpler compounds. For example, wood changes to charcoal (carbon, an element), some liquids (a mixture of simpler compounds) and some gases (a mixture of simpler compounds).

The elements in living materials

Many compounds that are found in living materials (or that are made from living materials) are made of only carbon, hydrogen and oxygen. Figure 2.9

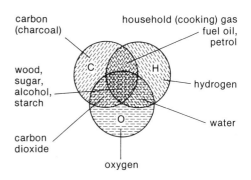

Figure 2.9

gives some of these. The substances in the central area are made of all three elements. Those in the areas where two circles overlap are made of only two of them.

Questions

1. Complete the following passage about blue copper sulphate by filling in the gaps with the words listed below.

anhydrous, crystals, hydrated, mass, water of crystallization

When blue copper sulphate .... are gently heated, there is a loss in .... because their .... is driven off. A white solid called .... copper sulphate is formed. If drops of water are added to this white solid, it goes back again to .... copper sulphate.

2. Cerium sulphate is a yellow solid. When it is heated, it turns orange and a colourless liquid forms on the sides of the test-tube.
  Explain what has happened using the terms *hydrated*, *anhydrous* and *water of crystallization*. How could the orange solid be changed back to the yellow solid?

3. Lead iodide is a solid compound made of two elements.
 (*a*) What are the names of these two elements?
 (*b*) Describe one way in which you might try to break down lead iodide.

4. People used to think that water was an element. What experiments can you think of to show that it is a compound?

5. Write down *three* changes that could happen during a chemical reaction. Two of the changes should concern what happens to a solid and the other should involve another kind of observation.

3 MIXING AND JOINING

Figure 3.1

More about chemical reactions

Sometimes when two substances are mixed together a chemical reaction takes place and sometimes it does not. Figure 3.1 explains a chemical reaction using shapes in a box rather than particles of the substances in a test-tube or beaker.

In the simple mixing process, the two kinds of shapes just get jumbled up but you can still see the separate shapes. In the 'chemical reaction', the two kinds of shapes join with one another to form a new kind of shape altogether. For real chemical reactions, these new 'shapes' are new particles that are different from the particles of the original substances.

Suppose that water is mixed with salt (sodium chloride) in the apparatus shown in Figure 3.2. The salt would get wet but there would be little or no change in the temperature shown on the thermometer. No new substance is formed in this experiment.

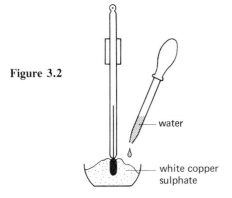

Figure 3.2

Now suppose that the solid in the crucible is white copper sulphate. This time the white solid would change to a dry blue solid and there would also be a large rise in the temperature shown on the thermometer. This is a chemical reaction in which water and white copper sulphate join together to form a new substance, blue copper sulphate.

Chemical reactions in which a rise in temperature takes place are said to be *exothermic*, that is the reaction mixture gives out heat to the surroundings. Though not all chemical reactions are exothermic, heat given out from a mixture is often a sign that the substances in it are joining with one another to form new substances.

FROM ELEMENTS TO COMPOUNDS

Compounds can be built up from their elements in a process called *synthesis*. This is the opposite of the breaking down (decomposition) of compounds by heat and electricity (page 9). Compounds can sometimes be made directly from their elements as shown in Figure 3.3, but often it is a longer and much more complicated process.

In the next experiment, you can look at the direct synthesis of another compound from its elements, iron and sulphur.

Experiment 1 Mixing and joining iron and sulphur

You are given a mixture of 7 g of iron filings and 4 g of powdered sulphur. Hold a magnet above the

Figure 3.3

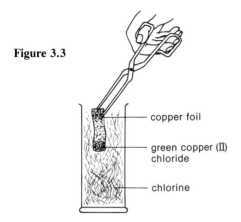

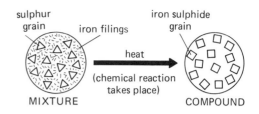

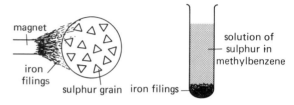

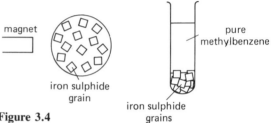

Figure 3.4

mixture and observe any changes.

- Which element is attracted by the magnet?

Add a spatula measure of the mixture to about 25 cm³ of methylbenzene (toluene) in a boiling tube, close the tube with a bung, and shake it. (WARNING: methylbenzene easily burns; do not have any burners alight when you are using it.) Filter the mixture and look at the residue and the liquid. Do this experiment again using only iron filings and then using only powdered sulphur.

Put the rest of the mixture in a hard-glass test-tube and clamp it in a horizontal position. Heat a *small* part of this mixture. Take away the Bunsen burner as soon as the mixture starts to glow.

- What happens to the mixture?

Look at the solid left after heating.

- What effect does a magnet have on it?

Write up your notes in a table like the one below.

Substance	Appearance	Effect of a magnet	Effect of methylbenzene

In the 'Substance' column you should write *iron filings, sulphur, a mixture of iron filings and sulphur,* and *the solid formed after heating.*

- How can you tell that a chemical reaction is taking place when a mixture of iron and sulphur is heated?

Some of the results of this experiment are given in Figure 3.4. The compound built up from iron and sulphur is called iron sulphide and the word equation for this chemical reaction is:

$$\text{iron} + \text{sulphur} \xrightarrow{\text{heat}} \text{iron sulphide}$$

The mixture of iron and sulphur has very different properties from iron sulphide. Both the mixture and the compound are made of atoms of iron and sulphur only. In the mixture the iron atoms and the sulphur atoms are still separate from one another. But in the compound they are joined together (Figure 3.5).

The chemical reaction between iron and sulphur changes the properties of the elements. By just looking at iron sulphide you would not be able to guess that it is made of either iron or sulphur. But it must have both these elements in it because it can be built up from them.

The above experiment starts with a mixture of 7 g of iron and 4 g of sulphur. This is because these are the masses of the two elements that just react with one another (this is also true for 14 g and 8 g, 21 g and 12 g, and so on). If 8 g of iron and 4 g of sulphur are heated together, the solid formed still has 1 g of iron that has not reacted.

Two important facts about compounds are shown by this experiment.

Compounds have very different properties from their elements.

Compounds are made of fixed proportions of their elements: they have a fixed composition.

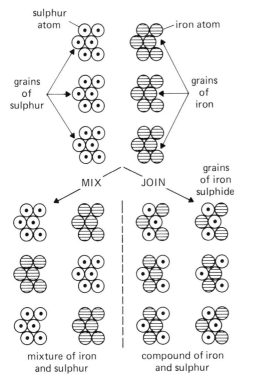

(The grains are magnified to show a few of the atoms in them)

Figure 3.5

The second fact means that a compound always has the same elements joined together in the same proportions. This is true how ever the compound is made. Chemists call this the *Law of Constant Composition*.

Questions

1. 1 part by mass of zinc reacts completely with 4 parts by mass of iodine to form zinc iodide only. Suppose that 4 g of zinc is made to react with 18 g of iodine.

(*a*) What important law of chemistry is illustrated by the first sentence in this question?
(*b*) Which of the two elements would not all react?
(*c*) What mass of this element would be left after the chemical reaction?
(*d*) What mass of zinc iodide would be formed?
(*e*) What important law of chemistry do you have to use in order to answer (*d*)?

2. Which of the following everyday processes do you think involve chemical reactions? Give your reasons for each answer. (You may need to talk about this question with your teacher before you start to answer it.)

(*a*) the setting of mortar (wet sand and cement).
(*b*) the rusting of iron
(*c*) using 'bicarbonate of soda' to cure acidity in the stomach
(*d*) milk going sour
(*e*) the decaying of food
(*f*) the burning of wood

3. Write down *two* important ways in which mixtures and compounds are different.

4 THE PARTICLE NATURE OF MATTER

There are three forms of matter—*solid, liquid* and *gas*. All matter is made up of tiny particles which are much too small to be seen directly. For some substances, they can be 'seen' using a scientific 'eye' called the electron microscope. Page 1 shows a photograph taken with this instrument. It shows the particles making up a protein. In this particular substance, the particles are collections of *atoms* called *molecules*. Many other substances are also made up of molecules. Some are made up of other kinds of particles which we will examine later (page 292).

The simple experiments which follow give evidence that (*i*) matter is made up of particles and (*ii*) particles of matter are continuously moving.

Crystals

Experiment 1 Growing crystals

a When a warm concentrated solution of a salt is cooled, crystals are formed. Place some hot concentrated salt solution on a microscope slide. Put the slide under a microscope. Observe the crystals as they form (Figure 4.1).

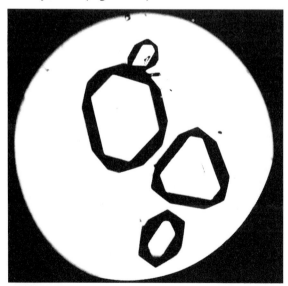

Figure 4.1 Crystals of potash alum (aluminium potassium sulphate) growing on a microscope slide.

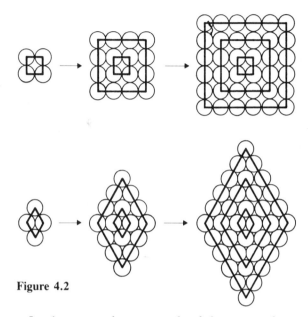

Figure 4.2

- Look at some large crystals of the same salt.

Note that each small crystal has a similar shape to the large crystal it grows into. 'Growth lines' sometimes show up within larger crystals. These lines are the original edges of the smaller crystals. The idea of particles can explain this regular growth. Particles in a crystal are packed together in a regular way. For a particular type of crystal, this packing is always in the same pattern. New particles added to the crystal always fit in with the pattern that is already there (Figure 4.2).

DIFFUSION

Smells, pleasant or otherwise, travel quickly and are caused by rapidly moving gas molecules. This spreading of gas is called *diffusion* (see also page 100).

Experiment 2 Observing diffusion

a Place a few drops of liquid bromine at the bottom of a gas jar. Cover this jar with another gas jar. After about 30 minutes, it can be seen that bromine vapour has travelled up the jars as shown

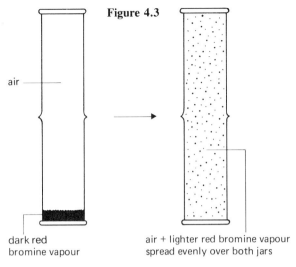

Figure 4.3

dark red bromine vapour

air + lighter red bromine vapour spread evenly over both jars

by Figure 4.3. The result is quite surprising when you realize that bromine vapour is more than five times denser (page 128) than air. In this experiment, the bromine vapour is not just spreading into a larger volume. It is also mixing with other gases in the air. The idea of particles explains this easily if it is assumed that there are large spaces between the molecules of a gas into which other molecules can move.

b Connect a porous pot containing air to a manometer. Pass hydrogen gas round the porous pot (Figure 4.4). The liquid in the manometer moves

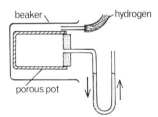

Figure 4.4

as shown by the arrows. This is due to the hydrogen molecules being lighter and moving faster than air molecules. They therefore diffuse into the pot faster than the heavier, slower moving air molecules can diffuse out.

- Why does the opposite happen if carbon dioxide surrounds the pot?

c Put a crystal of potassium dichromate at the bottom of a beaker. After a few hours, the orange colour caused by the crystal can be seen to have spread out through the liquid. Liquids must also be made of moving particles. Diffusion in liquids is much slower than in gases, so the particles in a liquid must be moving much more slowly than those in a gas.

Experiment 3 Brownian motion

The apparatus is shown in Figure 4.5. First light the paper drinking straw and blow it out so that smoke is produced. Fill the glass cell with smoke and replace the lid on the top. Put the cell on to the microscope platform. Switch on the lamp; the glass rod acts as a lens which focuses light on to the smoke.

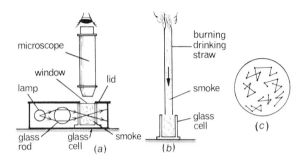

Figure 4.5

Carefully adjust the microscope until you see bright specks dancing around in a haphazard, or *random* way. These specks are smoke particles seen by reflected light. Their random motions are due to collisions with air molecules in the cell which knock them one way then another. This effect, called *Brownian motion*, also occurs with small particles suspended in a liquid. It was first seen by the botanist Robert Brown in 1827 when he was observing pollen grains in water. Notice that we would *not* expect to see Brownian motion if a large object is suspended in a liquid. In this situation, any motions caused by molecular bombardment would be too small to observe.

KINETIC THEORY OF MATTER

Solids, liquids and gases are made up of the same kinds of basic particles, but differ in several obvious ways. For example:

a Solids keep particular shapes, liquids take up the shape of the container, while gases spread out to fill all the space available.
b The densities (page 127) of solids and liquids are always much larger than the densities of gases.
c Only for gases is it possible to reduce the volume greatly by increasing the pressure, i.e. only gases are compressible.

From **a** we see that the forces acting between molecules are greatest in solids and least in gases. From **b** and **c** it is reasonable to assume that the molecules are much farther apart in gases than in the other two states. To these, we can add that the

15

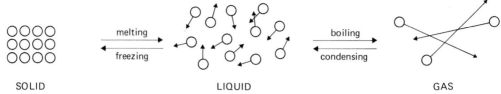

SOLID
Particles are arranged in a regular way, close together and held in position. The only motion possible is vibration about their fixed positions.

LIQUID
Particles are still close to each other, but can now move in between one another.

GAS
Particles are now much more widely separated, and are moving more rapidly than in a liquid.

Figure 4.6

molecules are in continuous motion.

Molecules exert strong electrical forces on one another when they are close together. These forces are both attractive and repulsive. Attractive forces hold the molecules together and resist stretching forces. Repulsive forces cause matter to resist being compressed. Using the *kinetic* (motion) theory, the existence of solid, liquid and gaseous states is explained as follows:

a *In solids* the atoms are close together and the attractive and repulsive forces between neighbouring atoms balance. Each molecule merely vibrates to and fro about one position in which it is more or less locked. Solids therefore have a regular, repeating molecular pattern, i.e. are crystalline, and their shape is definite.

b *In liquids* the molecules are usually slightly farther apart than in solids and, as well as vibrating, they can at the same time move rapidly over short distances. However, they are never near another molecule long enough to get trapped in a regular pattern, and a liquid can flow and takes the shape of the container.

c *In gases* the molecules are much farther apart than in solids or liquids (about ten times). They dash around at very high speeds (500 m/s for air molecules) in all the space available. It is only during the brief spells when they collide with other molecules or with the walls of the container that the molecular forces act. At other times the forces are so very small that they can be ignored.

Figure 4.6 shows a simple picture of how particles are probably arranged in the three states of matter.

Figure 4.7 shows models for demonstrating states of matter. If the tray (Figure 4.7a) is moved to and fro at increasing speeds, the motion of the marbles represents the motion of the particles in each state. Figure 4.7b is a model of a gas. If a polystyrene ball ('smoke particle') is dropped into the tube, it is bombarded by the metal spheres ('gas molecules') causing irregular motions (Brownian motion).

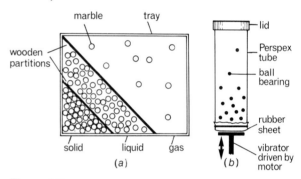

Figure 4.7

Facts and laws are discovered by experiments. To explain them and 'make sense' of science, we invent theories, like the kinetic theory, which enable us to understand and use many of the properties of matter.

Questions

1. How do we decide whether something is a solid, a liquid or a gas? Give an example of each.

2. A substance is most familiar in one particular form (solid, liquid or gas) but it can usually be changed from one form to another. Name the other two forms of water and say how they are produced.

3. What do the following tell us about matter?
(*a*) Crystals
(*b*) Diffusion

4. (*a*) What do we *see* when looking at Brownian motion?
(*b*) What *causes* what we see in Brownian motion?

5. How does the kinetic theory explain the existence of the solid, liquid and gaseous states of matter.

Air, Earth, Fire and Water

5 INVESTIGATING AIR

BURNING

When a substance burns, it takes part in a special kind of chemical reaction in which a great deal of heat is produced. Burning is an *exothermic* process (page 276).

When certain compounds are heated, they break down into simpler compounds and sometimes elements (page 10). Suppose an *element* like magnesium or sulphur is burned. As you can see from Figure 5.1, new substances are produced, but the new substances cannot be simpler than the original elements.

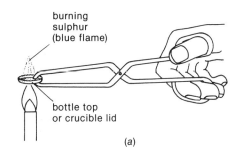

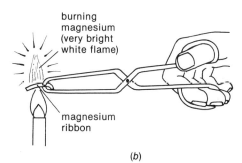

Figure 5.1

Here are the results of two experiments in which 0·30 g each of two metals are completely burned.

Magnesium
Mass of metal = 0·30 g
Mass of white powder produced = 0·50 g

Copper
Mass of metal = 0·30 g
Mass of black powder produced = 0·38 g

In both cases, there is an increase in mass. This must mean that the metal is reacting with another substance (labelled Q in Figure 5.2b). The substance produced is therefore a compound of metal and Q.

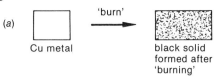

MASS OF COPPER does *NOT* equal MASS OF BLACK SOLID

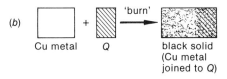

MASS OF (COPPER + Q) does equal MASS OF BLACK SOLID

Figure 5.2

By thinking about the results of the experiment shown in Figure 5.3 you can see that Q must be air. When the water in the tube is boiled, it turns to steam, and this pushes the air out of the test-tube.

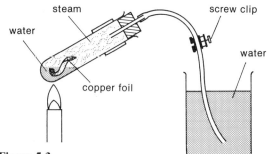

Figure 5.3

The copper foil does not change colour at all. The clip is then tightly closed and the rubber tube taken out of the water. The clip is opened and air gets into the test-tube. The hot copper foil now turns black as it 'burns' to a new substance. (In the case of copper, not enough heat is given out for the metal to burst into flames.)

What is air?

Air is not a pure substance: it is a mixture of several gases, some of which are elements and some of which are compounds. This will become clear if you read about, or carry out for yourself, the following experiments.

A burning candle is made to float on water and is then covered by a beaker held upside down (Figure 5.4). After a short time, the candle goes out and the water-level in the beaker goes up. The 'air' that

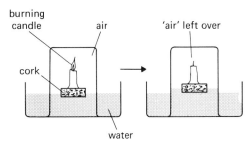

Figure 5.4

is left in the beaker does not let the candle carry on burning. There are two possible reasons for this:

(i) only part of the air is used for burning
(ii) the substances produced by burning make the candle go out.

Both these reasons, or just one of them, could be correct here.

To find out whether only part of the air is used for burning, a substance should be chosen that burns to give a new solid. This cannot then be the cause of the flame going out.

The syringes shown in Figure 5.5 can be used to show that only part of the air is used in burning. One of the syringes is filled with 50 cm³ of air. The copper wire is heated strongly. Next, using the plungers, the air can be passed to and fro over the hot copper. The copper turns black as it 'burns' in the air. When the chemical reaction has finished and the apparatus has cooled down, the final volume of 'air' is found to be 40 cm³. So only 10 cm³ out of 50 cm³ of air reacts with the copper.

Figure 5.5

One-fifth (20 per cent) of the air is 'active' while the other four-fifths (80 per cent) is 'inactive'. Air must be a mixture of at least two gases.

'Active' air

When a small piece of sodium is warmed on a bottle top, it melts and then catches fire. Like nearly all metals, sodium burns in the 'active' part of the air to form a new solid. This yellow solid must be made of sodium and the 'active' air. Very strong heating can break down the yellow solid and a colourless gas is formed. This will relight a glowing splint and is oxygen.

The experiment shown in Figure 5.6 demonstrates whether 'active' air is made of oxygen only. The potassium permanganate in the tube breaks down when heated. Oxygen is given off which then passes over hot copper wire. The changes seen are the same as those that take place when copper wire is heated in air, so oxygen is the same thing as 'active' air.

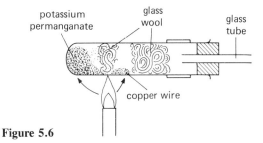

Figure 5.6

Experiment 1 Reacting oxygen with copper

Set up the apparatus shown in Figure 5.6. Heat the potassium permanganate until a glowing splint held close to the mouth of the glass tube relights. The test-tube is now full of oxygen and all the air has been pushed out.

Now heat the copper wire strongly until you see no more changes taking place.

- Is the chemical reaction exothermic?
- Does the remaining solid look like the solid left when copper wire is heated in air?

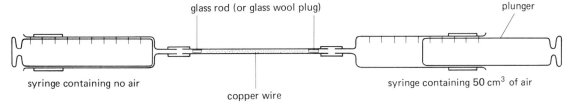

- Are there any differences between heating copper wire in oxygen and heating it in air?
- Do you think that the 'active' air is oxygen?

Other gases in the air

The two main gases in the air are both elements—oxygen (the 'active' part) and nitrogen, which is 'inactive'. There are other gases present in much smaller amounts. Two of these can be detected using the methods of the next experiment.

Experiment 2 Two more gases in the air

a Leave some clear lime water in the air on a watch-glass for a few hours.

- What happens on the surface of the lime water?
- How can you explain what you see?

b Put some white copper sulphate on a watch-glass and then find the total mass. Leave the solid standing in the air until the next science lesson.

- What happens to the colour of the solid?

c Find the new mass of the watch-glass and its contents.

- How can these results be explained?

Figure 5.7 shows what dry air is made of. The percentage of water vapour in the air varies

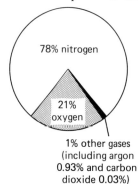

Figure 5.7

between nearly zero and 4 per cent depending on the weather conditions. The 1 per cent of dry air that is not oxygen or nitrogen is mainly made of a group of elements called the *noble gases*. These are like nitrogen in being very 'inactive'. Over 99 per cent of the noble gas mixture in air is argon.

MORE ABOUT OXYGEN

When substances are heated in air or pure oxygen and react, they form *oxides*.

Oxides are compounds made of oxygen and another element.

One of the best ways of making oxygen in the laboratory is from an unusual oxide called hydrogen peroxide. This is a solution with water that gives off oxygen when it is heated. But it gives off the gas more quickly if black manganese oxide is added to it. This substance acts as a *catalyst* (page 288) in the chemical reaction.

A catalyst is a substance that speeds up a chemical reaction without itself being used up.

Experiment 3 More experiments with oxygen

Use the apparatus shown in Figure 5.8. Add the

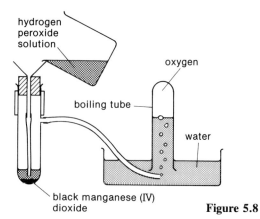

Figure 5.8

hydrogen peroxide solution to the black manganese oxide carefully, a few drops at a time. Collect five boiling tubes full of oxygen and close each tube with a bung.

- Do you think that oxygen can be very soluble in water?

Heat each of the following elements in a small combustion spoon until they start burning and then

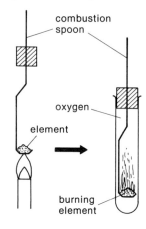

Figure 5.9

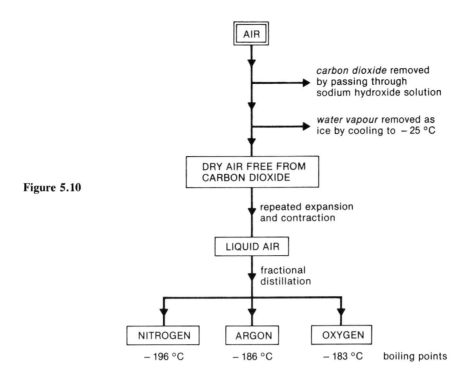

Figure 5.10

put the combustion spoon into a boiling tube full of oxygen (Figure 5.9): charcoal (carbon), iron filings or steel wool, clean magnesium ribbon, powdered sulphur, zinc powder.

- What happens to each element?

For further information on the oxides produced in this experiment see page 55.

Making oxygen in large amounts

The cheapest source of oxygen is the air. Figure 5.10 shows how pure oxygen can be separated from the other gases in the air.

First carbon dioxide and water vapour are removed from the air. The dry air, free from carbon dioxide, is then expanded quickly. This makes it cool down so much that it changes to a liquid, a mixture of liquid oxygen and liquid nitrogen (with small amounts of the noble gases). When the liquid mixture is distilled, the more volatile nitrogen (the gas with the lowest boiling point of the three in Figure 5.10) boils off first.

This separation of nitrogen from oxygen is like the part-separation of alcohol from water mentioned on page 4, but the temperatures of the two fractional distillations are very different indeed!

Uses of oxygen

Most of the oxygen produced as above is used by the steel industry for burning out impurities from molten steel (page 44). Other uses include: *burning* with a gas like hydrogen or acetylene to produce high temperatures used for welding and cutting torches; in *rockets* (page 280) where the thrust is produced by burning the oxygen in a fuel like kerosene or hydrogen; *as an aid to breathing*. Normally, the air contains enough oxygen for breathing, but for some patients in hospital, extra oxygen is needed. Both high altitude climbers and underwater divers carry an oxygen supply in cylinders.

Questions

1. What mass change would you expect when some powdered sulphur is burnt in air on a bottle top or crucible lid? Explain why your answer does not contradict the Law of Conservation of Mass.

2. A square of copper foil is folded several times and then hammered flat to make an 'envelope'. This 'envelope' is then strongly heated in air. It is finally cooled and opened out to make the original square.

Describe and explain what you would expect to see.

3. In the experiment shown in Figure 5.4, the fraction of 'air' left after the candle has stopped burning is about six-sevenths.

(a) What fraction of 'air' would you have expected to be left?
(b) Explain why the fraction in (a) is less than six-sevenths.

4. White phosphorus burns slowly in air without first being heated. It is always kept under water because of this property.

Describe an experiment in which you could use a small piece of white phosphorus to show that 20 per cent of the air is used in burning. Draw a diagram of the apparatus you choose.

5. (a) Design some apparatus that you could use to remove water vapour, carbon dioxide and oxygen from the air. What substances would you use to remove each of these gases?
   (b) What gases would be left at the end of your experiment? How would you collect these?

6. The 'air' we breathe out has about 17 per cent oxygen and 4 per cent carbon dioxide.

(a) How are these percentages different from those of the air we breathe in?

(b) Explain these differences and so say why it is essential for us to breathe in order to live.

7. When iron is left in the air, it slowly goes brown and powdery. This is because it rusts. Figure 5.11 shows an experiment on the rusting of iron.

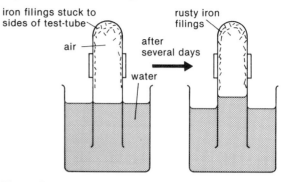

Figure 5.11

(a) Explain the results of this experiment.
(b) What has rusting got in common with burning?

6 MINERALS

Describing minerals

A *mineral* is a single substance that can be found in rocks. Rocks can be made of one mineral only or, more usually, several different minerals.

Different samples of the same mineral look very alike (except perhaps in colour) and they also behave in the same way when they are tested. Just as for the solids on the shelves of a laboratory, each mineral can usually be given a single chemical name. But all minerals have a common name as well.

The common name for the mineral in Figure 1.9 (page 5) is halite (or rock salt) and its chemical name is sodium chloride. Like many minerals it is often not completely pure when dug up and these impurities can affect its colour. So pure halite is colourless or white but it is often coloured brown or orange because of sand and clay impurities.

The crystals of some other minerals are shown in Figures 6.1 to 6.4. You can use the shapes of the crystals to identify them. But perfect specimens are not always available and other observations have to be made to make an identification.

Figure 6.1 Quartz

Figure 6.2 Copper Pyrites

Figure 6.3 Galena

Figure 6.4 Calcite

Table 6.1 Mohs' scale of hardness for minerals

	Hardness	Mineral	
↑	10	Diamond	
	9	Corundum	
	8	Topaz	
	7	Quartz	
increasing	6	Feldspar	← a penknife blade (hardness 5½)
hardness	5	Apatite	
	4	Fluorite	
	3	Calcite	← a 10p coin (hardness 3½)
	2	Gypsum	← a finger-nail (hardness 2–2½)
	1	Talc	

Hardness is a very useful test for a mineral. An order of hardness, called Mohs' scale, is given in Table 6.1. For each hardness number, one mineral is used as the standard. The hardness of three everyday materials are also shown.

The way the mineral splits (cleaves) is also a useful test. In Figure 6.5 a lump of calcium carbonate (calcite) is being hit in a certain direction with a sharp implement. This makes smaller bits of calcite with a definite shape, and this shape is made by three sets of *cleavage faces*.

Figure 6.5

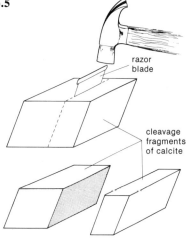

Sometimes the cleavage fragments have the same shape as the crystals of the mineral. This is true of halite. In other cases such as calcite, the cleavage fragments and the crystals look very different.

Some minerals do not split in a regular way: they have no cleavage. Despite this, they may still break in a way that can be used to identify them. Quartz is a good example: it breaks to form a curved and ribbed surface that looks like some kinds of sea shell.

Experiment 1 Comparing two minerals

For this experiment, choose two minerals that look alike, perhaps because they are both colourless or white. What you have to do is to find ways of telling them apart. Carry out each of the following tests on both minerals and put the results in a table with the headings shown below (Table 6.2).

a *Crystal shape and cleavage*: Use a hand-lens to look for flat faces which could be crystal faces or cleavage face, or both.

- Draw any crystals with a definite shape that you can see.
- Does the mineral show any cleavage faces? If so, how many sets of faces are there and what angles do they make to one another?
- Does the mineral show any irregular breaks?

b *Hardness*: See if you can scratch the mineral with your finger-nail, a coin or a penknife. Each time you try this, rub the 'scratch' with a wet finger and look at the mark under a hand-lens. Decide whether it is a proper scratch and not just a few loose grains of the mineral or a grey mark made by the coin or the knife.

- Use Table 6.1 to decide roughly where the two minerals come in Mohs' scale.
- Which mineral is likely to be able to scratch the other? Find out by a test if you are right in your prediction.

c *Colour and lustre*:

- What colour is the mineral?
- Is it transparent (can be seen through), translucent (lets light through but cannot be seen through), or opaque (cannot let light through)? Hold the mineral up to the light so that you can see its lustre (shine).
- Which one of these lustres (if any) does the mineral have: metallic, glassy, dull?

Table 6.2

Mineral	Crystal shape and cleavage	Hardness	Colour and lustre	Streak	Reaction with acid	Relative density
A						
B						

d *Streak*: Make a streak with the mineral by scraping it across a tile of white unglazed porcelain (a streak plate).

- If a mark is made, what is its colour?
- Is the colour of the streak the same as the colour of the whole mineral?

e *Reaction with acid*: Add a few drops of dilute hydrochloric acid to the mineral.

- Is there any bubbling (effervescence) which shows that a chemical reaction is taking place?

Of the common minerals, only calcite shows this property.

f *Density*: Find the density of the mineral using the method shown on page 127.

- Which of the two is the denser mineral?

This experiment shows how you can describe minerals and how you can tell two minerals from one another. To identify an unknown mineral, you need to obtain the results of tests like those above and then compare them with the detailed descriptions of a lot of minerals given in textbooks of mineralogy.

Ores

Most minerals are compounds made of several elements and one of these elements is often a metal.

Ores are minerals from which a metal can be extracted.

Two of the minerals shown in Figures 6.1 to 6.4 are ores. Copper pyrites is a copper ore and galena is a lead ore, and both are important sources of their metals.

It is sometimes possible to identify the metal in a mineral by a simple chemical test.

Experiment 2 The flame test for some ores

Use a powdered sample of the ore for this experiment.

Clean a nichrome wire by dipping it into concentrated hydrochloric acid. (CARE: do not let any acid come into contact with your clothes or your skin.) Heat it strongly in a non-luminous Bunsen flame until it no longer gives a colour to the flame. Then moisten the wire with the acid, dip it into the powdered ore and heat it strongly (Figure 6.6).

- What colour (if any) does the Bunsen flame go?
- Use the information in Table 6.3 to identify the metal in your ore.

Table 6.3

Compound of:	Flame colour
Barium	Green
Calcium	Brick red
Copper	Blue-green
Lead	Blue
Lithium	Very intense red
Potassium	Lilac (easily masked) by sodium impurities); 'crimson through blue glass'
Sodium	Persistent intense yellow
Strontium	Bright red

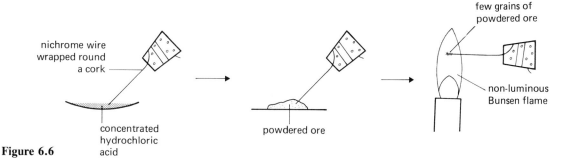

Figure 6.6

As a check on your conclusion, carry out the flame test again on a compound from your chemicals cupboard that you know (from its name) must contain this particular metal.

Questions

1. Complete the following passage about quartz by filling in the gaps with the words listed below.

cleavage, crystals, hardness, metal, mineral, ore

Quartz is a .... made of two elements, silicon and oxygen. It can be identified by the shape of its .... and by the fact that it has a .... of 7 on Mohs' scale. Also, it breaks unevenly when struck because it has no ....

Neither silicon nor oxygen is a .... and so quartz is not an example of an ....

2. Would you be able to scratch a lump of calcite with a lump of quartz? Explain your answer.

3. Explain how you would set about identifying an unknown white mineral using simple tests.

7 ROCKS

Layering in rocks

Rocks can be seen at the earth's surface where there is no soil or vegetation to hide them, for example, in a quarry, a crag or a sea cliff. In Figure 7.1 you can see a rock that shows a well-marked layering. Many rocks are layered in this way and they are nearly all examples of *sedimentary rocks*.

To get an idea of how sedimentary rocks are formed, first think about a simple experiment in which handfuls of differently coloured sand are added in turn to some water in a gas jar (Figure 7.2). Each handful settles into a layer at the bottom of the jar so that a succession of more or less flat layers is produced.

Many sedimentary rocks come from grains of material that have settled in layers under water. Huge amounts of sediment are brought by rivers into seas and lakes where they slowly settle into layers at the bottom. After millions of years, these layers 'harden' into layers of rocks which are lifted up above sea-level to form land.

Figure 7.2

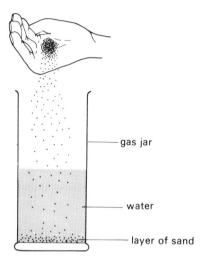

You can find out more about how layers of loose sand (the sediment) are made into layers of sandstone (the rock) by carrying out the next experiment.

Figure 7.1

Experiment 1 Making a 'rock' from sand

a Fill a small plastic syringe with damp sand (Figure 7.3a). Squeeze the sand by pushing it with the plunger against a thumb placed over the sawn-off end of the syringe (7.3b). Finally push the squeezed lump of damp sand out of the syringe on to a tile (7.3c).

- Does squeezing the sand make the grains stick together better than in loose sand? What would happen as the sand dried out?

b Do the experiment in **a** again, but this time use a damp mixture of sand and clay in the proportions (by volume) of 3 to 1. Leave the 'rock' for a few days and then look at it under a hand-lens.

- Does the clay seem to help the sand grains to stick together?

c Repeat **a**, but this time use a damp mixture of sand and plaster of Paris in the proportions (by volume) of 3 to 1. Leave the 'rock' for a few minutes and then look at it under a hand-lens.

- Arrange the 'rocks' formed in **a**, **b** and **c** in order of how well the grains stick together. (Put the *least* easily crumbled one first.)

The photograph (Figure 7.4) shows what sandstone looks like under a microscope. The rock is mostly made of sand grains, but there is also some material between the grains. This is the 'glue' or 'cement' that holds the grains together and makes it more difficult to crumble the rock. In your experiment, the 'cement' is clay or plaster of Paris, but in natural sandstones it is commonly calcium carbonate, calcium sulphate or various iron compounds.

Two things happen when sand changes to a sandstone:

(*i*) the sand grains are pressed more closely together
(*ii*) a 'cement' is formed between the grains.

In the experiment you used a syringe plunger to put pressure on the sand. In nature this pressure comes from layer after layer of other material which gets piled on top. This is, of course, far

Figure 7.3

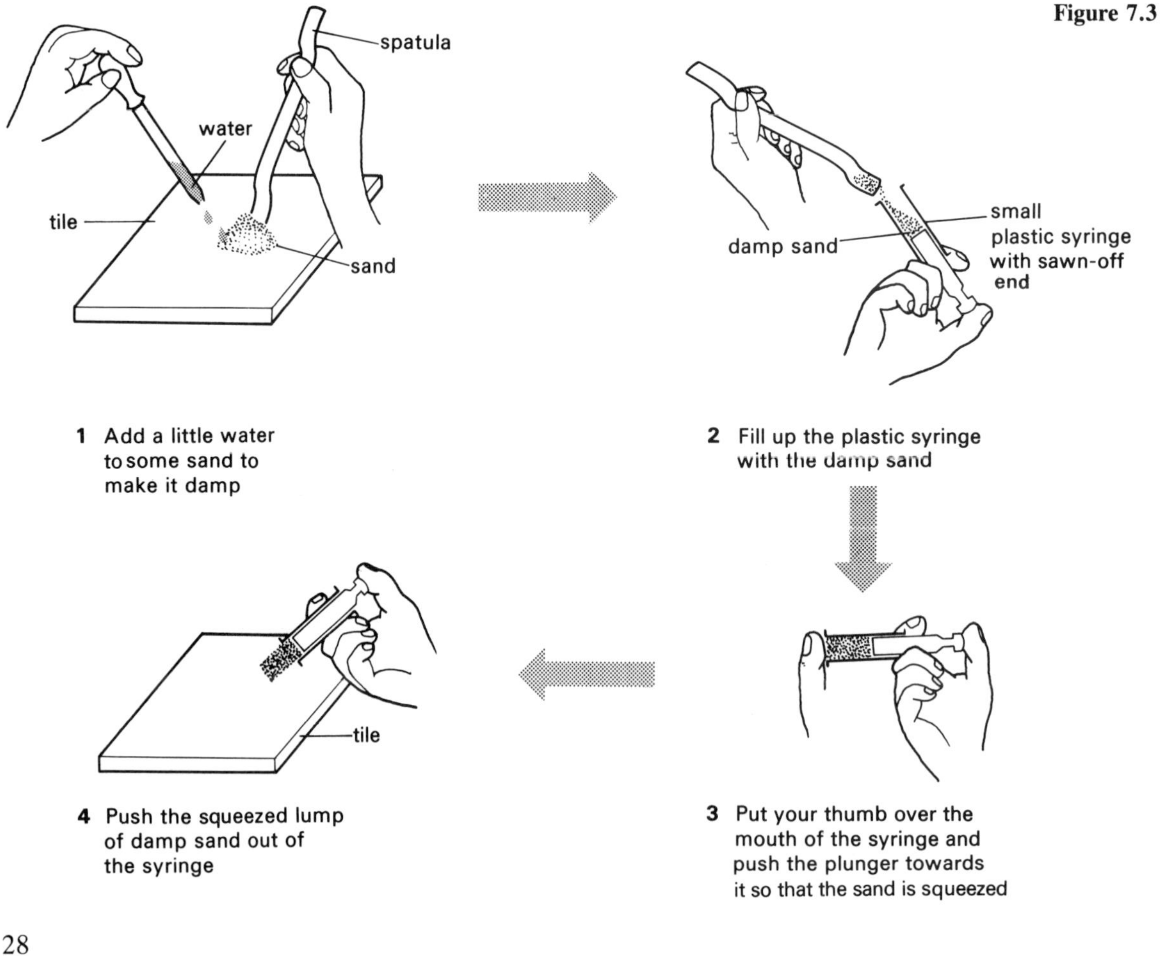

1 Add a little water to some sand to make it damp

2 Fill up the plastic syringe with the damp sand

3 Put your thumb over the mouth of the syringe and push the plunger towards it so that the sand is squeezed

4 Push the squeezed lump of damp sand out of the syringe

Figure 7.4

greater than any pressure you can apply using a syringe, and it happens over millions of years, not just over a few seconds.

Another piece of evidence that shows that many sedimentary rocks were formed under water comes from the discovery of fossils in them. Fossils found in the rock tell us something about life in the past. Some of the fossils in the limestone in Figure 7.5 look very like modern sea shells. This limestone was obviously formed from layers of shells on the sea-floor.

Figure 7.5

Rocks from inside the earth

The rock in the sea cliff in Figure 7.6 is nothing like the one in Figure 7.1. Although there are many vertical cracks (joints) in it, no horizontal layers can be seen at all. This rock is called granite and it is an example of a completely different class of rocks known as *igneous rocks*.

Figure 7.6

The close-up of a polished face of a granite in Figure 7.7 shows interlocking crystals rather than grains 'cemented' together. In the eighteenth century, some scientists thought that these crystals were formed from the sea as the water slowly evaporated. But we now know that this idea is wrong and that they were formed as molten (melted) materials slowly cooled deep inside the earth.

Figure 7.7

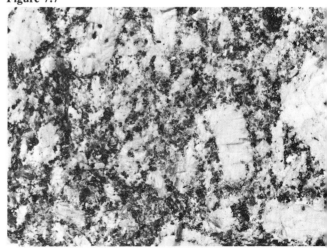

Experiment 2 Making crystals from molten material

Warm some salol (phenyl salicylate) in a test-tube placed in a water-bath kept at 50–60 °C until it has all melted. Quickly add one drop of the liquid to a microscope slide at room temperature and to one which has just been taken out of a refrigerator. Use a hand-lens to watch the crystals growing on the two slides.

- What is the difference between the size of the crystals on the slides?

The crystals formed on the cold slide are much smaller than those formed on the slide kept at room temperature. This is because the molten salol cools down more quickly on the cold slide: small crystals are formed by rapid cooling and large crystals by slow cooling.

The crystals in granite (Figure 7.7) are large and so granite must have been formed by the slow cooling of molten material. But the crystals in the igneous rock shown in Figure 7.8 are so small that they can only be seen under a microscope. Obviously, this rock must have been formed by very rapid cooling.

How quickly molten material cools down depends on where it is in the earth. Material that is deeply buried cools slowly because of the higher temperatures inside the earth. But material (lava) that rushes up to the surface through the vent of an active volcano (Figure 7.9) is cooled very quickly by the air. So granite was formed deep inside the earth but volcanic rocks were formed by cooling at the earth's surface.

Figure 7.9

Figure 7.8

Questions

1. The table below has brief descriptions of three rocks.

Granite	Limestone	Sandstone
Large interlocking crystals	Fossil shells stuck together by calcite crystals	Grains of sand stuck together by a 'cement'

(a) Limestone contains some crystals and yet is a sedimentary rock. How do we know this?
(b) What is the evidence that suggests that granite was not formed by the cooling of lava from a volcano?
(c) Would you expect any samples of (i) granite, and (ii) sandstone to contain fossils? Explain each of your answers.
(d) What would happen if you added some drops of dilute hydrochloric acid to a lump of limestone?

2. Explain how layering is formed in sedimentary rocks. Can you think of any ways in which material from a volcano might be formed in layers?

3. Granite is formed deep inside the earth but there are many places where this rock can be seen at the surface? How do you think this comes about?

8 SOIL

CONSTITUENTS OF SOIL

Soil is a mixture of (a) sand and clay particles, (b) humus, (c) water, (d) air, (e) dissolved salts, and (f) bacteria and other micro-organisms.

(a) *Sand and clay particles*: These are formed from rocks which have been weathered and broken down by the action of wind, rain, rivers and glaciers. Sand particles are larger than clay particles and have a different chemical composition. The sand is mainly *silicon oxide* while clay is a mixture of aluminium and silicon oxides. *Crumbs* of soil consist of groups of clay and sand particles stuck together by humus. The crumb structure of a soil contributes to its drainage, air content and general fertility.

(b) *Humus*: This is formed from decaying organic matter, e.g. dead remains of plants and animals and animal faeces. The name, humus, is popularly used to describe coarse organic material such as leaf mould, peat or compost. To the soil scientist, however, humus is a fine, structureless organic material which forms part of the soil crumbs. In either case, its origin is the same and its value to the soil is in providing a source of nitrates, phosphates, etc., improving the water-holding properties of the soil and 'glueing' the clay and sand particles together to form crumbs.

(c) *Water*: In a moist soil, the water penetrates the crumbs and forms a film over the particles (Figure 8.1). It is held to the particles by the force of *capillary attraction* (page 134). The root hairs of plants (page 103) come into close contact with the soil particles and have to overcome the capillary attraction in order to absorb water.

(d) *Air* occupies the spaces between the soil crumbs. The larger the crumbs, the bigger will be the air spaces. The air in these spaces supplies the oxygen necessary for the respiration of soil organisms and the roots of plants. In a water-logged soil, the air spaces are filled with water so there is not enough available oxygen.

(e) *Mineral salts*: Salts (page 65) of potassium, iron, and magnesium, as well as phosphates, sul-

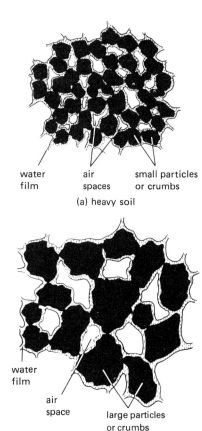

Figure 8.1
(a) heavy soil
(b) light soil

phates and nitrates are present in solution in the soil water. They originate from the action of bacteria on the organic matter, humus, in the soil (see 'Nitrogen Cycle' page 98). These salts are taken up by the roots of plants and used to build up their cell substances.

(f) *Bacteria*: The soil contains a wide variety of small animals such as mites, insects and worms. There is also a large population of micro-organisms such as fungi and bacteria. All these organisms influence the structure and properties of the soil but the bacteria are particularly important. They act on the organic remains of plants and animals, break them down to a fine humus, and release the salts that are needed by plants.

HEAVY AND LIGHT SOILS

Heavy soil

A heavy soil has a high proportion of clay particles. Partly because of their smaller size, the water film round the clay particles makes them stick together more strongly than sand particles. This makes the soil 'heavy' to dig or plough because it is sticky when wet and forms hard clods that are difficult to break up, when dry. The small spaces between clay particles reduce the amount of air available and slow down the drainage of water through the soil. A heavy soil with a good crumb structure does not necessarily show all these disadvantages because the crumbs have properties similar to the larger sand particles. In addition, the slower drainage allows more water to be retained in times of drought and prevents the soluble salts from being leached away by the rain.

A heavy soil can be made lighter and easier to work by adding humus or lime. The lime makes the clay particles clump together and so gives bigger air spaces between them.

Light soil

In a light soil there are more sand than clay particles. The large spaces between the particles give the soil good drainage and aeration. The sand particles are easy to separate so that the soil is 'light' for digging or ploughing. Adding humus improves the water-holding properties and provides a source of mineral salts.

Loam

This is the best type of soil for most agriculture. It contains a balanced mixture of clay and sand particles with abundant humus and a good crumb structure.

EXPERIMENTS ON SOIL

Experiment 1 Mass of water in soil

Place a sample of soil in a weighed evaporating dish and then reweigh it. Heat the dish in an oven at 100 °C for 2 days to drive off the soil water. Then reweigh the dish and soil. Carry on doing this until two weighings give identical results, showing that all the soil water has evaporated.

The difference between the first and final weighings gives the mass of water that was originally present. Temperatures higher than 100 °C must not be used as they will burn the humus in the soil and so give an additional loss of mass.

Sample calculation:

Mass of basin = 200 g
Mass of basin and moist soil = 250 g
∴ Mass of soil = 250 − 200 = 50 g
Final mass of basin and dry soil = 240 g
Loss in mass = 250 − 240 = 10 g

% water in moist soil = $\dfrac{10 \times 100}{50} = 20\%$

Experiment 2 Mass of humus in soil

Transfer the dry soil from the previous experiment to a metal tray and heat strongly over a Bunsen flame until smoke ceases to come off and the charred organic matter has disappeared, leaving only the grey or reddish mineral particles. Let the soil cool and then reweigh it. The loss in weight is due to the organic matter being burnt away to gases that escape (carbon dioxide and water).

Sample calculation:

Mass of dry soil = 40 g
Mass of 'burnt' soil = 38 g
∴ Mass of humus = 40 − 38 = 2 g

% humus in dry soil = $\dfrac{2 \times 100}{40} = 5\%$

Figure 8.2

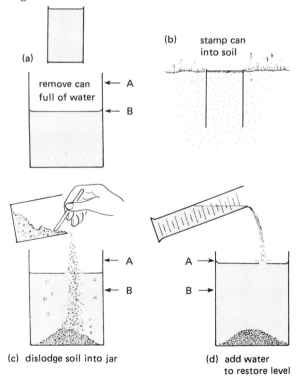

(a)
(b) stamp can into soil
remove can full of water ← A
← B
(c) dislodge soil into jar
← A
← B
(d) add water to restore level
A →
B →

Experiment 3 Volume of air in soil

Fill a large glass jar with water and mark the level, A. Remove a can full of water from the jar so that the level drops to B (Figure 8.2a) and measure the volume of water in the can by pouring it into a measuring cylinder. To obtain a can full of soil, stamp the can, open end down, into the soil until its base is level with the soil. Then dig the can out and cut the soil level with the top of the can (Figure 8.2b). The volume of soil in the can is the same as the volume of water removed from the glass jar. Loosen the soil from the can and tip it into the glass jar, breaking up all lumps at the same time (Figure 8.2c). Adding the soil to the jar will raise the water level again, but because air has escaped from the soil, the level will not have returned to A. To bring the level back to A add some water to the jar from a measuring cylinder (Figure 8.2d). The volume of water added is the same as the volume of air which escaped from the soil.

Sample calculation:

Volume of tin can = 200 cm³
Volume of water added to jar = 28 cm³
∴ Volume of air in 200 cm³ soil = 28 cm³
% air in soil = $\frac{28 \times 100}{200}$ = 14%

Experiment 4 Permeability of soil

Permeability measures the ability of soil to allow water to pass through it. Figure 8.3 shows an arrangement which can be used to compare the permeabilities of two soils. Plug two funnels with glass wool and put equal volumes of two soils into each. Cover both with water and keep the level the same throughout the experiment. This means that the water pressure is the same for both. Collect the water that runs through in a given time in a measuring cylinder. The sandy soil being more permeable to water than the clay soil, will allow far more water to run through in the fixed time.

Experiment 5 Capillarity of soil

Pack two glass tubes 1 cm or more wide and about 50 cm long with dry sand. One tube should contain fine sand and the other coarse sand. Plug the ends with glass wool and clamp the tubes upright in a beaker of water (Figure 8.4). The water travels up

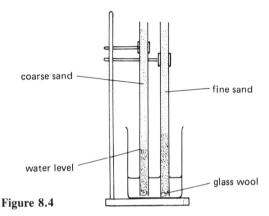

Figure 8.4

through the sand by capillary attraction (page 134) and its level can be seen by the darker colour of the sand. Measure the levels and compare them after one day.

The results show that water travels further in the fine sand. The smaller the particles in a soil, the greater is the capillary attraction.

Questions

1. What are the conditions in the soil which make it a suitable environment for microscopic plants and animals?

2. Write down the biological significance of each of the following:

(a) ploughing farmyard manure (animal faeces and straw) into the soil
(b) adding lime to the soil
(c) spreading sulphate of ammonia on the land

3. Why is it necessary to start with dry soil in order to measure the mass of humus in the soil?

4. What agricultural practices and climatic conditions might affect the amount of air in the soil?

5. What steps, in principle, could you take to improve the fertility of the soil?

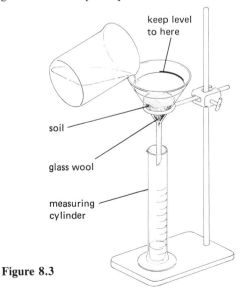

Figure 8.3

9 TWO CHEMICALS FROM WATER: OXYGEN AND HYDROGEN

WHAT IS WATER?

Water must be a compound because it can be broken down by electrolysis (page 58). A little dilute sulphuric acid is usually added to speed up the electrolysis, and the products are two gases, hydrogen and oxygen.

This experiment does not definitely prove that water is a compound of the elements hydrogen and oxygen only. Final proof comes from a successful synthesis of water from its elements.

Figure 9.1 shows that water can be synthesized by burning pure dry hydrogen in the oxygen of the air. One name for water is hydrogen oxide.

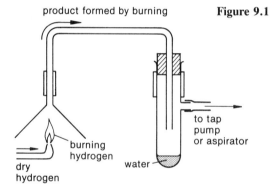

Figure 9.1

The word equation for this chemical reaction is:

hydrogen + oxygen ⟶ water

Hydrogen gas is made of molecules, each of which contains two hydrogen atoms joined together. Oxygen molecules also contain two oxygen atoms joined together. Water molecules contain two hydrogen atoms and one oxygen atom joined together.

Figure 9.2 shows how diagrams of the molecules can be used to explain the synthesis of water. Notice that two molecules of hydrogen are needed for every one molecule of oxygen to form two molecules of water. This is because the total number of hydrogen atoms on each side of the equation must be the same, and because the total number of oxygen atoms on each side must also be the same.

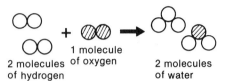

Figure 9.2

Figure 9.2 can be written as a *symbol equation* (page 238):

$$2H_2(g) + O_2(g) \longrightarrow H_2O(l)$$

where 'g' and 'l' stand for 'gas' and 'liquid'.

Competing for oxygen in water

Some metals react with water. Oxygen removed from the water molecules combines with the metal to form an oxide. The hydrogen remaining is released as a gas. The word equation for the reaction is:

metal + hydrogen ⟶ metal + hydrogen
        oxide   oxide

Some metal oxides react with water and form metal hydroxides.

Experiment 1 Reacting calcium with cold water

Using the apparatus shown in Figure 9.3, make a test-tube full of hydrogen. Hold a burning splint close to the mouth of the tube.
This is the test for hydrogen.

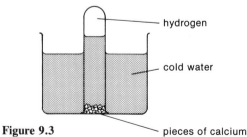

Figure 9.3

- What happens?
- What does hydrogen form when it burns explosively in air?

Magnesium only reacts very slowly with cold water, but it does react easily with steam (Figure 9.4). The word equation for this chemical reaction is:

magnesium + hydrogen ⟶ magnesium + hydrogen
oxide   oxide

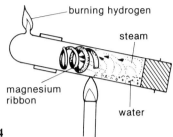

Figure 9.4

Written as a symbol equation, this becomes:

$$Mg(s) + H_2O(g) \longrightarrow MgO(s) + H_2(g)$$

where 's' stands for 'solid'.

Magnesium and hydrogen both react with oxygen. But magnesium is more reactive towards it than hydrogen is. In a competition for oxygen, magnesium wins and so hydrogen oxide loses its oxygen to magnesium.

OXIDATION AND REDUCTION

When a substance burns in oxygen or removes oxygen from an oxide, both *oxidation* and *reduction* are said to take place.

Oxidation in a chemical reaction takes place when oxygen is added to a substance.

Reduction in a chemical reaction takes place when oxygen is removed from a substance.

In the reaction between magnesium and steam, the magnesium gains oxygen from the steam: this is oxidation. Another way of saying the same thing is that the magnesium removes oxygen from the steam: this is reduction. Both oxidation and reduction take place at the same time in the same chemical reaction. The magnesium is called the *reducing agent* in this chemical reaction while the steam is called the *oxidizing agent*.

```
            ┌──── oxygen ────┐
            │                ▼
oxidizing agent + reducing agent ⟶ products
(gives up oxygen  (removes oxygen    of reaction
   to reducing      from oxidizing
     agent)             agent)
```

This word equation shows that oxidation and reduction always go hand in hand. Such chemical reactions are often called *redox* reactions (*red*uction–*ox*idation reactions).

A CLOSER LOOK AT HYDROGEN

All acids are compounds of hydrogen and one or more other elements (page 61). Sometimes the hydrogen can be 'pushed out' of acids, for example, when acids are added to some metals.

Experiment 2 Making and testing hydrogen

Set up the apparatus shown in Figure 9.5. Collect at least four test-tubes full of hydrogen and close each tube with a bung.

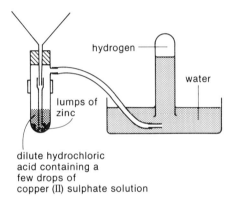

Figure 9.5

CARE: no Bunsen burners should be alight during this part of the experiment. Only when all the sets of apparatus have been emptied of their contents should any testing of the hydrogen be carried out.

Test the gas with:

(*i*) a few cm³ of lime water
(*ii*) a burning splint held near to the mouth of the test-tube.

Try also to show that hydrogen has a much lower density than air (air is about fifteen times as dense as hydrogen).

Hydrogen can remove oxygen from the oxides of some metals. It does this if it is more reactive towards oxygen than the metal is.

Figure 9.6 shows dry hydrogen being reacted with hot black copper oxide. The black solid changes to pink copper metal and drops of water appear to the right of it in the combustion tube. The word equation and the symbol equation for this chemical reaction are:

black copper + hydrogen ⟶ copper + hydrogen
oxide   oxide
$$CuO(s) + H_2(s) \longrightarrow Cu(s) + H_2O(l)$$

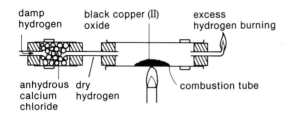

Figure 9.6

This is another example of a redox reaction. The black copper oxide is the oxidizing agent and gives up its oxygen to the hydrogen which is the reducing agent.

Questions

1. Heated zinc reacts with steam to form hydrogen.

(a) Draw a diagram of the apparatus you would use to make and collect some hydrogen from this chemical reaction.

(b) How would you prove that the gas collected is hydrogen?

(c) Write a word equation and a symbol equation for this reaction. (Write zinc oxide as ZnO.)

(d) Name the oxidizing agent and the reducing agent and so explain what happens to the oxygen in this reaction.

2. Explain why the experiment shown in Figure 5.3 (page 18) would not be successful if magnesium ribbon were used instead of the copper foil.

3. When hydrogen is passed over yellow lead oxide (PbO), lead metal and water are formed.

(a) Write down the word equation and the symbol equation for this chemical reaction.

(b) Which substance, hydrogen or yellow lead oxide, acts as an oxidizing agent in this reaction, and which substance acts as a reducing agent? Explain your answers.

(c) What would you *see* happening to the solid (yellow lead oxide) during the experiment?

Classifying Elements and Compounds

10 METALS AND NON-METALS

CLASSIFYING ELEMENTS

Here is a list of elements that may be found in a school laboratory:

aluminium, bromine, calcium, carbon, copper, hydrogen, iodine, iron, lead, magnesium, mercury, nitrogen, oxygen, phosphorus, potassium, sodium, sulphur, zinc.

One obvious way of classifying these is into solids, liquids and gases. Linked with this idea is a classification based on melting points or boiling points. But Figure 10.1 shows that the elements have very widely different boiling points which make it difficult to sort them into several classes. At one extreme, aluminium, copper and iron have very high boiling points while at the other extreme, hydrogen, nitrogen and oxygen have very low boiling points. But there are a lot of elements with boiling points in between these extremes.

An easier and more important way of classifying elements is into *metals* and *non-metals*. The next experiment helps us to sort some elements into these classes.

Experiment 1 Some properties of metals and non-metals

Use the following elements: carbon (a rod of graphite), sulphur (a lump), copper, lead, zinc and aluminium (all foils), iodine (a crystal).

- Put all the results in a table with the headings given below.

Element	What is its colour?	Is it shiny or dull?	Does it bend easily or break?	Is it a good conductor of electricity?

To find out whether each element conducts electricity, set up the apparatus shown in Figure 10.2.

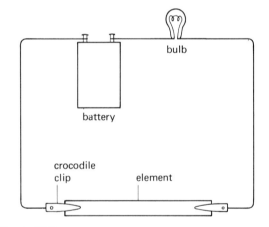

Figure 10.2

- What happens if the element does conduct?
- Which of the elements can be called metals and which can be called non-metals?

Figure 10.1

C 3730
(carbon goes directly from a solid to a gas at this temperature)

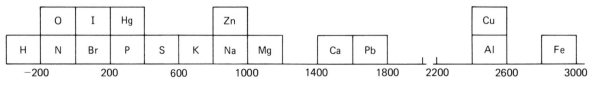

boiling point in °C

Table 10.1

Element	Appearance	Hardness	Conduction of electricity
Red phosphorus	Red powder with no shine	Easily breaks	Does not conduct
Sodium	Freshly cut surface is shiny and grey coloured	Easily cut with a knife	Conducts well

From your results, make a list of the main properties of metals and non-metals.

Some results for two elements that are too dangerous for you to use are given in Table 10.1. Red phosphorus is obviously a non-metal. Sodium is not quite so easy to classify because metals are usually not soft enough to cut with a knife. But sodium must be a metal in view of its shiny surface and its conduction of electricity.

Table 10.2 gives some of the differences in properties between metals and non-metals. There are other differences, especially those connected with chemical reactions. For example, metals often react with dilute acid (page 42) whereas non-metals do not; and metals and non-metals burn to form different kinds of oxides (page 55).

CARBON—AN UNUSUAL NON-METAL

A carbon rod is a good conductor of electricity, so good that it is often used to carry an electric current from the wires into a substance that is being electrolysed (page 58). Yet in other respects, carbon is clearly a non-metal.

The carbon in a carbon rod is called graphite. Like many non-metals, it is a brittle substance and

Figure 10.3

can also be easily scratched by a knife. But another form of pure carbon is completely different in appearance from graphite. This is diamond (Figure 10.3).

Table 10.2

Property	Metals	Non-metals
Melting-point and boiling point	Usually high	Usually low
Appearance	Usually shiny (metallic lustre); often 'ring' when struck	Solids do not have characteristic appearance and are often dull; do not 'ring' when struck
Conduction of electricity	Good (decreases as the temperature rises)	Poor
Conduction of heat	Good	Poor

Table 10.3

Property	Diamond	Graphite
Appearance	White or colourless with 'glassy' lustre	Grey with metallic lustre
Conduction of electricity	Poor	Good
Hardness on Mohs' scale	10	1–2
Density in g/cm^3	3·51	2·25
Temperature at which burning starts in °C	800–900	about 700

Carbon is said to show *allotropy* or to be allotropic. Diamond and graphite are the different allotropic forms, or *allotropes*. Table 10.3 shows some of the main differences in the properties of the two allotropes.

These properties are so different that you may think it is extremely unlikely that diamond and graphite could possibly be made of the same atoms. But they have the same chemical reactions, for example, both burn in oxygen to form the same oxide, carbon dioxide. The differences between them come about because of the different ways that the carbon atoms are arranged in the crystals.

Questions

1. Sodium is soft and yet it is a metal. Write down *two* metallic properties that sodium shares with all other metals.

2. Here are some melting points of a selection of solid elements:

aluminium	660 °C	potassium	64 °C
carbon	3730 °C	sodium	98 °C
copper	1083 °C	sulphur	113 °C
iron	1535 °C	zinc	420 °C
phosphorus	44 °C		

Is there any link between the melting point of the element and whether it is a metal or a non-metal? Which melting points surprise you most?

3. (*a*) List *three* ways in which the properties of diamond are different from those of graphite.
 (*b*) Explain the meaning of the term *allotropy*.

11 REACTING METALS

A REACTIVITY SERIES FOR METALS

Some metals give out a lot of heat and light energy when they burn in air or oxygen. Others react in a much less spectacular way. This means that some metals are more 'keen' to react with oxygen than others. They are said to be more *reactive* towards oxygen or, to put it in another way, they have a greater *affinity* (liking) for oxygen.

Think of the reaction between magnesium and air or oxygen. Once the magnesium ribbon has been heated to a high temperature, its reaction with oxygen is rapid and a lot of heat and light energy is produced. The heat given out when the first part of the magnesium burns must be enough to heat the next part of the metal to the high temperature required for it to react rapidly. The large amount of heat energy given out shows that magnesium has a high affinity for oxygen, that is, a high reactivity towards oxygen.

Now think of the reaction between copper and air or oxygen. This is slow and shows no obvious signs of heat energy being given out. A lot of heating is needed here because copper has a low affinity for oxygen, that is, it has a low reactivity towards oxygen. Any heat energy given out by the reaction of one part of the copper with oxygen is not enough to make the next part react.

Figure 11.1 shows an experiment which can be used to compare the reactivities of some metals towards oxygen. To make a fair comparison, all the metals must be in the same form; in this case, it is powder form. The oxygen is produced by the heating of the potassium permanganate and is then passed over the hot metal powder. The reactivity of each metal can be judged not only by how much the metal glows but also by how quickly the soft-glass test-tube starts to bend.

Experiments like this help to build up a *reactivity series* for metals in which the most reactive metal is put at the top and the least reactive one at the bottom. A short reactivity series is given below.

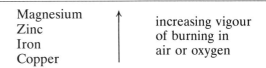

Other metals such as sodium and gold could also be put in this series. Sodium burns very vigorously indeed in air. It has to be stored in oil to prevent it from burning in air, even without first being heated. But gold does not burn at all in air, even at a very high temperature. So sodium can be placed above magnesium in the reactivity series and gold can be placed below copper.

REACTING METALS WITH WATER AND ACIDS

The idea of a reactivity series for metals also works for their reactions with substances other than air or oxygen.

Table 11.1 lists the observations that can be made during the reaction of some metals with water or steam. You should see that the order is very like the one for the reactions of the metals with air or oxygen.

Experiment 1 A reactivity series for metals in their reactions with acids

You will need about 25 cm³ of each of the following acids:

(*i*) very dilute hydrochloric acid
(*ii*) dilute hydrochloric acid*

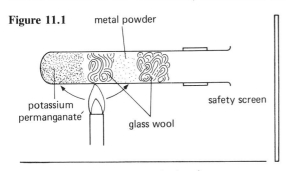

Figure 11.1

*The acid should be of normal 'bench' concentration (i.e. 2 M).

Table 11.1

	Metal	Reaction with water or steam
↑	Sodium	Dangerous reaction with cold water; a large amount of heat energy is given out
	Calcium	Exothermic reaction with cold water (less vigorous than the reaction of sodium with water)
increasing reactivity	Magnesium	Only just reacts with cold water; reaction between hot metal and steam gives out a lot of heat and light energy
	Zinc / Iron	Metals may glow a little as some heat energy is given out during the reactions with steam; do not react with cold water
	Copper / Gold	Do not react with either water or steam

(*iii*) moderately concentrated hydrochloric acid.

CARE: any acid you may spill on to your skin or clothes must be washed straight away with a large amount of water.

Use 10 mm squares (or strips) of magnesium ribbon, and copper, lead and zinc foils. A small nail can be used as your sample of iron.

Add a few cm³ of each acid to each metal in separate test-tubes.

- How can you tell whether or not a reaction is taking place?

If a reaction does take place with the cold acid, you can assume that there would also be a reaction with the warm acid. *For the dilute acids only*, if no reaction takes place, warm the test-tube gently *with its mouth pointing well away from yourself or anyone else.*

- Does a reaction take place now?

Do not confuse boiling with a gas (hydrogen) being given off at the surface of the metal.

- Fill in a table, like the one given below, putting a tick whenever you think a reaction has taken place. One line is filled in to show you how to do this.

Metal	Acid (*a*)		Acid (*b*)		Acid (*c*)
	Cold	Warm	Cold	Warm	Cold
Magnesium	✓	✓	✓	✓	✓

- Which metals are the most reactive towards acids, those with a lot of ticks or those with only a few?
- Arrange the metals in a reactivity series, putting the most reactive first.
- Is the order here the same as the order for the reactivities of the metals towards oxygen and water?

GETTING METALS FROM SOLUTIONS

When an iron nail is placed in copper sulphate solution, the nail gets coated with a red solid (Figure 11.2). This solid is copper metal and it has been

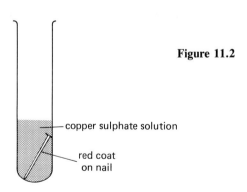

Figure 11.2

'pushed out' (displaced) from copper sulphate by the iron. The word and symbol equations for this reaction are:

iron + copper ⟶ copper + iron
sulphate   sulphate
$Fe(s) + CuSO_4(aq) \longrightarrow Cu(s) + FeSO_4(aq)$

where 'aq' stands for 'aqueous' and shows that the copper sulphate and iron sulphate are dissolved in water.

Reactions like this are called *displacement reactions*. It is possible to predict if a metal is likely to displace another metal from a solution of one of its compounds by knowing the reactivity series for metals. For example, iron comes higher than copper in the series and so it can displace copper from a solution of copper sulphate.

Experiment 2 Investigating some displacement reactions

Use the metals and solutions listed in Table 11.2. In each case, add a small piece of cleaned metal (foil or nail) to a few cm³ of the solution in a test-tube.

Table 11.2

Solution / Metal	Copper sulphate	Iron sulphate	Lead nitrate	Magnesium sulphate	Zinc sulphate
Copper		×			
Iron	✓				
Lead					
Magnesium					
Zinc					

- Can you see any new deposit ('coat') on the surface of the metals?

If no change takes place straight away, leave the test-tube standing in a rack for a few minutes and then look at the metal again.

- Copy Table 11.2 and put a tick in the appropriate box if you think a reaction has taken place, and a cross if you think it has not. (One tick and one cross are already filled in for you.)
- What is the order of reactivity for the metals in these displacement reactions?
- Does this order agree with the order of reactivity of metals towards acids?

Questions

1. Suppose you were given a sample of tin metal. Describe how you would decide its position in the reactivity series. Your answer should include details of *several* different experiments.

2. Here are some facts about lead.

(a) When heated strongly, lead reacts slowly with air to form an oxide.
(b) Lead reacts with moderately concentrated hydrochloric acid when it is boiled, but it does not react with the dilute acid.

Say whether you think that lead is more or less reactive than (*i*) magnesium, (*ii*) copper, and (*iii*) iron. Give reasons for your answers.

If you were given samples of lead foil, lead nitrate solution, copper foil, copper sulphate solution, an iron nail, iron sulphate solution, magnesium ribbon and magnesium sulphate solution, how would you set about checking your answers to this question?

3. (a) What are the metals gold and silver used for?
(b) Where do you think that these metals come in the reactivity series?
(c) Explain the connection between the answers to (a) and (b).

4. Describe an experiment you could use to show that magnesium powder is more reactive than iron filings towards sulphuric acid.

12 USING METALS

Metals and alloys

Metals are good conductors of heat and electricity. Many are also strong and can be bent into different shapes without breaking. You should be able to link these properties with the three uses of metals shown in Figures 12.1 to 12.3.

For some uses, metals need to be very pure. For example, the copper in copper wire has to be very pure (up to 99.98 per cent) so that it conducts electricity as well as possible. But for many purposes, mixtures of two or more metals are used. These mixtures (called *alloys*) are made by melting the metals together and letting them cool and solidify. Alloys have different properties from the separate metals, and nowadays scientists can make alloys with the properties they want for a particular job.

Generally, alloys are harder than the separate metals and have lower melting points. They may also be more resistant to *corrosion* (see below).

Table 12.1 shows the properties and uses of some common alloys. Some gaps have been left in the 'Uses' column for you to fill in after you have copied the table into your notebook.

The best-known 'alloy' is steel, which is made from iron and a non-metal (carbon) rather than from two metals. For steel making, the molten iron is contained in a furnace and, in a modern process, oxygen is blown on to its surface. The impurities

Figure 12.1

Figure 12.2

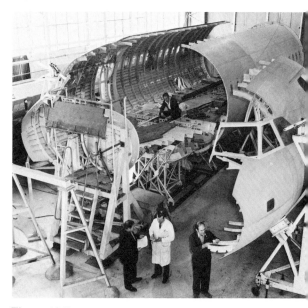

Figure 12.3

are oxidized and either they are given off as gaseous oxides or else they combine with calcium oxide (from added limestone) to form a slag which floats on the surface of the iron. When the purity of the iron has reached the right level, the furnace is tapped. If an 'alloy steel' is required (stainless steel, invar steel, and so on), then the required metals are added at this stage.

The corrosion of metals

Everyone knows that iron forms a layer of brown crumbly rust on its surface when it is exposed to the 'weather'. Rusting is caused by the reaction of iron with oxygen and water in the air.

Many other metals corrode in air. You can tell this because they lose their shine and sometimes change colour as well. The surfaces of copper water-tanks and bronze statues start off copper-coloured but slowly go brownish, grey and then green as they corrode.

The layer formed on most metals by corrosion is an oxide or a carbonate of the metal. Unlike rust itself, this layer does not usually flake off quickly. Instead it stays there and helps to stop the air from attacking the fresh metal underneath. So these metals have their own 'do-it-yourself' protection. This is why some metals like aluminium resist corrosion better than iron even though they are higher in the reactivity series than iron.

It is very difficult to stop iron and ordinary steel from rusting, though regular painting or greasing does help. One of the best ways is to make an alloy of iron and another metal: stainless steel (Table 12.1) is a good example.

But steel alloys can be much more expensive than ordinary steel. No one would think of making very long underground pipelines or large ships out of stainless steel; and yet the ordinary steel in these things could rust very quickly because acids in the soil and salt in the sea speed up the process a lot.

Ordinary steel can be protected from rusting in another way by using a more reactive metal, for example zinc, in conjunction with it. The next experiment will help you to find out why.

Experiment 1 Investigating the rusting of iron nails

Use three shiny iron nails in this experiment. Wrap a small piece of zinc foil tightly round a 1 cm length of one nail, and do the same to another nail with a small piece of copper foil. (CARE: the edges of the

Table 12.1

Alloy	Percentage composition	Important properties	Uses
Brass	Cu 60 Zn 40	Harder than copper but just as corrosion-resistant	
Bronze	Cu 85 Sn 15	As for brass	Statues, plaques, ships' propellers
Cupro-nickel	Cu 75 Ni 25	As for brass, looks silvery	'Silver' coins
Duralumin	Al 95 Cu 4 Traces of Mn and Mg	Density just as low as aluminium or magnesium but much stronger and more corrosion-resistant	Aircraft construction
Magnalium	Al 70 Mg 30		
Solder	Pb 67 Sn 33	Harder than lead and has an even lower melting point	
Steel	Fe 98 C 1 Traces of other elements	Much stronger than iron	
Invar steel	Fe 63 Ni 36 Trace of C	Much lower coefficient of expansion than iron	In thermostats and clock pendulums
Stainless steel	Fe 85 Cr 14 Ni 1	Harder than ordinary steel and much more corrosion-resistant	

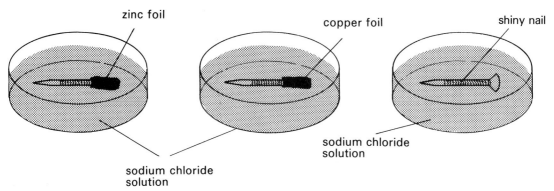

Figure 12.4

foils may be sharp.) Leave the third nail as it is.

Put each nail into a separate Petri dish and cover it with sodium chloride solution (Figure 12.4). Rusting can take place in the solution because it contains dissolved air as well as sodium chloride.

Leave the nails for at least a day. After this, look for signs of brown rust on each nail.

- Which nail has rusted the most, and which one has rusted the least?

The order of reactivity towards oxygen of the three metals is (most reactive first): zinc, iron, copper. When a metal that is more reactive than iron is wrapped round an iron nail, it is this that reacts with the oxygen in the dissolved air. So the iron wrapped in zinc foil does not rust. But if the metal wrapped round the nail is less reactive than iron, the iron rusts quickly while the unreactive metal (in this case, copper) does not change at all.

Questions

1. (a) What is an *alloy*?
   (b) Why are alloys often better than the pure metals for a whole variety of uses in the home and in industry?
   (c) State *one* reason why some alloys are not widely used despite having very useful properties.

2. Type-metal is an alloy with the percentage composition lead (Pb) 80, antimony (Sb) 15, tin (Sn) 5. It is used in printing because it is hard, melts at a low temperature (240 °C) and expands on solidifying.

   (a) What are the differences in composition between type-metal and solder?
   (b) Give one property which type-metal and solder have in common.
   (c) Would you expect the melting point of pure lead to be higher or lower than 240 °C? Give a reason for your answer.

3. (a) Find as many ways as you can in which iron or steel can be protected from rusting, and make a list of these.
   (b) Why do the following articles made of iron or steel not rust easily (if at all): cutlery, the bumpers of cars, a can for storing food, a metal bridge?

4. The processes of burning and rusting have some features in common. Make a list of their differences and a list of their similarities.

13 FAMILIES OF ELEMENTS

Some unusual metals

Sodium and potassium are quite like one another and very unlike most other metals. They are very soft and they quickly burst into flames when they are heated. They also have to be stored in oil so that they cannot come into contact with oxygen or water vapour in the air.

The reaction of sodium with water is quite dramatic. A molten ball of metal darts across the water surface and very quickly disappears. The experiment shown in Figure 13.1 proves that hydrogen is given off. A blue colour is formed when you add some drops of universal indicator solution to the water after all the sodium has gone, and this shows that the water is now alkaline (page 53). Pure water is neutral, so there must be something dissolved in it that makes it alkaline. This solution is sodium hydroxide, and it is being made as the sodium metal disappears into the water.

The word equation for this reaction is:

sodium + water ⟶ sodium + hydrogen
hydroxide

Potassium reacts with water even more dramatically than sodium. A *very* small piece splutters vigorously and quickly disappears. Hydrogen is given off and again the solution left over goes blue when some drops of universal indicator solution are added.

Lithium is a metal that is similar in appearance to sodium and potassium. In the next experiment you can find out how similar it is in its chemical reactions.

Experiment 1 Two reactions of lithium

a Observe what happens when a very small piece of lithium (no bigger than a rice grain) is heated on a tin lid covered by a gauze (Figure 13.2).

Figure 13.1

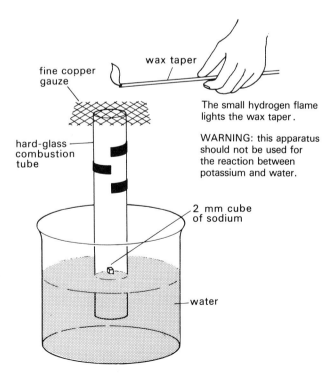

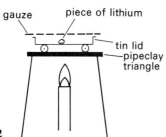

Figure 13.2

b Add a very small piece of lithium (no bigger than a rice grain) to some cold water in a beaker. (CARE: lithium is a reactive metal.)

• What can you see?

Test the final solution with a few drops of universal indicator solution.

• Is the final solution acidic or alkaline?
• Is lithium more reactive or less reactive towards water than sodium?

Lithium, sodium and potassium belong to a chemical 'family' called the *alkali metals*. They can be put in the same family because they have a lot in common and the differences between them are not very great. As with the members of some human families, everyone can tell almost at a glance that they are closely related.

A family of non-metals

Chlorine, bromine and iodine are all non-metals (Table 13.1). You might think that these three non-metals are not much alike. But they do have something in common: they can exist in the form of coloured gases with nasty smells.

Chlorine is used in the water in indoor swimming pools because it helps to kill the germs. Bromine also behaves in this way, and 'tincture of iodine' is a solution of iodine in methylated spirit used as an antiseptic to stop cuts from getting infected. Here again, these non-metals have something in common.

Table 13.1

Name of element	Appearance
Chlorine	Greenish yellow GAS
Bromine	Dark red LIQUID which forms a red-brown gas
Iodine	Grey shiny SOLID which forms a violet vapour

Three experiments using chlorine are shown in Figure 13.3. When you have looked at these, study the experiments shown in Figure 13.4 for bromine. You should decide that chlorine and bromine show similar chemical reactions but that chlorine is the more reactive element.

Experiment 2 Is iodine like chlorine and bromine?

a Use some tongs to add a small crystal of iodine to a test-tube half full of water. Shake the tube well.

• Has the iodine dissolved?

Now warm the water.

• Has any more iodine dissolved?

Add a piece of blue litmus paper and a piece of red litmus paper (page 52) to the water.

• Is there any sign of bleaching?

b Use some tongs to add a small crystal of iodine to a test-tube half full of sodium hydroxide solution. Warm it gently *with the mouth of the tube pointing well away from yourself or anyone else.*

• What happens?

c Use the apparatus in Figure 13.5 and heat the iron wool very strongly. Then move the Bunsen flame under the iodine crystal so that some iodine vapour passes through the iron wool.

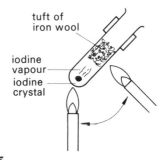

Figure 13.5

• What happens?
• What is left on the iron wool at the end of the experiment?

Chlorine, bromine and iodine are closely related and belong to a family of non-metals called the *halogens*. They are a family even though at room temperature one is a gas, one is a liquid and one is a solid. The halogens do not look so alike as the alkali metals, but the family likenesses show up in some of their reactions.

Figure 13.3

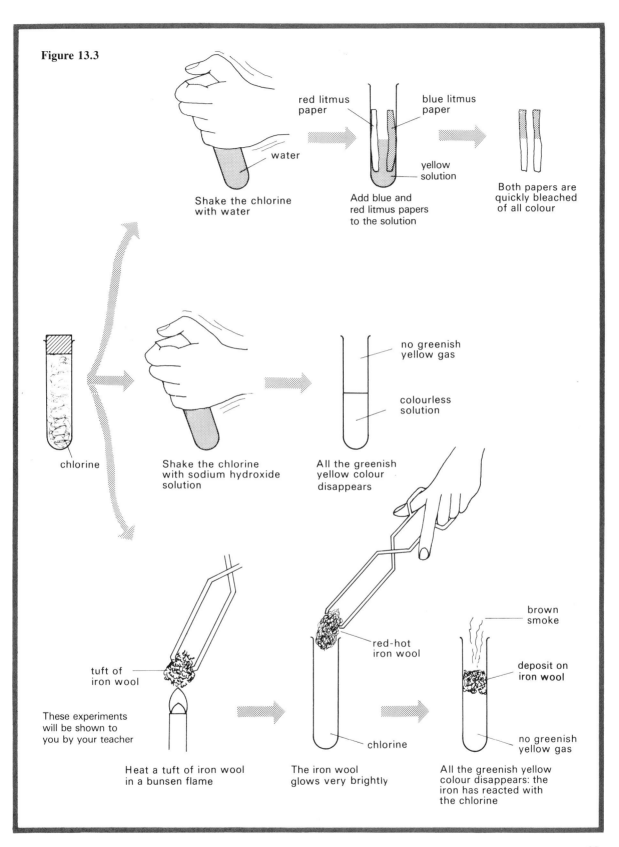

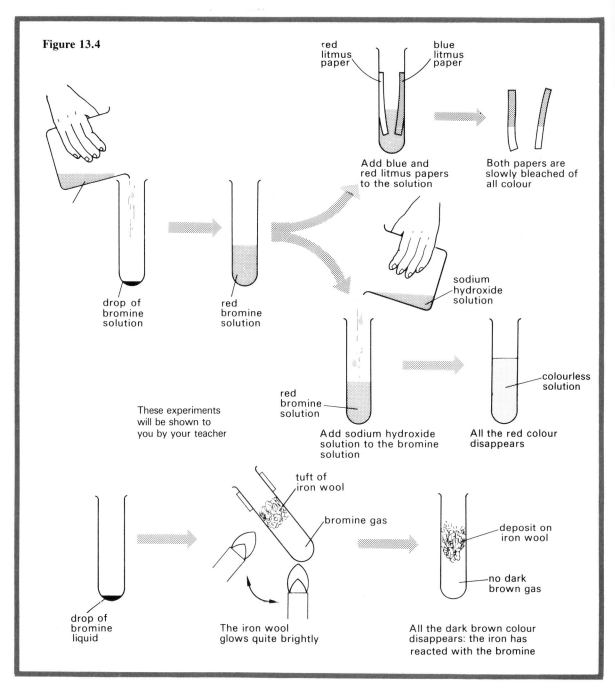

Figure 13.4

Just as with some human families, the members do not look much alike at first sight. But when you look more closely at particular features, you may see some of the less obvious family likenesses.

The alkaline earth metals

Magnesium and calcium are two metals that look alike. Both are grey, rather soft, and shiny when fresh; but both get coated with a dull layer of white or grey oxide when exposed to the air. This last fact suggests that they are reactive metals. But they are not stored under oil so they are obviously not quite as reactive as the alkali metals.

Both metals burn in air and react with water or steam (Table 11.1, page 42), though in each case not as vigorously as members of the alkali metal family. Magnesium and calcium belong to the family of *alkaline earth metals*.

THE PERIODIC TABLE

Scientists are always trying to find out more about the patterns in the world around us. One of the most important patterns in chemistry is the *periodic table* (inside back cover).

This is a way of arranging all the elements so that patterns in their properties are highlighted by the positions of the elements in the table. It so happens that the order in which the elements occur in the periodic table is, with one or two special exceptions, the order of their relative atomic masses (atomic weights, inside back cover).

You should try to find the chemical families that are 'hidden' in the table. These are arranged in vertical *groups*. The alkali metals and the alkaline earth metals are in groups I and II on the left-hand side of the table, and the halogens are in group VII on the right-hand side. You may also notice the general pattern that metals are found on the left-hand side or in the middle while non-metals are found on the right-hand side.

Questions

1. This is a question about an element called rubidium (Rb).

(*a*) To which family of elements does rubidium belong?
(*b*) What would you expect rubidium to look like?
(*c*) What is likely to happen when a piece of rubidium is added to cold water? What would you expect to see when some drops of red litmus solution are added to the solution that is formed?

2. Write down the evidence you know that shows that chlorine is a more reactive element than iodine.

3. Look at the position of fluorine (F) in the periodic table.

(*a*) To which family of elements does fluorine belong?
(*b*) Is fluorine likely to be a gas, a liquid or a solid at room temperature?
(*c*) Name two substances you would expect fluorine to react with.

14 ACIDS AND ALKALIS

Indicators

Citrus fruits such as oranges, lemons and limes have a sharp taste. This is caused by an *acid* (citric acid) in the juices. If the juice of a citrus fruit is squeezed on to blue litmus paper, the paper turns red. This happens because litmus is a substance that changes colour when acids are added to it: litmus is an *indicator* of acids.

Many other substances are acidic and turn blue litmus red (Table 14.1), and there will be some in your own laboratory.

Table 14.1 Some common acids

Acid	Comments
Carbolic acid	Its chemical name is phenol; it is used as a disinfectant
Carbonic acid	Soda water is made by dissolving carbon dioxide in water; rain water is a very dilute solution of this acid
Citric acid	The acid in citrus fruits
Ethanoic (acetic) acid	Vinegar is a very dilute solution of this acid; wines form ethanoic acid when they go sour
Lactic acid	The acid in sour milk, and in human muscles after prolonged exercise
Tartaric acid	A substance made from this acid (called 'cream of tartar') is used as the acid part of baking powder
Hydrochloric acid Nitric acid Sulphuric acid	The three important laboratory acids

The opposite of an acid is an *alkali*. This is a substance that turns red litmus blue. Again, there will be alkaline substances in your own laboratory: some of them have a name ending in *hydroxide* (Table 14.2).

Laboratory acids and alkalis must be used with a lot of care. Although a few, such as citric acid, are harmless, most are very dangerous and should *never* be tasted or allowed to touch the skin.

Table 14.2 Some common alkalis

Alkali	Comments
Ammonia solution	Used in some household cleaners
Calcium hydroxide	Its solution is known as lime water
Magnesium hydroxide	'Milk of magnesia'
Potassium hydroxide	Often called caustic potash
Sodium carbonate	Used as washing soda
Sodium hydrogen-carbonate	Used as stomach powder
Sodium hydroxide	Often called caustic soda

Litmus comes from a kind of lichen (a plant) that grows in West Africa. Many of the coloured dyes in plants, vegetables and flowers are also good indicators.

Experiment 1 Making and testing an indicator

Tear the petals off some blue or red flowers. Put them into a mortar and add a few cm³ of ethanol (alcohol). Grind up the petals with a pestle until the ethanol becomes deeply coloured. (The ethanol acts as a solvent for the dye in the petals.) Now pour off the solution into a test-tube.

Find out the colour of your indicator in acids and its colour in alkalis.

Indicators can be used to test for gases. Some, like carbon dioxide, are acidic and turn *damp* blue litmus paper red. Some are *neutral*, like oxygen and hydrogen, and have no effect on either damp blue litmus paper or damp red litmus paper. One common gas, ammonia (page 271), is alkaline and turns damp red litmus paper blue.

It is important always to use damp litmus paper in these tests. Only when there is water present can the colour change take place.

Neutralization

Acids and alkalis have the opposite effect on litmus. In the next experiment, you can find out what effect they have on each other.

Experiment 2 Mixing acids and alkalis

Pour about 2 cm³ of dilute sulphuric acid into a test-tube and add a few drops of litmus solution. Put a −10 °C to 110 °C thermometer in the mixture and note the temperature. Then, using a dropper, add dilute sodium hydroxide solution a small amount at a time until no further colour change can be seen.

- What happens to the colour of the litmus?
- Can you detect any change in the temperature of the mixture as the reaction takes place?

Do the experiment again, this time using 2 cm³ of dilute hydrochloric acid in the test-tube and adding dilute calcium hydroxide from the dropper.

- Can you make the same two observations for this second reaction as you did for the first?

Alkalis counteract acids in a chemical reaction called *neutralization*. The solution finally formed in the above experiment is usually alkaline rather than neutral because it is very difficult to stop adding the alkali at exactly the moment when the neutral point has been reached. The products of the reaction are a *salt* (page 65) and water so that the general word equation for all neutralization reactions is:

$$\text{acid} + \text{alkali} \longrightarrow \text{salt} + \text{water}$$

pH values

Universal indicator is a mixture of several indicators and shows, by its colour, how acidic or alkaline a substance is. Scientists have found it easier to use a scale of numbers to describe how acidic or how alkaline a substance is, rather than to use colours. This scale is called the *pH scale* and the numbers are *pH values*.

The colours of a universal indicator and the pH scale are shown in Figure 14.1. Strong acids have pH values of 0 or 1 while weak acids have pH values in the range 4 to 6. Strong alkalis have pH values of 13 to 14 and weak alkalis have pH values in the range 8 to 10. Neutral solutions, in which the acidity and alkalinity cancel each other out, have a pH value of 7.

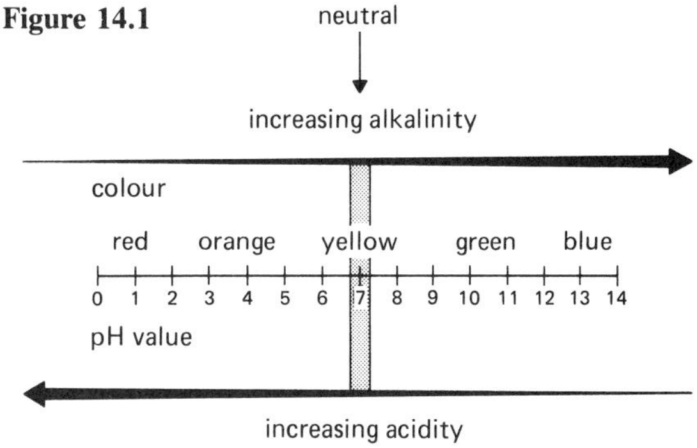

Figure 14.2 shows how you can find the pH value of a solid. Everyday substances you could use include sugar, salt, soil, wood ash or charcoal ash, and soap. You may also want to test the pH value

Figure 14.1

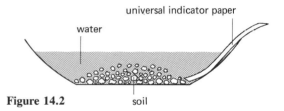

Figure 14.2

of some everyday liquids and gases. If you use a gas, remember first to make the paper damp with a little water, which must be neutral itself. But you should always ask your teacher which substances are safe to test before you try the experiment either at home or at school.

Some results you might get are given in Figure 14.3. Soils mostly have a pH value fairly close to 7: for many plants, the most suitable value is 6·5, though some plants prefer a more acidic soil than this and others like more alkaline soils.

The fairly high pH value of the ash made by burning plant material, such as wood or seaweed, is due to the presence of a substance called 'potash'. In the past, the fact that potash is alkaline was made use of in the manufacture of soap.

Soap was first made in Mediterranean countries by heating animal fat or vegetable oil with an alkali called 'soda' which is found naturally in the Nile Valley. But it is known from the writings of the day that the Romans made soap from goat's tallow (fat) and the ashes of beechwood. Modern soap making uses vegetable oil and sodium hydroxide.

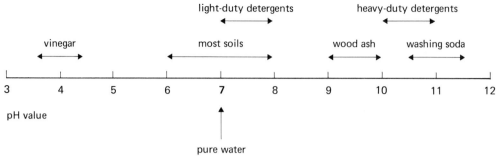

Figure 14.3

Some detergents have a very high pH value and are called heavy-duty detergents. They are able to clean very dirty clothes but they unfortunately remove some dyes and are harmful to the skin. Other detergents have much lower pH values and are called light-duty detergents. These are not quite as good at getting dirt out of heavily soiled clothes but they have the advantage of not removing the dyes from them and they do not cause irritation to the skin.

The pH value of fresh cow's milk is about 6·5. When milk goes sour, its pH value drops to about 5·5. This shows that an acid must be formed during this chemical reaction.

Questions

1. Describe what happens when a dilute acid is slowly dropped into a solution of an alkali to which a few drops of universal indicator solution have been added.

2. Why does acid that has been accidentally spilled on a book often cause the colour of the book cover and the ink to change?

3. What substances are used to counteract acidity in soils? What is this process of counteracting acidity called?

4. Here is a list of well-known gases: ammonia, carbon dioxide, hydrogen, nitrogen, oxygen.

Classify these as acidic, alkaline or neutral towards damp litmus paper.

15 OXIDES

Acidic and alkaline oxides

Oxides are formed whenever an element (or a compound) burns in air or oxygen. They can be classified according to how they behave towards damp indicator paper.

Experiment 1 Testing oxides with indicator paper

Prepare five boiling tubes full of oxygen by using either the apparatus shown in Figure 5.8 (page 20) or an oxygen cylinder. Close each tube with a bung. To get the oxide, burn each of the following elements as described in the experiment on page 20: powdered charcoal (carbon), iron filings or steel wool, clean magnesium ribbon, powdered sulphur, zinc powder.

As soon as the burning has stopped, test the oxide formed with damp universal indicator paper. Figure 15.1 shows what you should do if the oxide is a gas or a solid.

- Put all the results in a table with the headings given below.

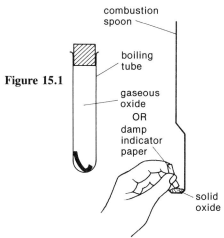

Figure 15.1

Name of element	Is its oxide a gas or a solid?	Is its oxide acidic, alkaline or neutral?

In Table 15.1 there are some details about the oxides of some more elements.

From these results, it seems that oxides can be sorted out into three classes. Some oxides (these are usually gases) dissolve in water to form an acidic solution. Some oxides (these are usually solids) dissolve in water to form an alkaline solution. The rest (these are also usually solids) are neutral towards damp indicator paper.

There is a link between the class of element being burnt and the class of oxide that is formed (Figure 15.2). Some metals form alkaline oxides, though some form oxides that have no effect on damp indicator paper. Some non-metals form acidic oxides, though some also form oxides that have no effect on damp indicator paper.

When oxides dissolve in water to form acidic or alkaline solutions, a chemical reaction is always taking place.

Table 15.1

Element	Oxide	Effect on damp indicator paper
METAL	Calcium oxide Potassium oxide Sodium oxide	All dissolve in water and have an alkaline reaction to damp indicator paper
	Copper oxide	Does not dissolve in water and has no effect on damp indicator paper
NON-METAL	Phosphorus oxide	Dissolve in water and has an acidic reaction to damp indicator paper

Figure 15.2

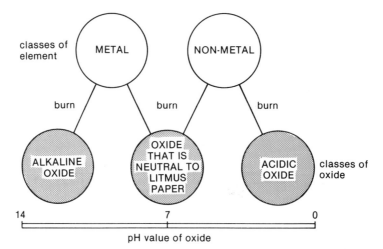

Experiment 2 Reacting calcium oxide with water

Put a small lump of calcium oxide on a watch-glass and add water drop by drop. Watch carefully what happens and feel the watch-glass underneath the solid.

- What do you observe which suggests that a chemical reaction is taking place?

You would get a similar result if you added water to any other alkaline oxide such as magnesium oxide or sodium oxide. They all react with water to form metal hydroxides, and these metal hydroxides are alkalis. So the word and symbol equations for the reaction of calcium oxide with water are:

calcium oxide + water ⟶ calcium hydroxide
$$CaO(s) + H_2O(l) \longrightarrow Ca(OH)_2(s)$$

Basic oxides

Many metal oxides, like copper oxide and iron oxide, do not react with water to form alkalis and so they are neutral towards damp indicator paper. But they do react in another way, and this reaction is also given by alkaline oxides.

Experiment 3 Reacting metal oxides with acids

Warm a few cm³ of dilute hydrochloric acid in a boiling tube. (CARE: *do not boil this solution, and do not point the mouth of the tube towards yourself or towards anyone else.*) Then add a very small amount of black copper oxide on the tip of a spatula to the warm acid. Stir the mixture with a glass rod.

- What happens to the solid?
- What colour change takes place in the solution?
- Copy Table 15.2 into your notebook and fill in the results for this experiment in the appropriate box.

Do the experiment again, this time using each of the metal oxide–acid pairs shown in Table 15.2.

Table 15.2

Acid \ Metal oxide	Black Copper oxide	Magnesium oxide
Dilute hydrochloric acid		
Dilute nitric acid		
Dilute sulphuric acid		

- Does a chemical reaction take place in each case?

Oxides that react with acids are called *basic oxides*. Basic oxides are always made of a *metal* and oxygen.

Some metal oxides, such as magnesium oxide, are basic oxides but they also react with water to form alkalis. So alkaline oxides are a special kind of basic oxide (Figure 15.3).

Figure 15.3

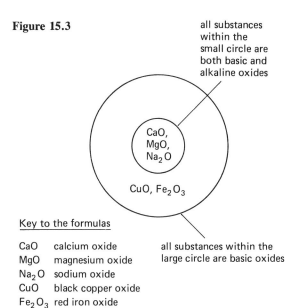

Key to the formulas

CaO   calcium oxide
MgO   magnesium oxide
Na_2O sodium oxide
CuO   black copper oxide
Fe_2O_3 red iron oxide

Metal oxides with a difference

Just to complicate matters a little, some metal oxides are not just basic oxides, as the next experiment shows.

Experiment 4 Reacting some metal oxides with acids and alkalis

a Warm a small amount of yellow lead oxide (enough to cover the tip of a spatula) with a few cm³ of dilute acid in a boiling tube.

• Does the oxide react?

Do the experiment again, this time using zinc oxide and dilute nitric acid.

b Carry out the two experiments in **a** using the same two oxides but warm them with dilute sodium hydroxide solution instead of the acid.

• Does each oxide react?

These two oxides react with both acids and alkalis. When they react with acids, they are behaving like basic oxides; and when they react with alkalis, they are behaving like acidic oxides. They can do two completely different things depending on what they are mixed with.

Oxides that can behave in both basic and acidic ways are said to be *amphoteric*.

Oxides and the reactivity series for metals

There are some links between the class of metal oxide and the position of the metal in the reactivity series (page 41).

Very reactive metals like sodium and calcium form alkaline oxides. Metals lower in the reactivity series can form basic oxides or, in some cases, amphoteric oxides.

Questions

1. Explain carefully what is wrong with each of these statements.

(*a*) Some, but not all, alkaline oxides are also basic oxides.
(*b*) Non-metals usually form basic oxides when they burn in air or oxygen.
(*c*) Sodium oxide dissolves in water to form a solution: no chemical reaction occurs between the oxide and the water.
(*d*) Metal oxides all belong to the class of oxides called basic oxides.

2. What can be observed when water is added drop by drop to a lump of calcium oxide that suggests a chemical reaction is taking place?

3. (*a*) Describe an experiment you could perform to show that zinc oxide is an amphoteric oxide.
(*b*) What important difference in properties is there between a basic oxide and an amphoteric oxide?

16 EFFECTS OF ELECTRICITY ON COMPOUNDS

Conduction in solids and liquids

Though some solid elements (metals and graphite) conduct electricity well, all solid compounds are non-conductors (insulators).

One liquid that conducts electricity is the element mercury. Mercury is a metal and so we expect it to show this property. The aim of the next experiment is to find out whether any pure compounds that are liquids, or any solutions, are also good conductors.

Experiment 1 Looking at the effect of electricity on some liquids and solutions

Set up the apparatus shown in Figure 16.1. The carbon rods are used to carry the electricity into the liquid: they are called *electrodes*. If the liquid is

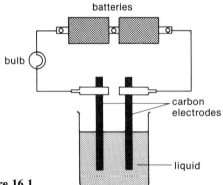

Figure 16.1

a conductor, then electricity passes between the electrodes so completing the circuit. The bulb then lights up.

Use each of these liquids, one after the other in the beaker: tap water, distilled water, ethanol (alcohol).

- Do any of these make the bulb light up?

b Now try some solutions of solid compounds in water. Suitable solids include green copper chloride, lead bromide, potassium iodide, sodium chloride (salt) and sugar. Dissolve as much solid as you can in about $50 \, cm^3$ of water and add the solution to the beaker. (Note that lead bromide is only slightly soluble in water.)

- Do any of the solutions make the bulb light up?
- What happens in these solutions if electricity does pass through them?

c Lastly, find out whether dilute acids and dilute alkalis conduct electricity.

- What happens in the solutions?

Water on its own is a very poor conductor of electricity. But some solutions made by dissolving a substance in water are good or fairly good conductors.

All these substances are called *electrolytes*. The water in rivers, seas and lakes, as well as the water from a tap, conducts electricity because it always contains dissolved solids. This means that it can be dangerous to handle electrical appliances with wet hands.

Electrolytes do not just let electricity pass through them: something happens in each solution as well. Sometimes bubbles of gas can be seen at one or both electrodes, and sometimes a solid or a liquid appears around an electrode. New substances are obviously being formed.

Electrolytes let electricity pass through them and, at the same time, are chemically changed by it.

This kind of chemical reaction is called *electrolysis*.

Electrolysis is the breaking down of a compound by electricity.

Figure 16.2 shows some apparatus that can be used to collect any gases given off at the electrodes during the electrolysis of a solution. One electrode is connected to the negative end of a battery: this is the *cathode*. The other electrode is connected to the positive end of a battery: this is the *anode*.

Some results for the electrolysis of solutions are given in Table 16.1. For some electrolytes, the new substances formed at the electrodes come from the dissolved solid. So the copper and the chlorine formed by the electrolysis of copper chloride solution must come from the copper chloride itself.

Table 16.1

Solution	Product at:	
	Cathode	Anode
Potassium iodide	Hydrogen	Iodine
Sodium chloride	Hydrogen	Chlorine and/or oxygen
Magnesium nitrate	Hydrogen	Oxygen
Zinc sulphate	Zinc and hydrogen	Oxygen
Lead bromide	Lead	Bromine
Copper chloride	Copper	Chlorine and/or oxygen

Note: Oxygen cannot be detected at a *carbon* anode. Instead, carbon dioxide is formed by the reaction of the oxygen with the carbon. So some of these results can only be obtained with another kind of anode, for example, a platinum one.

But the electrolysis of some solutions gives hydrogen at the cathode and oxygen at the anode, even though the solid dissolved in the water might contain only one of these two elements or neither of them. In these cases, the electricity must be breaking down the water into its elements.

The electrolytes in Table 16.1 are arranged according to the position of the metal in the reactivity series, with those containing the most reactive metals at the top. A pattern can be seen in the elements that are formed at the cathode. At the top, the cathode product is always hydrogen while at the bottom, it is the metal. In between, both the metal and hydrogen are produced.

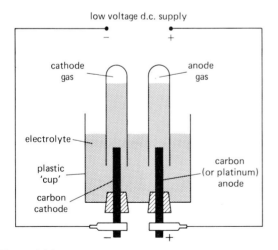

Figure 16.2

Conduction in fused (melted) substances

In the next experiment, you can investigate which solid compounds conduct in the molten state.

Experiment 2 Looking at the effect of electricity on some molten substances

Use some of the following solids: green copper chloride, lead bromide, potassium iodide, sodium chloride (salt), sugar, wax. Put a small amount of one of the solids between two carbon electrodes placed very close together (Figure 16.3). Heat the solid strongly until it melts.

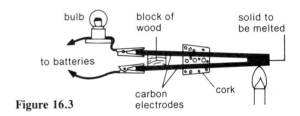

Figure 16.3

- Does the bulb light up now?
- Is the solid a conductor or a non-conductor when molten?

Do the experiment again with some of the other solids.

- Make a list of solids that conduct electricity when molten and another of solids that do not conduct when molten.

Those solids that are electrolytes when dissolved in water are also electrolytes when molten. All these solids belong to a class of compounds called *salts* (page 65).

It is not easy to detect any of the new substances formed during the electrolysis of a molten electrolyte using the apparatus shown in Figure 16.3. But any compound made of two elements is always broken down when molten to give one of the elements (the metal) at the cathode and the other (the

59

Figure 16.4

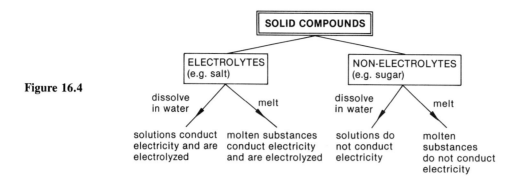

non-metal) at the anode. A good example is the electrolysis of molten copper chloride mentioned on page 9.

A SUMMARY

Figure 16.4 shows the main ideas that are dealt with in this unit.

Questions

1. Mercury is a conductor but not an electrolyte. Molten sodium chloride is a conductor and an electrolyte. Explain these two statements.

2. What are the likely products at the anode and at the cathode when these solutions are electrolyzed:

(*a*) copper bromide solution
(*b*) potassium chloride solution

3. Here are some results for passing electricity through solutions of salts in water.

Salt	Substance formed at cathode
Lead nitrate	Lead
Magnesium nitrate	Hydrogen
Silver nitrate	Silver
Sodium nitrate	Hydrogen

(*a*) Sort out the salts into a class that forms a metal at the cathode and a class that forms hydrogen.
(*b*) Where do the metals that form at the cathode come in the reactivity series for metals?
(*c*) What is likely to be formed at the cathode when electricity is passed through lithium nitrate solution? Explain your answer.

17 MORE ABOUT ACIDS

Acidic properties

Acids are important chemicals and their reactions are dealt with in several units of this book. Here is a summary of the main properties of a typical acid. In each case, the acid is dissolved in water: no acid can show its acidic properties if it is completely free from water.

A typical acid:
—*has a pH value of less than 7* (page 53)
—*reacts with many metals to give hydrogen* (page 42)
—*reacts with alkalis to form new compounds called salts* (page 53)
—*reacts with carbonates and hydrogencarbonates to give off carbon dioxide* (page 66)
—*reacts with basic oxides to form salts* (page 66)
—*is an electrolyte* (page 58)

Metals are elements. Since hydrogen is often given off when metals are added to acids, it is obvious that acids must be made of hydrogen joined with one or more other elements (Table 17.1).

Table 17.1

Name of acid	Elements in acid	Formula
Hydrochloric acid	Hydrogen and chlorine	HCl
Nitric acid	Hydrogen, nitrogen and oxygen	HNO_3
Sulphuric acid	Hydrogen, sulphur and oxygen	H_2SO_4

OXIDIZING PROPERTIES

Of the three common dilute acids found in the school laboratory, dilute nitric acid is a little unusual. Whereas dilute hydrochloric acid and dilute sulphuric acid react with many metals to give off hydrogen, dilute nitric acid forms another gas altogether.

Experiment 1 Reacting dilute nitric acid with copper and iron

Add a few cm³ of dilute nitric acid to some copper turnings in a boiling tube. If no reaction takes place, or if the reaction is slow, warm the mixture gently *with the mouth of the tube pointing well away from yourself or anyone else.*

Test the gas given off with:

(*i*) a burning splint held close to the mouth of the tube
(*ii*) damp universal indicator paper.

Also look closely at the contents of the tube to see if there are any colour changes taking place, (it helps to view the tube with a piece of white paper held behind it).

- What is the evidence that the gas given off is not hydrogen?

Do the experiment again, this time using some iron filings rather than copper turnings.

- Is the gas given off the same as the one given off when dilute nitric acid is added to copper?

Dilute hydrochloric acid and dilute sulphuric acid only react with metals that are found either around the middle of the reactivity series or even higher than this. So they react with iron or magnesium, giving off hydrogen, but they do not react with copper.

Dilute nitric acid reacts with most metals, including metals like copper which are found towards the bottom of the reactivity series. But the gas that is formed is nearly always not hydrogen.

Figure 17.1 shows what happens in the reaction between copper and dilute nitric acid. If you look closely at the boiling tube, you may notice that the gas only gets coloured some distance above the level of the acid. This is because the gas produced at the surface is colourless nitrogen monoxide (NO) which reacts with oxygen from the air in the upper part of the tube to produce yellowish brown nitrogen dioxide (NO_2).

A clue as to what might be happening in this reaction of dilute nitric acid comes from the

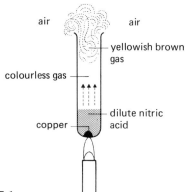

Figure 17.1

The concentrated nitric acid is the oxidizing agent and the wood is the reducing agent. Some of the oxygen from the acid is added to the wood and this causes the wood to break down into carbon dioxide and water. The nitric acid is reduced to nitrogen dioxide.

Whenever either dilute nitric acid or concentrated nitric acid reacts to give nitrogen dioxide, it means that the acid is acting as an oxidizing agent. So in its reactions with copper and iron in the above experiment, dilute nitric acid is showing oxidizing properties, and not the acidic properties of a typical acid.

Table 17.2 gives a summary of some of the main properties shown by the three common laboratory acids.

Besides being oxidized by dilute nitric acid, copper (and many other metals) are also oxidized by concentrated nitric acid and concentrated sulphuric acid. The word equations for these reactions are:

copper + nitric acid $\longrightarrow$ copper + nitrogen + water
(concentrated)   nitrate   dioxide

copper + sulphuric acid $\xrightarrow{\text{heat}}$ copper + sulphur + water
(concentrated)   sulphate   dioxide

Again, in the case of nitric acid, nitrogen dioxide is formed as the concentrated acid releases some of its oxygen to the copper. Concentrated sulphuric acid is made of three elements, hydrogen, sulphur and oxygen. This acid also releases some of its oxygen to the copper and is reduced itself to an acidic gas called sulphur dioxide (page 266).

experiment shown in Figure 17.2. Here concentrated nitric acid is being broken down by heat with the help of the glass wool which acts as a catalyst (page 288). Brown fumes of nitrogen dioxide can be seen just past the glass wool and oxygen collects in the inverted boiling tube. The word equation for the reaction is:

nitric acid $\xrightarrow{\text{heat}}$ nitrogen + oxygen + water
(concentrated)   dioxide

Nitric acid is made of the three elements, hydrogen, nitrogen and oxygen. Under certain conditions, it can release some of this oxygen to another substance, so acting as an oxidizing agent (page 35).

Your teacher may show you just how powerful an oxidizing agent nitric acid is by carefully adding a few drops of the concentrated acid to some warm sawdust (a compound of carbon, hydrogen and oxygen). The sawdust bursts into flames straight away and a lot of brown fumes of nitrogen dioxide are given off. The word equation for this reaction can be written as:

nitric acid + wood $\longrightarrow$ nitrogen + carbon + water
(concentrated)   dioxide   dioxide

Dehydrating properties

It is extremely dangerous to pour water on to concentrated sulphuric acid. The correct way to dilute some of the concentrated acid is to add the acid

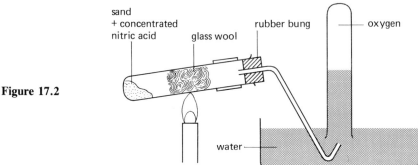

Figure 17.2

Table 17.2

Acid		Hydrogen-producer with fairly reactive or reactive metals	Oxidizing agent
Hydrochloric acid	Dilute	Yes	No
	Concentrated	Yes	No
Nitric acid	Dilute	Occasionally gives hydrogen with reactive metals	Oxidizes metals
	Concentrated	No	Oxidizes metals, carbon and carbon compounds (e.g. wood), and sulphur
Sulphuric acid	Dilute	Yes	No
	Concentrated	No	Oxidizes metals, carbon and sulphur

slowly and carefully to a large amount of water. The mixing of these two substances is very exothermic and, if done wrongly, can produce a violent reaction.

Concentrated sulphuric acid has a strong attraction for water. If a few drops are added to blue copper sulphate crystals, the solid soon goes white as the acid takes away the water of crystallization (page 9). This process of removing water (or the elements of water, hydrogen and oxygen) from a compound is called *dehydration*: concentrated sulphuric acid is a *dehydrating agent*. Neither of the other two common laboratory acids, dilute or concentrated, acts as a dehydrating agent.

One use of the dehydrating properties of concentrated sulphuric acid is in the drying of damp gases during their preparation and collection in the laboratory.

Questions

1. Explain carefully what is wrong with each of these statements.

(*a*) All acids contain oxygen.
(*b*) Every solution which, during electrolysis, gives off hydrogen at the cathode, must contain an acid.
(*c*) All metals react with all dilute acids to give off hydrogen.

2. Name the gas given off when the following pairs of substances are mixed:

(*a*) calcium carbonate and dilute hydrochloric acid
(*b*) copper and concentrated nitric acid
(*c*) zinc and dilute sulphuric acid
(*d*) sodium hydrogencarbonate and dilute nitric acid
(*e*) copper and concentrated sulphuric acid

3. When water is added to white (anhydrous) copper sulphate, the solid turns blue. This process can be called *hydration* and it is a chemical reaction in which the water gets 'locked up' inside the solid as *water of crystallization*.

(*a*) What is the opposite reaction called in which the water of crystallization is removed from blue copper sulphate?
(*b*) What substance could be added to the blue copper sulphate to make this reaction occur?
(*c*) Describe one other way in which blue copper sulphate can be changed to white copper sulphate.

4. 'Both dilute and concentrated nitric acid can act as oxidizing agents.'

Explain this statement, giving some examples of their reactions.

18 MAKING SALTS

Metals and salts

When a metal reacts with dilute hydrochloric acid or dilute sulphuric acid, hydrogen is given off and the metal dissolves to form a solution. In the next experiment you can investigate this solution further. The reaction chosen is the one between magnesium and dilute sulphuric acid.

Experiment 1 Investigating a metal–acid reaction

a Add some strips of clean magnesium ribbon to a few cm³ of dilute sulphuric acid in a test-tube. Wait until the reaction has stopped. If there is no magnesium ribbon left over, add some more strips and wait again until the reaction has stopped. Carry on doing this until there is an excess of the ribbon left over in the mixture.

Now pour the solution into a small evaporating dish, leaving any extra magnesium in the test-tube. Evaporate the solution to half its volume (Figure 18.1).

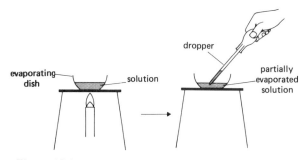

Figure 18.1

Use a dropper to transfer a small amount of the hot solution to a microscope slide. Look at the solution under a microscope.

- What kind of crystals can you see? Are they anything like the ones in the photograph (Figure 18.2)?

b Your teacher will give you some crystals that have been made by a method like the one you used in **a**. Dissolve these crystals in a little distilled water in a test-tube. Shake or stir it well so that all the

Figure 18.2

solid dissolves. Pour about half the solution into another test-tube.

To the first tube, add two or three drops of dilute hydrochloric acid and then a 1 cm depth of barium chloride solution. (CARE: barium chloride is poisonous.)

- What can you see now?
- What does this result tell you about the dissolved crystals (Table 18.1)?

To the second tube, add two or three drops of magneson reagent followed by two drops of sodium hydroxide solution.

- What can you see now?
- What does this result tell you about the dissolved crystals (Table 18.1)?

Table 18.1

	Test for:	Observation
CHLORIDE	Add dilute nitric acid followed by silver nitrate solution	White cloudiness (proves CHLORIDE)
SULPHATE	Add dilute hydrochloric acid followed by barium chloride solution	White cloudiness (proves SULPHATE)
MAGNESIUM	Add magneson reagent followed by sodium hydroxide solution	Blue colour (proves MAGNESIUM)

Magnesium reacts with dilute sulphuric acid to form hydrogen and a solution of magnesium sulphate. The magnesium metal 'pushes out' (displaces) all the hydrogen from the acid (Figure 18.3). The other part of the acid (in this case, the

Figure 18.3

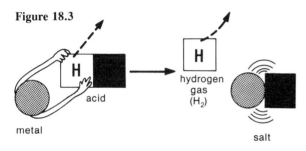

sulphate part) becomes linked to the magnesium.

The word and symbol equations for this reaction are:

magnesium + sulphuric ⟶ magnesium + hydrogen
              acid   sulphate
            (dilute)

$Mg(s) + H_2SO_4(aq) \longrightarrow MgSO_4(aq) + H_2(g)$

What are salts?

Magnesium sulphate is an example of a class of compounds called *salts*. Most salts are made of a metal part and an acid part. The crystals of each salt have a particular shape. But the usual way of finding out which salt is which is by the kind of *chemical analysis* carried out in part **b** of the above experiment.

Salts formed from sulphuric acid are called *sulphates*; salts formed from hydrochloric acid are called *chlorides*; and salts formed from nitric acid are called *nitrates*.

So the acid part of a salt's name sometimes ends in -*ide* and sometimes in -*ate*. The -*ide* ending tells you that there is only one element in the acid part of the salt. For example, sodium chloride is made of the elements sodium and chlorine only. But the -*ate* ending means that the salt also contains oxygen. For example, sodium sulphate is made of the elements sodium, sulphur and oxygen.

Figure 18.4 shows that there are several ways of making salts but that one of the two substances you start with is always an acid. Another way of saying this is that a salt is formed whenever an acid is neutralized by a member of one of the classes of substances in Figure 18.4.

Figure 18.4

If you look back to Chapter 16, you will find that a lot of the substances used in the experiments on electrolysis are salts. All salts are electrolytes when they are dissolved in water or in the molten state.

Other ways of making salts

Not all salts can be made by using a metal to displace hydrogen from the acid. For one thing, there is no reaction at all when some metals are added to a dilute acid. For another, it is very dangerous to add a very reactive metal to an acid. This is why the best-known salt, common salt (sodium chloride), should never be made by adding sodium metal to dilute hydrochloric acid.

In the next experiment, you can make some

sodium chloride in a safe way using the neutralization of an acid by an alkali (page 53).

Experiment 2 Making a salt by an acid–alkali reaction

Add about 20 cm³ of dilute sodium hydroxide solution to an evaporating dish on a white tile. Slowly add dilute hydrochloric acid to the alkali, stirring the mixture all the time with a glass rod (Figure 18.5).

Keep on taking the glass rod out of the mixture and spotting it on to a piece of blue litmus paper (or universal indicator paper) on the white tile next to the dish. Stop adding the acid after a drop of the solution has *just* turned the indicator paper red.

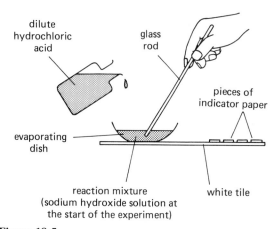

Figure 18.5

Now evaporate some of the water from this solution and test it in the same way as the magnesium sulphate solution in part **a** of Experiment 1.

- What shapes are the crystals under the microscope?

You should see that they are very like the crystals of rock salt shown in Figure 1.9 (page 5).

Quite a few salts can be made by the neutralization of an acid by an alkali. You can make another one in this way if you carry out the preparation of a fertilizer as described on page 273.

Basic oxides and salts

Basic oxides react with acids to form a solution, but no gas is given off. The general word equation for this reaction is:

basic oxide + acid ⟶ salt + water

An experiment using basic oxides and acids is described on page 56. To get crystals of the salt from the solution formed in the reaction, you need to filter off any excess (extra) basic oxide and then partially evaporate the filtrate: crystals are formed as this solution cools.

A basic oxide is made of a metal part and an oxygen part. When the oxide reacts with an acid, the oxygen part joins with the hydrogen part of the acid to make water (Figure 18.6). The metal part of the oxide joins with the other part of the acid to make a salt.

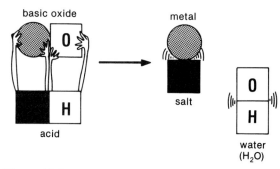

Figure 18.6

For example, the word and symbol equations for the preparation of copper sulphate from black copper oxide and dilute sulphuric acid are:

copper + sulphuric acid ⟶ copper + water
oxide   (dilute)   sulphate
$CuO(s) + H_2SO_4(aq) \longrightarrow CuSO_4(aq) + H_2O(l)$

Carbonates and salts

Experiment 3 Reacting carbonates with acids

a Add half a spatula measure of copper carbonate to a few cm³ of dilute sulphuric acid in a test-tube.

- What happens to the mixture in the test-tube?

Now show that the gas being given off is carbon dioxide by carrying out the test using lime water that is shown in Figure 2.4 (page 8).

- What happens to the lime water?
- Copy Table 18.2 into your notebook and fill in the gaps next to copper carbonate and sulphuric acid.

b Do the experiment five more times, each time using one of the carbonate–acid pairs shown in Table 18.2.

- Does a reaction take place in each case?

Table 18.2

Carbonate	Dilute acid	Is carbon dioxide given off? (✓ or ✗)	Colour of solution formed
Copper carbonate	Sulphuric acid		
Copper carbonate	Nitric acid		
Nickel carbonate	Sulphuric acid		
Nickel carbonate	Nitric acid		
Zinc carbonate	Sulphuric acid		
Zinc carbonate	Nitric acid		

The general word equation for the reaction of carbonates with acids is:

carbonate + acid ⟶ salt + carbon + water
                                   dioxide

A metal carbonate can be thought of as a metal oxide part joined to a carbon dioxide part (Figure 18.7). When an acid is mixed with the carbonate,

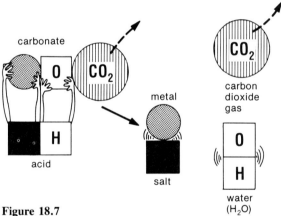

Figure 18.7

the metal oxide part makes a salt and water, as in Figure 18.6. The carbon dioxide is then set free and comes off as a gas.

For example, the word and symbol equations for the preparation of copper sulphate from copper carbonate and dilute sulphuric acid are:

copper   + sulphuric acid ⟶ copper   + carbon + water
carbonate   (dilute)   sulphate   dioxide
$CuCO_3(s) + H_2SO_4(aq) \longrightarrow CuSO_4(aq) + CO_2(g) + H_2O(l)$

You should know how to get crystals of copper sulphate from the solution formed in this reaction.

Questions

1. Explain carefully what is wrong with each of these statements.

(a) Copper chloride can be made by mixing copper and dilute hydrochloric acid.
(b) Lead and nitrogen are the only elements in lead nitrate.
(c) Lead chloride cannot be electrolysed because it does not dissolve in water.

2. (a) Suppose that you wanted to make some crystals of copper sulphate starting from copper carbonate and dilute sulphuric acid. Explain how you would do this in two parts—first how you would make a solution of copper sulphate, and then how you would make some crystals from this solution.
(b) You again want to make some copper sulphate crystals, but this time you are given some powdered copper and dilute sulphuric acid. No other laboratory chemicals are available. Explain in outline how you would carry out the preparation, especially pointing out where it would be different from the preparation in (a).

3. Fill in the gaps in the following word equations:

(a) zinc + sulphuric acid ⟶ .... + ...
(b) lead oxide + nitric acid ⟶ .... + water
                (dilute)
(c) .... + hydrochloric acid ⟶
        (dilute)   magnesium + carbon + water
                        chloride   dioxide
(d) potassium + ... ⟶ potassium + water
    hydroxide   sulphate

19 SOLUBLE AND INSOLUBLE SALTS

Adding salts to water

A lot of salts dissolve in water. Table 19.1 shows that all nitrates are soluble whereas there are just a few chlorides and sulphates that are insoluble.

Table 19.1

	Nitrates	Chlorides	Sulphates
Soluble	All	Most	Most
Insoluble	None	Lead chloride, Silver chloride	Barium sulphate, Calcium sulphate, Lead sulphate

Experiment 1 Dissolving salts in hot and cold water

a Add flat spatula measures of potassium nitrate to about 5 cm³ of cold water in a boiling tube, stirring the mixture with a glass rod, until no more solid dissolves. There should now be a very small amount of solid left undissolved at the bottom of the tube.

Set up the apparatus shown in Figure 19.1, and heat the water in the water-bath to a temperature of about 40 °C. Add one more flat spatula measure

Figure 19.1

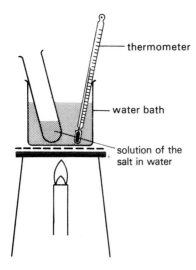

of potassium nitrate to the solution in the tube and stir the mixture until it dissolves. Keep on adding more flat spatula measures until no more solid dissolves, counting the number as you go along.

- How much better does potassium nitrate dissolve in water at 40 °C than in water at room temperature?

Now heat up the water in the water-bath to a temperature of about 60 °C, and find out how much better the solid dissolves in water at 60 °C than in water at 40 °C.

b Carry out the whole experiment again, this time using sodium chloride instead of potassium nitrate.

- Does sodium chloride dissolve better in hot water or in cold water?

Salts vary in how well they dissolve in water at different temperatures. Just a few of them dissolve better as the temperature of the water goes down. But most salts are like potassium nitrate, though the difference between the dissolving at low and high temperatures may not be quite so great (see Figure 19.2).

Experiment 2 Adding lead bromide to water

CARE: lead bromide is poisonous.

Add a *very small* amount of lead bromide or lead iodide on the tip of a spatula to about 15 cm³ of cold water in a boiling tube. Shake the tube and then heat the mixture strongly so that the water nearly boils. Let the mixture cool and look at it against a light background.

- Is lead bromide soluble or insoluble in (*i*) cold water and (*ii*) hot water?
- What can you see as the hot solution cools?
- Explain what you see.

Solubility

When a solid is being added to water, a stage is always reached when no more of it can be dissolved because the solution has become *saturated*: the solution contains the maximum possible amount of

Figure 19.2

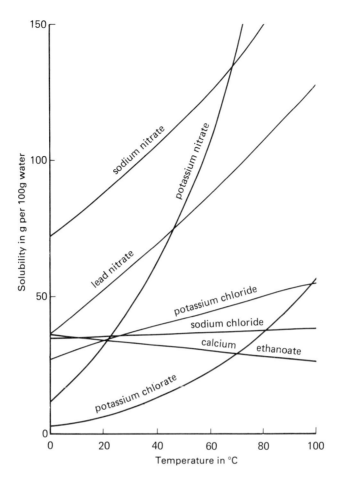

dissolved solid. Any excess of solid stays undissolved and settles at the bottom of the container.

The proportion of solid in the solution is known as the *solubility*. For example, at 20 °C, 36 g of sodium chloride dissolves in 100 g of water, and the solubility of sodium chloride is 36 g per 100 g of water.

If the experiment is done again at a different temperature, a different solubility is found. The way in which the solubilities of some salts, including sodium chloride, vary with temperature is shown in Figure 19.2. You should compare the results shown here for potassium nitrate and sodium chloride with your own results for the first experiment in this chapter.

Mixing solutions of salts

When pairs of soluble salts are mixed together in water, no reaction usually takes place. But some pairs of soluble salts do react when they are mixed together.

Experiment 3 Reacting calcium chloride and sodium sulphate solutions

Add a few cm³ of calcium chloride solution to a few cm³ of sodium sulphate solution in a test-tube.

• What can you see?

This experiment shows that a reaction can take place between two soluble salts if an insoluble salt is formed from them. The insoluble salt then appears as a *precipitate* in the solution, that is, grains of undissolved solid which may be scattered all through the liquid or which may collect at the bottom of the container. When a precipitate is formed in a reaction, the reaction mixture goes cloudy.

The precipitate formed in the above reaction is calcium sulphate. The two soluble salts have 'swapped partners', the sulphate from sodium sulphate switching to the calcium, and the chloride from the calcium switching to the sodium (Figure 19.3). Calcium sulphate appears as a precipitate because it is an insoluble salt. On the other hand,

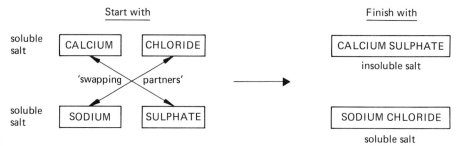

Figure 19.3

sodium chloride is a soluble salt and so stays in solution.

Precipitation reactions are useful in chemical analysis. Mixing magnesium sulphate solution and barium chloride solution (in the first experiment in Chapter 18) gives a white precipitate. 'Swapping partners' produces magnesium chloride, which is soluble and stays in solution, and barium sulphate, which is insoluble and forms a precipitate. The word equation for this reaction is:

magnesium sulphate + barium chloride ⟶ magnesium chloride + barium sulphate

Any soluble sulphate, not just magnesium sulphate, gives a white precipitate with barium chloride solution. So barium chloride can be used to test for a sulphate.

Also, silver nitrate solution can be used to test for a chloride (Table 18.1, page 65). This is because 'swapping partners' here produces silver chloride, and this salt is insoluble in water. So mixing silver nitrate solution and a solution of a soluble chloride always gives a white precipitate of silver chloride. For example, the word equation for the reaction of sodium chloride and silver nitrate solution is:

sodium chloride + silver nitrate ⟶ sodium nitrate + silver chloride

Salts in natural waters

Complete evaporation of samples of sea water and tap water gives solid residues. The amount of the residue is much greater in the case of sea water because it contains a higher concentration of soluble salts than tap water.

Precipitation reactions can be used to find out whether any of the dissolved salts in sea water and tap water are sulphates or chlorides. This analysis works even though the concentrations of salts in natural waters are not as high as in many laboratory solutions.

Experiment 4 Analysing sea water and tap water

Use the instructions given in Table 18.1 (page 65) to test separate samples of sea water and tap water (and perhaps also river or stream water) for sulphates and chlorides.

- Do you know which salt forms about three-quarters of all the dissolved salts in sea water?
- Would you expect this salt to be present in tap water?

Questions

1. Explain why some mixtures of two soluble salts in solution produce a precipitate while some other mixtures do not.

2. Lead sulphate is insoluble in water whereas lead nitrate is soluble. Suppose two mixtures are made: (*a*) lead carbonate and dilute sulphuric acid, and (*b*) lead carbonate and dilute nitric acid.

Describe what you would expect to happen in each case. In which mixture would a solution of a lead salt be formed?

3. Using the patterns in solubility given in Table 19.1 (page 68), predict whether or not a precipitate of an insoluble salt is likely to form when the following pairs of solutions are mixed:

(*a*) calcium nitrate and potassium sulphate solutions.
(*b*) lead nitrate and sodium sulphate solutions
(*c*) magnesium chloride and zinc sulphate solutions
(*d*) silver nitrate and dilute hydrochloric acid
(*e*) barium chloride and copper nitrate solutions

If you think that a precipitate will form, then name it, and write the word equation for the reaction.

4. Describe *two* precipitation reactions which are used in chemical analysis. Write a word equation for one example of each reaction.

20 BREAKING DOWN SALTS

Many salts break down on heating to give several simpler compounds and sometimes even elements. In this chapter, you will find that the ease with which this breakdown takes place is linked with the position of the metal in the reactivity series.

Heating carbonates

Carbonates are salts of an acid called carbonic acid (page 52). They react with dilute acids to give carbon dioxide and another salt (page 67), and they also release carbon dioxide when they are heated. But some carbonates release carbon dioxide only at very high temperatures whereas others release it at temperatures which can be easily reached in a school laboratory.

Experiment 1 Carbonates and the reactivity series

Use copper carbonate, anhydrous sodium carbonate and zinc carbonate in this experiment. Heat a small amount of each carbonate using the same Bunsen flame and the apparatus shown in Figure 2.4a (page 8).
Find out which two of the carbonates gives off carbon dioxide and which of these gives it off more easily.

- Write down the three carbonates in order of the ease with which they are broken down by heat, starting with the one that is not broken down at all.

- What is the link between this order and the order of the three metals (copper, sodium and zinc) in the reactivity series?

This experiment shows that carbonates of reactive metals are much more difficult to break down than carbonates of less reactive metals. Reactive metals must be bound much more strongly to the other elements in a carbonate than less reactive metals are.

Heating nitrates

You can now find out whether there is the same kind of link between the reactivity series and the ease with which members of another class of salts are broken down by heat. Two gases are given off by some nitrates: these are nitrogen dioxide (which is a brown acidic gas) and oxygen (which relights a glowing splint). A few nitrates only give off oxygen when they are heated.

Experiment 2 Nitrates and the reactivity series

Use copper nitrate, sodium nitrate and zinc nitrate in this experiment. Put a few crystals of each nitrate in a small test-tube and heat each tube strongly in turn with the same Bunsen flame. (CARE: carry out this experiment in a fume cupboard and do not inhale the gases given off.)
Test the gases with:

(i) a glowing splint.
(ii) pieces of damp indicator paper.

- Which crystals give off water of crystallization on heating?

- Which two nitrates give off two gases when heated and which one gives off only one gas?

- Write down the three nitrates in order of the ease with which they are broken down by heat.

- Is this order linked to the order of the metals in the reactivity series?

Heating oxides

It is generally the case that, for all compounds of metals and not just salts, it is easier to break them down by heat if they are made from a metal which has a low position in the reactivity series. Oxides are a good example of this. Most oxides cannot easily be decomposed by heat but the oxides of mercury and silver can be easily broken down to the metal and oxygen (see Figure 2.6, page 9).

A SUMMARY

Table 20.1 gives the results of a large number of experiments on the heating of carbonates, nitrates and oxides. Notice that the compounds of very unreactive metals like silver are decomposed by heat straight to the metal.

Table 20.1

The reactivity series for metals	Action of heat on:		
	Oxide	Carbonate	Nitrate
Potassium	Do not decompose on heating	Do not decompose on heating	Decompose to metal nitrite and oxygen
Sodium Calcium Magnesium Aluminium Zinc Iron Lead Copper		Decompose to metal oxide and carbon dioxide	Decompose to metal oxide, oxygen and nitrogen dioxide
Mercury Silver	Decompose to metal and oxygen	Decompose to metal, oxygen and carbon dioxide	Decompose to metal, oxygen and nitrogen dioxide

Note: These results are those you would obtain by heating the solids to a temperature no higher than 800 °C. Many chemical reactions that do not take place at this temperature may well take place at higher temperatures.

Questions

1. The ease with which a metal carbonate is broken down by heat to an oxide and carbon dioxide depends on the position of the metal in the reactivity series.
   (a) Arrange these carbonates in the probable order of the ease of decomposition, putting the one that is most difficult to break down first: copper carbonate, lead carbonate, magnesium carbonate, sodium carbonate, zinc carbonate.
   (b) Suggest a way in which you could test your order in (a) by experiment. Explain carefully what you would look for.

2. When green iron sulphate crystals are gently heated, a white residue is formed and a colourless liquid appears on the cooler parts of the test-tube. This liquid turns white copper sulphate powder blue.
   If heating is continued, this time more strongly, the white solid turns red and a gas is given off that has an acidic reaction with damp indicator paper. The red solid reacts with dilute sulphuric acid to form a solution of a salt, without giving off a gas.
   (a) Explain why you know from the above description that green iron sulphate is a hydrated salt.
   (b) What is the chemical name for the white residue?
   (c) What happens to the white residue when it is heated strongly?
   (d) To what class of compounds does the red solid belong?
   (e) What could the acidic gas be?

3. Fill in the gaps in the following word equations:
   (a) lead nitrate $\xrightarrow{heat}$ .... + ... + ...
   (b) zinc carbonate $\xrightarrow{heat}$ .... + ...
   (c) mercury oxide $\xrightarrow{heat}$ .... + ...

Living Organisms

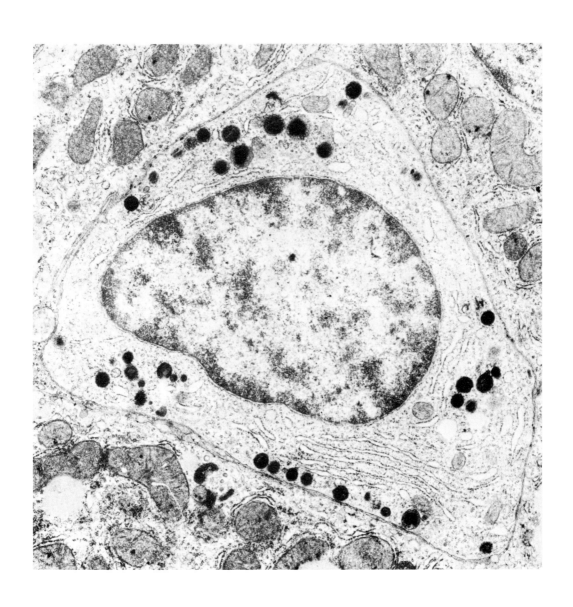

21 CHARACTERISTICS OF LIVING ORGANISMS

Biology is the study of life (Greek *bios* = life, *logos* = knowledge) or, in other words, the study of living organisms. Most animals move about, feed, have young, respond to changes in their surroundings and are thus *seen* to be alive. The living processes in plants are less obvious. When dealing with micro-organisms like bacteria and viruses, the distinction between living and non-living can often be made only by a trained biologist.

The main differences between living organisms and non-living objects can be summarized as follows.

RESPIRATION

This is the production of energy as a result of chemical reactions within the organism. The commonest of these is the chemical decomposition of food usually by combining it with oxygen. This is not a particularly obvious occurrence but it is fairly easy to show that living creatures take in air, remove some of the oxygen from it and increase its content of carbon dioxide. Experiments on these lines are described on pages 80–3. Put more simply, we can say that living organisms take in oxygen and give out carbon dioxide. Sometimes this takes place with obvious breathing movements (see page 307). Respiration also results in a rise in temperature.

EXCRETION

The process of living involves a great many chemical reactions, including respiration. Many of these reactions produce substances that are poisonous when moderately concentrated. The removal of these poisons from the body of the organism is called excretion. Excretion in man is explained in Chapter 74.

FEEDING

This is an essential preliminary to respiration, since energy comes first from food. The feeding of a plant is much less obvious than that of an animal, which moves actively in search of food. Feeding in plants involves a process called *photosynthesis* which is dealt with in Chapter 24. Feeding in animals requires a digestive system (see Chapter 71).

GROWTH

Strictly speaking, growth is simply an increase in size, but usually the organism also becomes more complicated and more efficient as it grows. Experiments on the growth of roots and shoots are given on pages 121–2.

MOVEMENT

An animal can move its whole body, whereas the movements of plants are usually restricted to certain parts, such as the opening and closing of petals or changes in the direction of growth. Chapter 77 gives an account of how the skeleton and muscles are able to produce movement in animals.

REPRODUCTION

No organism can live for ever. But although individual organisms must die sooner or later, their life is handed on to new individuals by reproduction. This results in the continued existence of the species. For reproduction in plants see Chapter 29. Reproduction in man is described in Chapter 76.

SENSITIVITY

This is the ability to respond to a stimulus. Obvious signs of sensitivity are the movements made by animals as a result of noises, on being touched or on seeing an enemy. If we try the same stimuli on fully grown plants they do not seem to make any response. The growing parts of seedlings can respond in ways that can be seen. Roots, for example, will change their direction of growth to grow towards the pull of gravity. Growing shoots respond to gravity by growing away from its directional pull, and respond to light by growing towards it.

Human sense organs are described in Chapter 78 and reflex responses on page 334.

22 CELLS

CELL STRUCTURE

If almost any part of a living organism is cut into thin slices and examined under a microscope, it appears to be made up of distinct units called *cells*. In plants, these look like miniature boxes (Figure 22.1), closely packed together. In animals, the cells are less distinct but can still be recognized easily in certain tissues (Figure 22.2). Thousands of these cells, arranged in an orderly manner, make up the tissues of living organisms; the skin, bone and muscle of an animal or the leaf and stem tissue of a plant. Cells may be of very different shapes and sizes. Muscle cells and nerve cells, for example, are very long (see Figure 79.3, page 333). However, all cells consist of a *cell membrane* enclosing a living substance called *cytoplasm* in which is embedded a *nucleus*.

Figure 22.1

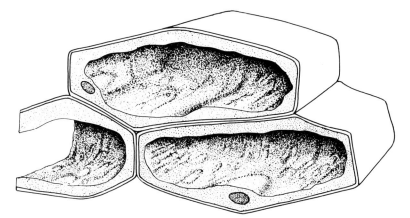

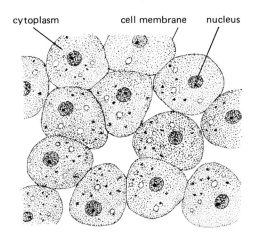

Figure 22.2

Cell membrane

This is a flexible layer, thought to be only a few molecules thick, surrounding the cell. Although it is so thin, it keeps the cell contents from escaping and mixing with the surroundings. It also exercises control over the substances which are allowed to enter and leave the cell.

Cytoplasm

Seen under an ordinary microscope, this has the appearance of a rather granular, jelly-like fluid. The electron microscope, however, reveals complicated patterns of membranes running through it. It contains enzymes (page 79), which control the chemical reactions inside it, and a variety of miniature structures which play a part in building up new cytoplasm, producing substances to be used outside the cell or processing food material in order to produce energy. The cytoplasm may also contain food reserves such as starch grains or oil droplets.

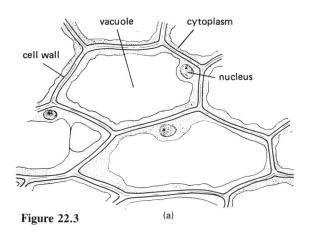
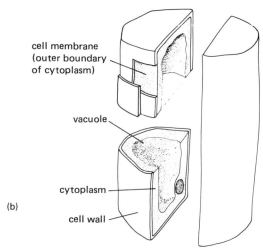

Figure 22.3

Nucleus

Each cell has one nucleus and this is usually an almost spherical structure contained in, but distinct from, the cytoplasm. The job of the nucleus is to control and direct the chemical reactions which take place in the cytoplasm. It controls, for example, what kinds of protein the cell will make and when they are produced.

The nucleus also begins cell division, as shown in Figure 22.5 (page 77). A cell without a nucleus cannot reproduce. Once a cell has ceased dividing, chemicals from the nucleus will determine what kind of cell it will become, for example, a blood cell or a liver cell.

Plant cells

Plant cells (Figure 22.3) differ from animal cells in several ways. They all have a cell wall outside the cell membrane. This is a non-living layer of transparent, fairly rigid *cellulose*, which allows liquids and dissolved substances to pass through in either direction. Under the microscope, it makes plant cells quite distinct and easy to see (Figure 22.4). Most plant cells also have a large, fluid-filled space called a *vacuole*. This vacuole contains *cell sap*, a watery solution of sugars, salts and sometimes pigments. This large central vacuole restricts the cytoplasm to a thin lining just inside the cell wall. It is the outward pressure of the vacuole on the cytoplasm and cell wall which makes plant cells and their tissues firm (see page 102).

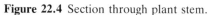

Figure 22.4 Section through plant stem.

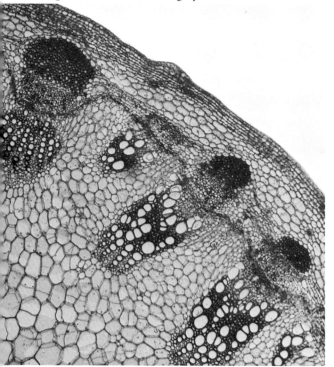

CELL DIVISION

When plants and animals grow, their cells divide in the growing regions. Typical growing regions are the ends of bones, layers of cells in the skin, root tips and buds. Each cell divides into two. Both daughter cells may divide again but usually one of the cells grows, changes its shape, becomes specialized to do a particular job and loses the ability to divide and continue the growth of the tissue. Growth is therefore the result of cell division and cell enlargement.

(a) A plant cell about to divide has a large nucleus and no vacuole

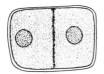
(b) The nucleus divides first

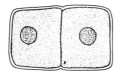
(c) The middle lamella develops and separates the two cells

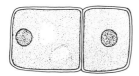
(d) The cytoplasm forms layers of cellulose on each side of the middle lamella

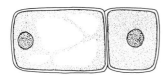
(e) One of the cells develops a vacuole and enlarges. The other cell retains the ability to divide again

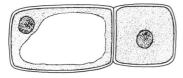
(f) The cell with the vacuole may change further and take on a special function

Figure 22.5

Figure 22.5 shows the process of cell division in a plant cell. The events in an animal cell are similar but since there is no cell wall, the cytoplasm simply constricts in the middle to separate the two nuclei, and no vacuolation takes place.

RELATIONSHIP OF CELLS TO THE ORGANISM

There are some microscopic organisms that consist of a single cell and they can carry out all the processes necessary for their existence. The individual cells of larger plants and animals, however, cannot survive on their own. A muscle cell could not obtain its own food or oxygen. Other specialized cells in a muscle would have to provide the food and oxygen for the muscle cell to live. Unless individual cells are grouped together in large numbers and they work together, they cannot exist for long.

Tissue such as bone, nerve or muscle in animals, and epidermis, phloem or pith in plants, is made up of many hundreds of cells of one or a few types. Each type has a similar structure and function so that the tissue also has a particular function, for example, nerves conduct impulses, phloem carries food in plants. Figure 22.6 shows how cells may form tissues.

Organs consist of several tissues grouped together making a functional unit. For example, a muscle is an organ containing long muscle cells held together with connective tissue and penetrated by blood vessels and nerves. The arrival of a nerve impulse causes the muscle fibres to contract, using the food and oxygen brought by the blood vessels to provide the energy needed. In a plant, the roots, stems and leaves are the organs.

System usually refers to a series of organs whose functions are co-ordinated to produce effective action in the organism. For example, the heart and blood vessels make up the circulatory system; the brain, spinal cord and nerves make up the nervous system.

An organism results from the efficient co-ordination of the organs and systems to produce an individual capable of separate existence and able to reproduce its own kind.

Specialization in cells

Cells in certain tissues often become specialized to carry out particular jobs. This often means that they have a characteristic shape and specialized chemistry in their cytoplasm. Nerve cells are very long and chemical reactions in their cytoplasm enable them to conduct electrical impulses very efficiently. Muscle cells, too, are elongated and their cytoplasm is able to contract. Question 5 overleaf refers you to pages where examples of specialized cells are drawn or described.

Figure 22.6

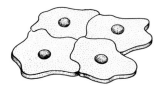

(a) Cells forming an epithelium, a thin layer of tissue, e.g. that lining the mouth cavity

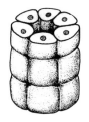

(b) Cells forming a fine tube, e.g. a kidney tubule (see Figure 74.3 on page 311)

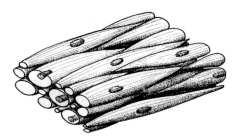

(c) Muscle cells forming a sheet of muscle tissue. Blood vessels, nerve fibres and connective tissue will also be present

(d) Cells forming part of a gland. The cells make chemicals which are released into the central space and are carried away by a tubule like the one shown in (b)

Questions

1. In what respects is a plant cell (*a*) similar to, (*b*) different from an animal cell?

2. In what way does the red blood cell described on page 301 differ from most other animal cells?

3. In Figure 25.5 (page 93), what features of plant cells have been omitted from the drawing?

4. Classify the following into cells, tissues and organs: lung (page 306), skin (page 313), root hair (page 103), mesophyll (page 93), multipolar neurone (page 332).

5. Study the illustrations on the following pages and name some of the cells which appear to be specialized for a particular function. Say what the function is and how the cell structure appears to be suited to that function. Pages 87, 93, 103, 301, 315, 333.

23 ACTIVITIES IN CELLS

The activities which make an organism alive take place inside its cells. In a muscle, for example, it is the contraction of muscle cells which results in movement. Electrical impulses are carried by the individual cells of a nerve. Each cell in the liver is engaged in chemical reactions such as turning glucose into glycogen for storage. All these processes need *energy* and they take place rapidly because the cells contain *enzymes*.

ENZYMES

Enzymes are chemical compounds made in the cells. There are many different kinds of enzyme but they are all *proteins* (page 292). Some are built into the cell membrane systems and others are in solution in the cytoplasm. All enzymes act as catalysts (page 288) and speed up chemical reactions in the cell without being used up themselves. A chemical reaction which would take hours or days to happen in a test-tube, takes only seconds or minutes when the right enzyme is present. Reactions in which complex substances are built up from simple substances probably would not take place at all without an enzyme. In a leaf cell, for example, glucose molecules are built up into the much larger starch molecule only by means of an enzyme. Figure 23.1 shows the way in which such a reaction is thought to work. The substance on which the enzyme acts, in this case glucose, is called the *substrate*. The enzyme combines temporarily with the substrate and makes it more reactive. The reactive substrate–enzyme complex readily combines with another substrate molecule and the enzyme is set free to take part again in the same reaction.

Classes of enzyme

The reaction described above is a synthesis, i.e. one in which substances are built up. An enzyme which acts as a catalyst for such a reaction might be

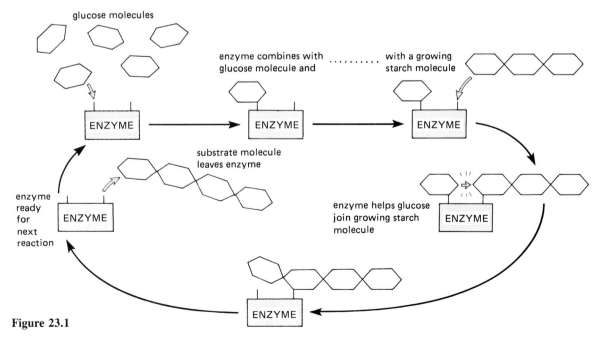

Figure 23.1

called a *synthetase*. There are also enzymes which speed up the breakdown of compounds to simpler substances. For example, in most cells, sugar molecules are broken down in the presence of oxygen molecules to carbon dioxide and water, so releasing energy. The enzymes which catalyse such an oxidation are called *oxidases*. The names of most enzymes end in *-ase* and describe the kind of reaction that they catalyse.

Most enzymes catalyse reactions inside cells. These are called *intra-cellular* enzymes. A few enzymes, however, are made in the cells and released or *secreted* outside the cell to do their work. These are called *extra-cellular* enzymes. The digestive enzymes described on pages 295–9 are examples of extra-cellular enzymes. They are made in the cells of digestive glands but secreted into the alimentary canal to digest the food when it arrives.

Characteristics of enzymes

(a) *Temperature.* A rise in temperature speeds up most reactions, including those catalysed by enzymes. Because enzymes are proteins, however, their chemical structure is damaged (*denatured*) by temperatures above 50 °C. This is one reason why most organisms are killed by temperatures of this level. Their enzymes are destroyed and their chemical reactions cease or go on too slowly to maintain life.

(b) *Acidity and alkalinity.* Each enzyme works best at a particular degree of acidity or alkalinity (pH, page 53). Inside most cells, the pH is about 7, i.e. neutral. In the stomach, however, the conditions are acid (pH = 2). This acid atmosphere is favourable for the stomach enzyme *pepsin*, but stops the action of the enzyme in saliva, *amylase*.

(c) *Specificity.* This means that a particular enzyme will act only on one specific substrate. For example, an enzyme which acts on starch, will not act on proteins. Most enzymes are even more specific than this. Pepsin acts on only certain parts of a protein molecule and breaks it down to peptides. A different enzyme is needed to break the peptides down into amino acids (see page 292).

ENERGY FROM RESPIRATION

Most of the processes taking place in cells need energy. Contraction of muscle cells, electrical activity in nerve cells and chemical synthesis in liver cells are all examples of energy-requiring processes. This energy is supplied by the food taken in by the cells and made available by the process of respiration.

Respiration is the process by which food is broken down usually in the presence of oxygen to release energy ('aerobic respiration').

The food always contains carbon and hydrogen (see page 292), and so the waste products of this oxidation are carbon dioxide and water. The foods most commonly oxidized in this way are carbohydrates and fats. The reaction can be summarized by the equation below:

glucose + oxygen ⟶ carbon dioxide + water + energy

$C_6H_{12}O_6 + 6O_2 \longrightarrow 6CO_2 + 6H_2O + 2830 \text{ kJ}$

The 2830 kilojoules (page 159) is the amount of energy that can be obtained by oxidizing one mole (180 g, see page 234) of glucose completely to carbon dioxide and water. In the cell, the energy is not released all at once. The reaction takes place in a number of small steps, each one being catalysed by its specific enzyme. Although the energy is used to drive the kinds of reactions listed above, some of it always appears as heat. In 'warm-blooded' animals, some of this heat is retained to keep the body temperature at a level above that of the surroundings. In 'cold-blooded' animals and in plants, the heat is lost to the atmosphere almost as fast as it is produced.

EXPERIMENTS: RESPIRATION

Figure 23.2 shows the equation for respiration in a way that indicates how the process may be investigated experimentally. If an organism can be shown to be taking in oxygen, giving out carbon dioxide, using up food or producing heat, it can be assumed that respiration is taking place. Production of water vapour is not good evidence of respiration because evaporation of water may take place from non-living material, e.g. clothes drying on a line.

Experiment 1 Decrease in dry weight (using up food)

If living material is changing food into carbon dioxide, which escapes into the air, its weight will decrease. However, it is the dry weight which must be measured because any material, living or non-living, may lose weight by the evaporation of water.

Soak one hundred seeds in water for 12 hours. Then kill half of them by boiling. These dead seeds are used as *controls*. Place the fifty living seeds in one dish with moist cotton wool and the fifty dead

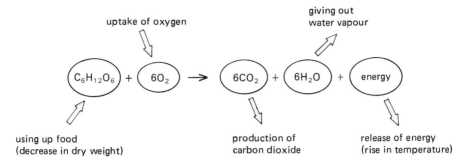

Figure 23.2

seeds in identical conditions. Every day, for 5 days, take ten seeds or seedlings from each dish and heat in an oven at 120 °C for 12 hours to evaporate all the water. Then weigh the two samples of seeds. In this way only the solid matter in the seeds is weighed. If the seeds are respiring, the solids in their food reserves should be decreasing as some of the food is oxidized to carbon dioxide.

Experiment 2 Uptake of oxygen

The apparatus is arranged as shown in Figure 23.3. After 5 minutes the tubes will be at the temperature of the water in the beaker, and the screw clips are closed. If the seeds are respiring they will give out carbon dioxide as fast as they take in oxygen and there will be no change in the volume of gas in the tube. However, soda lime will absorb all the carbon dioxide produced so that any volume change is due to the uptake of oxygen. If oxygen is absorbed by the living seeds, the level of liquid in the capillary will rise within 20 minutes or so. Any change in the temperature of the tubes will cause the air in them to expand or contract and alter the level of liquid in the capillary tube. This could be confused with the movements due to oxygen uptake. The water in the beaker, however, reduces temperature changes. Since any temperature change affects both tubes to the same extent, the change in volume due to oxygen uptake alone can

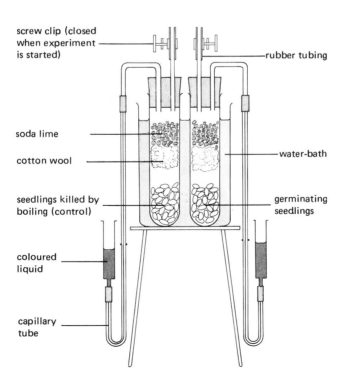

Figure 23.3

be seen by *comparing* the levels of liquid in the experiment and the control. (See page 87 for an explanation of 'control'.) The control thus allows for changes due to temperature variation and also serves to show that oxygen uptake results from a living process in germinating seeds and is not merely due to physical absorption by the seeds.

Experiment 3 Production of carbon dioxide (small organisms)

Put some blowfly maggots or germinating wheat grains in a large test-tube and seal the mouth of the tube with aluminium foil. Set up a control with a tube containing either seeds killed by boiling or an equal volume of glass beads. After 15–20 minutes extract a sample of air from the tube with living organisms, as shown in Figure 23.4, and bubble it through a small quantity of lime water in a test-tube. Repeat the test for the control tube.

Result. Air from the control tube will not affect the lime water, but air from the tube with living organisms will turn it milky. This proves the presence of carbon dioxide.

Interpretation. The carbon dioxide must have been produced by the living organisms and therefore they are respiring. The control shows that the carbon dioxide does not come from the atmospheric air or from anything else in the test-tube.

Experiment 4 Production of carbon dioxide (larger organisms)

Place the animal in the vessel C (Figure 23.5). By means of a filter pump at E draw a stream of air

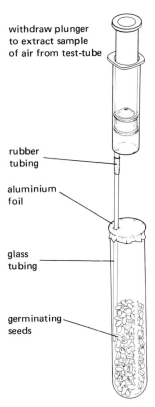

Figure 23.4

slowly through the apparatus. In A, soda lime absorbs the carbon dioxide from the incoming air. The lime water in B should stay clear and so prove that carbon dioxide is absent from the air going into vessel C. If carbon dioxide is given out by the organism, the lime water in D will go milky after a time. A control experiment would be to run the

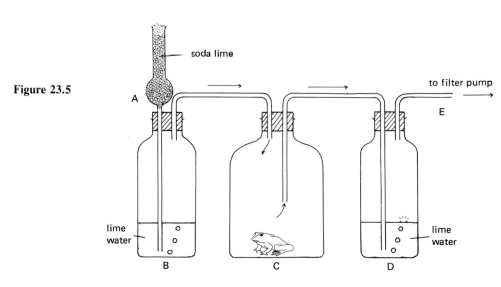

Figure 23.5

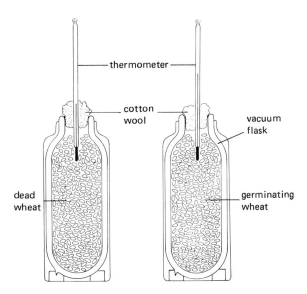

Figure 23.6

filter pump for the same length of time but with no animal present in C. The lime water in D should stay clear.

Experiment 5 Release of energy in germinating seeds

Heat production is a good indication of energy release. Measure out sufficient wheat grains to fill two small vacuum flasks. Soak the grains in water for 24 hours and then kill half of them by boiling for 10 minutes. Then soak both lots of wheat in a solution of sodium chlorate (I) for 15 minutes to kill any moulds growing on the seeds. After rinsing in water, place the living grains in one flask and the dead grains in the other. Insert thermometers and plug the mouths of the flasks with cotton wool (Figure 23.6).

Result. After a few days, the temperature in the flask with living wheat will be considerably higher than in the control.

Interpretation. During the germination of wheat, heat energy is released. This is what we would expect if respiration is taking place in the seeds.

Questions

1. There are cells in the salivary glands which make an extra-cellular enzyme, amylase. Would you expect these cells to make intra-cellular enzymes as well? Explain your answer.

2. It is thought that saliva contains an enzyme which digests starch. An experiment is designed by mixing starch and saliva. The starch disappears in 5 minutes. A control experiment is set up using saliva which has been boiled. The failure of the boiled saliva to digest starch is taken as evidence that the first reaction was due to an enzyme. Explain the reasoning behind this conclusion.

3. What name would you give to enzymes which (*a*) added phosphate to another molecule, (*b*) removed hydrogen from a molecule.

4. What would you accept as evidence that respiration was taking place in a tissue?

5. In what parts of a living organism does respiration take place?

6. Respiration needs a supply of glucose and oxygen. How does a mammal obtain these substances?

7. What is the difference between 'respiration' and 'breathing'?

8. In Experiment 2 (page 81), the seeds take in 5 cm^3 oxygen and give out 7 cm^3 carbon dioxide. What change in volume will take place (*a*) if no soda lime is present, (*b*) if soda lime is present?

9. The seeds in Experiment 5 will release the same amount of heat whether they are in a beaker or a vacuum flask. Why then is it necessary to use a vacuum flask for this experiment?

10. If the seeds, after completing Experiment 5, were found to have gone mouldy, why would this throw doubt on the results?

Reagents

Benedict's solution (1 litre)
Dissolve 173 g sodium citrate crystals and 100 g sodium carbonate crystals in 800 cm³ warm distilled water. Dissolve separately 17·3 g copper sulphate crystals in 200 cm³ cold distilled water. Add the copper sulphate solution to the first solution with constant stirring.

Note. (1) Benedict's solution is preferable to Fehling's solution as it is less caustic and does not deteriorate on keeping.

(2) The red deposit of copper(I) oxide that coats the inside of test-tubes used for the sugar tests can be removed with dilute hydrochloric acid.

Iodine solution (1 litre)
Dissolve 10 g iodine and 10 g potassium iodide in 1 litre distilled water by grinding the two solids in a mortar while adding successive portions of the water. The solution should be further diluted for class use, e.g. 5 cm³ in 100 cm³ water.

Studying Plants

24 PHOTOSYNTHESIS AND NUTRITION IN PLANTS

All living things (organisms) need food. They need it as a source of raw materials from which to build new cells and tissues in order to grow. They also need it as a source of energy, a kind of 'fuel', with which to drive essential living processes and chemical changes (see pages 80 and 292).

Animals take in food, digest it, and use the digested products to build their tissues or to produce energy. Plants, on the other hand, first *make* the food they need and then use it for energy and growth. The process by which plants make their food is called *photosynthesis* (Greek *photos* = light, *synthesis* = building up).

PHOTOSYNTHESIS

From carbon dioxide in the air, and water from the soil, a plant can build *sugars*. A sugar molecule contains the elements carbon, hydrogen and oxygen (e.g. glucose $C_6H_{12}O_6$). The carbon dioxide molecules provide the carbon and oxygen; the water molecules provide the hydrogen. From these simple compounds, the plant is able to build sugar molecules. For this process it needs *enzymes*, which are present in its cells, and *energy* which it obtains from sunlight. The process takes place mainly in the cells of the leaves and is summarized in Figure 24.1. Water is absorbed from the soil by the roots and carried in the microscopic water vessels of the 'veins' up the stem to the leaf. Carbon dioxide is absorbed from the air through the stomata. In the leaf cells the carbon dioxide and water are combined to make sugar; the energy for this reaction comes from sunlight which has been absorbed by the green pigment, *chlorophyll*. The chlorophyll is present in the chloroplasts of the leaf cells and it is inside the chloroplasts that the reaction takes place. Chlorophyll is the substance which gives leaves and stems their green colour. It has the property of being able to absorb the energy from light and make it available for the chemical reactions which convert water and carbon dioxide into sugar.

Photosynthesis is the build up of sugars from carbon dioxide and water by green plants, using energy from sunlight which is absorbed by chlorophyll.

A chemical equation for photosynthesis is:

$$\text{carbon dioxide} + \text{water} \longrightarrow \text{glucose} + \text{oxygen}$$

$$6CO_2 + 6H_2O \xrightarrow[\text{energy}]{\text{light}} C_6H_{12}O_6 + 6O_2$$

though this really represents only the beginning and end of the process, and cannot show the many intermediate steps involved.

You can see that a 'waste' product of photosynthesis is oxygen. In daylight, therefore, green plants are photosynthesizing—taking in carbon dioxide and giving out oxygen. This gaseous exchange is the opposite of that resulting from respiration (page 80). But it must not be thought that green plants do not respire. The energy they need for all their living processes, apart from photosynthesis, comes from respiration which is going on all the time, using up oxygen and giving out carbon dioxide.

During the daylight hours, plants are photosynthesizing as well as respiring so that all the carbon dioxide produced by respiration is used for photosynthesis and all the oxygen needed for respiration is provided by photosynthesis. Only when the rate of photosynthesis is faster than the rate of respiration will carbon dioxide be taken in and the excess oxygen given out.

Adaptation of the leaf for photosynthesis

The detailed structure of the leaf is described on page 92. Although there are wide variations in leaf shape, we can see that most leaves are like each other in the following ways:

a Their broad, flat shape offers a *large surface area* for absorption of sunlight and carbon dioxide.
b Most *leaves are thin* so the distances across which carbon dioxide has to diffuse to reach the mesophyll cells from the stomata are very short.
c The *large spaces between cells* in the mesophyll provide an easy passage through which carbon dioxide can diffuse.

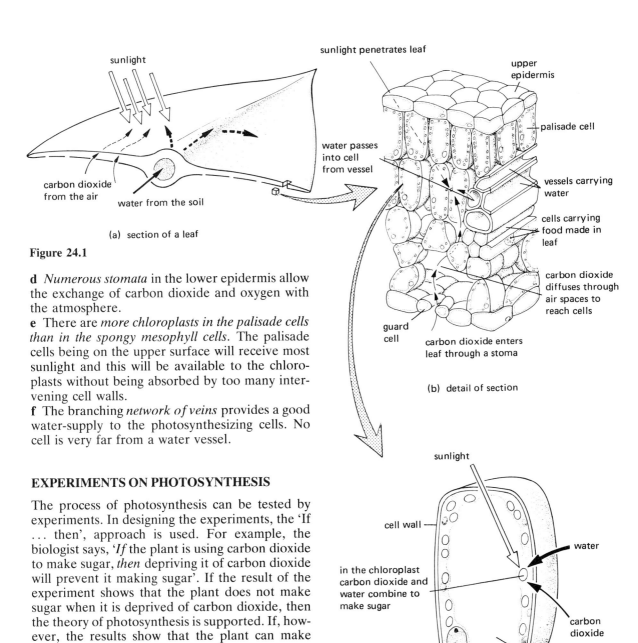

Figure 24.1

d *Numerous stomata* in the lower epidermis allow the exchange of carbon dioxide and oxygen with the atmosphere.

e There are *more chloroplasts in the palisade cells than in the spongy mesophyll cells*. The palisade cells being on the upper surface will receive most sunlight and this will be available to the chloroplasts without being absorbed by too many intervening cell walls.

f The branching *network of veins* provides a good water-supply to the photosynthesizing cells. No cell is very far from a water vessel.

EXPERIMENTS ON PHOTOSYNTHESIS

The process of photosynthesis can be tested by experiments. In designing the experiments, the 'If ... then', approach is used. For example, the biologist says, '*If* the plant is using carbon dioxide to make sugar, *then* depriving it of carbon dioxide will prevent it making sugar'. If the result of the experiment shows that the plant does not make sugar when it is deprived of carbon dioxide, then the theory of photosynthesis is supported. If, however, the results show that the plant can make sugar without carbon dioxide, the theory is wrong and must be altered.

The experiments described below, deprive the leaf of light, chlorophyll and carbon dioxide in turn, to see whether the leaf can still produce sugar. In designing the experiments it is very important to make sure that only one condition is altered. Suppose, for example, the method of keeping light away from a leaf *also* stopped carbon dioxide from reaching it. It is then impossible to decide whether it is the lack of light *or* lack of carbon dioxide which stops the production of sugar. To make sure that the experimental design does not interfere with more than one condition, a *control* is set up. A control is an identical situation to the experiment except that the condition missing from the experiment, e.g. light, carbon dioxide or chlorophyll, is present.

Although sugar is one of the first carbohydrates to be made by photosynthesis, the leaf quickly converts this into starch. It is easier to test a leaf for starch than for sugar, and so the presence of starch in a leaf can be regarded as evidence that photosynthesis has occurred, provided that no starch was present in the leaf to start with.

Destarching a plant. To ensure that leaves are free from starch at the start of an experiment the plant must be *destarched*. Potted plants are destarched by leaving them in a dark cupboard for 2 or 3 days. Experiments on plants in the open are set up the day before; during the night most of the starch will be removed from the leaves. Better still is to wrap the leaves to be tested in aluminium foil for 2 days and then test one of the leaves to see that no starch is present.

Testing a leaf for starch.

a Detach the leaf and dip it into boiling water for half a minute. This kills the cytoplasm by destroying the enzymes in it, and so prevents any further chemical changes. It also makes the cell more permeable to iodine solution.

b Boil the leaf in methylated spirit, using a waterbath (Figure 24.2), until all the chlorophyll is dissolved out. This leaves a white leaf and makes

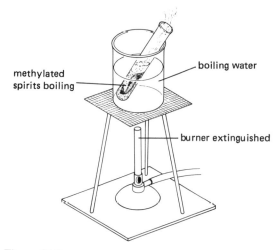

Figure 24.2

colour changes, caused by interaction of starch and iodine, easier to see.

c Alcohol makes the leaf brittle and hard, but it can be softened by dipping it once more into boiling water. Then spread it flat on a white surface such as a glazed tile.

d Place some iodine solution on the leaf. Any parts which turn blue have starch in them. If no starch is present, the leaf is merely stained yellow or brown by the iodine.

Experiment 1 Is chlorophyll necessary for photosynthesis?

It is not possible to remove chlorophyll from a leaf without killing it, and so a leaf, or part of a leaf which has chlorophyll only in patches is used. A *variegated* leaf of this kind is shown in Figure 24.3*a*. The white part of the leaf serves as the experiment, while the green part is the control.

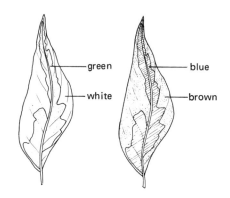

(a) variegated leaf   (b) after testing for starch

Figure 24.3

After destarching, expose the leaf, still on the plant, to daylight for a few hours. Remove the leaf and draw it carefully to show the distribution of chlorophyll. Test for starch as described above.

Result. Only the parts that were previously green turn blue with iodine. The parts that were white stain brown (Figure 24.3*b*).

Interpretation. Since starch is present only in the parts which originally contained chlorophyll it seems reasonable to suppose that chlorophyll is needed for photosynthesis. It must be remembered, however, that there are other possible interpretations which this experiment has not ruled out, e.g. starch could be made in the green parts and sugar in the white parts. Such alternative explanations could be tested by further experiments.

Experiment 2 Is light necessary for photosynthesis?

Cut out a simple shape from a piece of aluminium foil to make a stencil and attach it to a destarched leaf (Figure 24.4). After 4 to 6 hours of daylight, remove the leaf and test it for starch. Find out if only the areas which received light go blue with iodine.

Interpretation. If starch has not formed in the areas

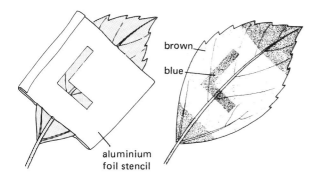

(a) leaf still attached to tree   (b) after testing for starch

Figure 24.4

without light, it may be assumed that light plays an essential part in starch formation and, thus, in photosynthesis. You might think, however, that the aluminium foil prevented carbon dioxide from reaching the leaf so it was shortage of this gas rather than absence of light which prevented photosynthesis. Against this it can be argued that a leaf produces carbon dioxide by its own respiration. Nevertheless, a further control could be designed using a transparent material instead of the aluminium foil stencil.

Experiment 3 Is carbon dioxide needed for photosynthesis?

Water two destarched potted plants and enclose the shoots in polythene bags. One (the experiment) should contain soda lime to absorb carbon dioxide from the air, and the other (the control) sodium hydrogencarbonate solution to produce carbon dioxide (Figure 24.5). Leave the plants in light for several hours and then detach a leaf from each and test for starch.

Result. The leaf deprived of carbon dioxide will not turn blue, while that from the atmosphere containing carbon dioxide will turn blue.

Interpretation. The fact that no starch is made in the leaf deprived of carbon dioxide suggests that this gas must be necessary for photosynthesis. The control rules out the possibility that high humidity or temperature in the plastic bag prevents normal photosynthesis.

Experiment 4 Is oxygen produced during photosynthesis?

Place a short-stemmed funnel over some Canadian pondweed in a beaker of water. Invert a water-filled test-tube over the funnel stem (Figure 24.6). Raise the funnel above the bottom of the beaker to

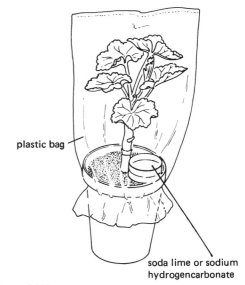

Figure 24.5

allow free circulation of water and place the apparatus in sunlight.

Result. Bubbles of gas soon appear from the cut stems, rise and collect in the test-tube. When sufficient gas has collected, remove the test-tube and insert a glowing splint. The splint bursting into flame shows that the gas is rich in oxygen (page 8). A control experiment should be set up in a similar way but placed in a dark cupboard. Little or no gas should collect.

Interpretation. The relighting of a glowing splint does not prove that the gas collected is *pure* oxygen. It does show that, in the light, this particular

Figure 24.6

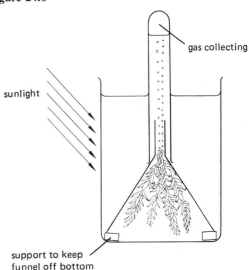

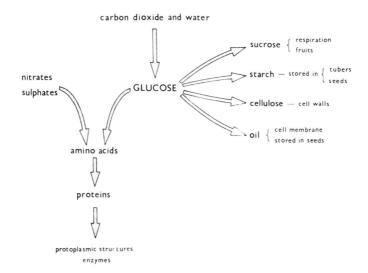

Figure 24.7

plant has given off a gas which is considerably richer in oxygen than ordinary air.

USE OF PHOTOSYNTHETIC PRODUCTS

So far, only the production of sugars such as glucose has been described but, in addition, the plant needs cellulose for its cell walls, fatty substances for the cell membranes, proteins for its cytoplasm and enzymes, pigments for its flower petals, etc. *All* these substances are built up from the sugar molecules produced in photosynthesis.

By joining hundreds of glucose molecules together, the long chain molecules of cellulose are built up and added to the cell walls. The production of amino acids (page 292) also depends on the formation of sugar molecules. These amino acids are then joined together to make proteins. The nitrogen for this synthesis comes from nitrates (page 98) which are absorbed from the soil by the roots. Similarly, other mineral salts such as sulphates and phosphates are taken in from the soil water and built into molecules essential for cell structures and processes. The chlorophyll molecule, for example, needs magnesium; this element is obtained from the soil. Some of the sugar is converted to starch and stored in the roots or stems, as in the case of potatoes, and some is converted into sucrose and stored in the fruits.

All these chemical processes need energy to make them happen. This energy is provided by respiration (page 80) in which some of the sugar is oxidized to carbon dioxide and water. Figure 24.7 summarizes the uses to which the glucose, made during photosynthesis, is put.

Questions

1. What substance does a green plant need to take in to make (*a*) sugar, (*b*) proteins? What materials must be present in the cells to make reactions, (*a*) and (*b*) work?

2. (*a*) Give two examples of processes in a plant cell which need a supply of energy.
   (*b*) For what purpose can sunlight be used by green plants as a source of energy?
   (*c*) What process provides energy for most of the reactions in the plant cell?

3. What is the essential difference between plants and animals in the way they obtain food?

4. In very bright conditions, some plants close their stomata, thus cutting off their supply of carbon dioxide. Discuss whether photosynthesis will come to a halt.

5. Suggest a possible advantage of the elongated shape of palisade cells in photosynthesis.

6. The leaves of some green plants never give a positive reaction when tested with iodine and yet there is other evidence to show that photosynthesis is taking place. Suggest reasons for this.

7. Discuss whether the method suggested for destarching a plant takes for granted the results of Experiment 2 (page 88).

8. In Experiment 2, a control was suggested to check whether the experimental design interfered with the uptake of carbon dioxide by the leaf. Describe how you would conduct this control experiment and say what results you would expect (*a*) if the stencil did prevent carbon dioxide reaching the leaf, and (*b*) if it did not.

9. Why do you think a pondweed rather than a land plant is used for Experiment 4 (page 89)? In what way might this choice limit the usefulness of the results?

25 PLANT STRUCTURE

The part of the plant above the ground is the *shoot*. It consists of stem, leaves, buds and flowers. Below ground is the *root system* (Figure 25.1).

SHOOT

Stem

Most stems are upright, supporting the leaves and flowers. Some, however, grow along the ground, for example, strawberry runners. Others may even grow horizontally under the ground as in couch grass and bracken. The functions of all stems are (*i*) to carry water and salts from the soil to the leaves, (*ii*) to carry food made in the leaves to other parts of the plant. In addition, upright stems (*iii*) support and space out the leaves so that they receive enough sunlight and (*iv*) hold the flowers above the ground which helps pollination by insects or wind (page 113) and seed dispersal (page 114).

Figure 25.2 is a diagram of a stem, showing its internal structure. A single layer of cells on the outside forms the *epidermis*. This helps to keep the shape of the stem and reduces evaporation.

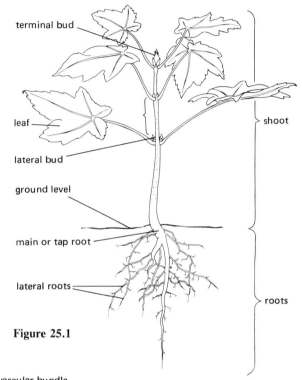

Figure 25.1

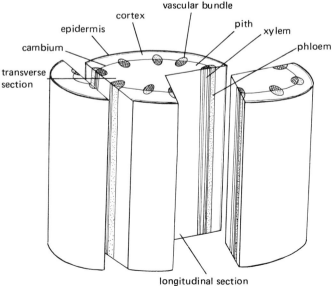

Figure 25.2

Running the length of the stem are a number of *vascular bundles* or 'veins'. These carry food, water and salts up or down the stem. The inner part of each vascular bundle consists of *xylem*. The xylem contains a large number of fine tubes called *vessels*. These vessels carry water and dissolved salts from the roots to the leaves. The outer part of each vascular bundle is the *phloem*. In the phloem are elongated cells which carry food made in the leaves, up or down the stem to other parts of the plant.

The region of tissue between the epidermis and the vascular bundles is called the *cortex*. The tissue in the centre of the stem is called the *pith*. Separating the cortex and pith is a layer of cells called the *cambium*. When the cambium cells divide, they increase the thickness of the stem.

Leaf

A typical leaf has a broad, thin *blade* supported by a network of *veins*. There is often a main vein, called the *midrib*, running down the centre of the

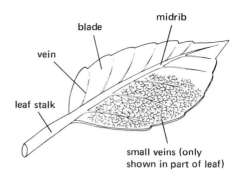

Figure 25.3

leaf. The midrib is continuous with the *leaf stalk*, which attaches the leaf to the stem (Figure 25.3). Running through the leaf stalk, midrib and veins are vascular bundles which connect with those in the stem and carry water and food to and from the leaf.

Figure 25.4 shows the shape and distribution of cells which form the leaf blade. The *epidermis* consists of a single layer of cells which helps the leaf to

Figure 25.4

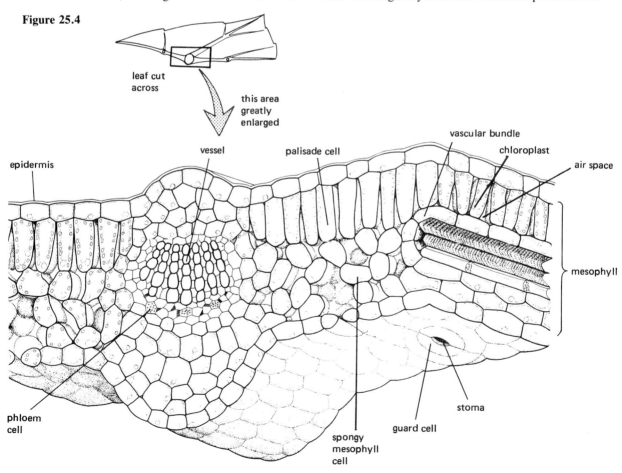

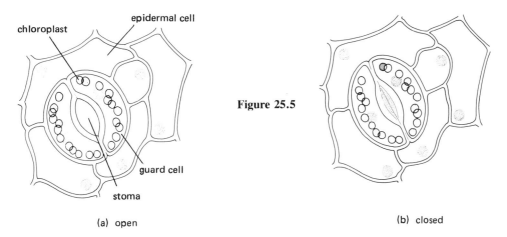

Figure 25.5 (a) open (b) closed

keep its shape, reduces evaporation and prevents entry of harmful organisms. In the lower epidermis there are openings called *stomata* (singular = stoma), which allow the exchange of oxygen, carbon dioxide and water vapour with the air outside the leaf. Each stoma has a pair of guard cells (Figure 25.5) which, by changing their shape, can open or close the stoma. Generally speaking, the stomata are open in the daytime and closed at night.

The tissue between the upper and lower epidermis is called *mesophyll* and consists of two zones, the *palisade mesophyll* and the *spongy mesophyll*. The palisade cells are usually elongated and contain structures called chloroplasts (Figure 25.6). The function of the chloroplasts is to absorb sunlight and make food by a process called *photosynthesis*. The spongy mesophyll cells vary in shape and fit together only loosely, leaving many large air spaces, *intercellular spaces*, between them. These intercellular spaces allow oxygen, carbon dioxide and water vapour to diffuse to and from the mesophyll cells. The spongy mesophyll cells also contain chloroplasts and make food. The xylem

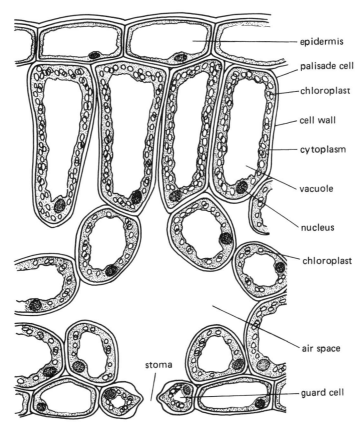

Figure 25.6

vessels in the vascular bundles bring water to the mesophyll cells for photosynthesis and the phloem carries away the food they make.

The function of the leaf is to absorb sunlight and make food by photosynthesis. This process, and the way the leaf structure is suited to it, are described on page 86.

Buds

Buds which occur at the tip of a shoot are called *terminal buds*. Others grow from the side of the stem at a point just above a leaf attachment; these are *lateral buds*. In either case, they consist of tightly packed, overlapping leaves on a short stem. The outer leaves, called *bud scales*, are often thick and tough, protecting the inner leaves from drying out and from attack by insects. When a bud sprouts, the bud scales fall off, the stem elongates, spacing out the leaves, the leaves expand and turn green (Figure 25.7). Terminal buds, when they sprout, continue the stem's growth in length. Lateral buds produce branches. Either type of bud may produce flowers.

Roots

Two types of root system are shown in Figure 25.8. If a single main root is seen, the system is a *tap root* (a). If all the roots are more or less equal in size, they form a *fibrous root* system (b). In either case, *lateral roots* grow out sideways from the roots which are growing vertically downwards.

Roots anchor the plant in the soil and prevent it from falling or being blown over. They also absorb water and salts from the soil and pass them to the rest of the plant. This process is described more fully on page 103. The root system is constantly growing and each root tip is protected by a *root cap* of cells which are continually worn away and

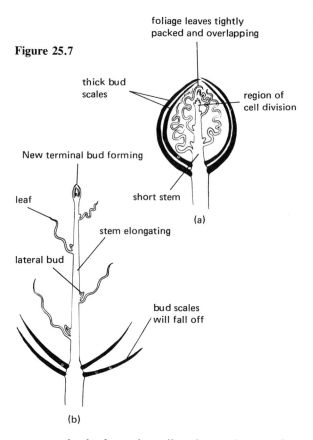

Figure 25.7

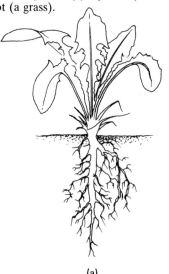

Figure 25.8 Types of root system: (*a*) tap root (dandelion); (*b*) fibrous root (a grass).

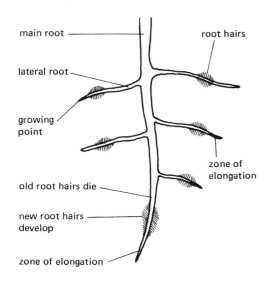

Figure 25.9

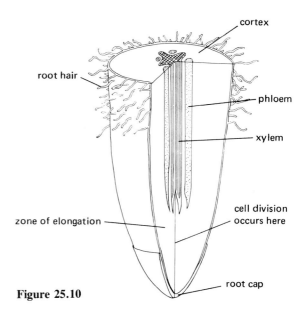

Figure 25.10

replaced as the root pushes through the soil. Just behind the root cap is a region where the cells are dividing rapidly and beyond this region the cells absorb water and expand (see page 102). When elongation has finished, the outer cells of the root develop microscopic *root hairs* (see page 103), which absorb water and salts from the soil (Figure 25.9). The vascular bundle is in the centre of the root (Figure 25.10). The xylem carries water and salts to the stem, while the phloem delivers food, made in the leaves, to the root.

Questions

1. State briefly the function of each of the following: xylem, palisade cell, root hair, intercellular space, root cap, stoma.

2. If you were shown a short cylindrical structure cut from part of a plant, how could you tell whether it was a piece of stem or piece of root?

3. Suppose that the terminal bud in Figure 25.7 (page 94) is cut off. Make a drawing to show how the shoot might look at the end of next year's growth period.

4. Describe the paths that a carbon dioxide molecule in the air and a water molecule in the soil would have to follow in order to meet in a palisade cell of a leaf.

5. Why do you think that root hairs grow only on the parts of the root system that have stopped growing?

6. Discuss whether you would expect to find a vascular bundle in a flower petal.

7. Look at Figure 25.6 (page 93). Why do you think photosynthesis does not take place in the epidermal cells?

26 THE INTERDEPENDENCE OF LIVING ORGANISMS

Food chains

Many animals feed on plants. Such animals are called *herbivores*, e.g. rabbits. Animals called *carnivores* prey on other animals. The predators, e.g. foxes, kill and eat their prey, e.g. rabbits. Scavengers eat dead remains of animals killed by predators. Whatever their diets, all animals ultimately depend on plants for their food. Foxes may eat rabbits but rabbits feed on grass. A hawk eats a lizard, the lizard has just eaten a grasshopper, but the grasshopper has fed on a grass blade. This relationship is called a *food chain*. Another example is:

lettuce ⟶ snail ⟶ thrush ⟶ sparrow-hawk

The organisms at the beginning of a food chain are usually very numerous but small in size. The animals at the end of the chain are often large and few in number. The *food pyramid* in Figure 26.1 shows this relationship. There will be millions of microscopic, single-celled green plants in a pond. These will be eaten by the larger but less numerous water-fleas, which in turn will fall prey to small fish like minnow and stickleback. The hundreds of small fish may be able to support only four or five large carnivores like pike or perch.

In reality, food chains are rarely as simple as described here because animals eat a variety of organisms. A fox, for example, does not feed entirely on rabbits but takes beetles, rats and even blackberries in its diet. To express these relationships more accurately a *food web* can be drawn up (Figure 26.2).

By taking the idea of a food chain one step further, we can see that all living organisms depend on sunlight and photosynthesis (page 86). Green plants make their food by photosynthesis for which sunlight is necessary. Since all animals depend, ultimately on plants for their food, they are therefore indirectly dependent on sunlight. Given below are a few examples of man's dependence on photosynthesis.

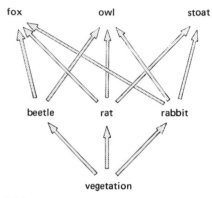

Figure 26.2

bread ⟵ flour ⟵ wheat grains ⟵ wheat grows by photosynthesis

cheese ⟵ milk ⟵ cow ⟵ grass ⟵ grows by photosynthesis

honey ⟵ bees ⟵ nectar ⟵ flowers ⟵ plants grow by photosynthesis

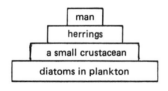
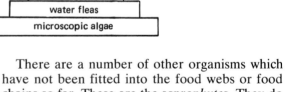

Figure 26.1

There are a number of other organisms which have not been fitted into the food webs or food chains so far. These are the *saprophytes*. They do not photosynthesize nor do they kill and eat living animals or plants. Instead they feed on dead and decaying matter such as dead leaves in the soil or rotting tree trunks. The most numerous examples are the fungi such as mushrooms, toadstools or moulds, and the bacteria—particularly those which live in the soil. They produce extra-cellular enzymes (page 80) which digest the decaying organic matter and then they absorb the soluble products back into their cells. In so doing, they remove the dead remains of plants and animals which would otherwise accumulate on the earth's surface. They also break these remains down into substances which can be used by other organisms. Some bacteria, for example, break down the protein of dead plants and animals and release nitrates. The nitrates are then taken up by plant roots and there built into new amino acids and proteins. This use and re-use of materials in the living world is called *recycling*.

THE CARBON CYCLE

Carbon is an essential element in all compounds of living organisms. The source of carbon for plants is the carbon dioxide in the atmosphere. Animals get their carbon from plants. The carbon cycle, therefore, is mainly concerned with what happens to carbon dioxide (Figure 26.3).

Removal of carbon dioxide from the atmosphere

Green plants remove carbon dioxide from the atmosphere by photosynthesis. The carbon of the carbon dioxide is built firstly into carbohydrates such as sugar or starch and then into the cellulose of cell walls and the proteins, pigments and other organic compounds of a plant. When the plants are eaten by animals, the organic plant matter is digested, absorbed and built into compounds making the animals' tissues. Thus, the carbon atoms from the plant become part of the animal.

Addition of carbon dioxide to the atmosphere

(*a*) *Respiration.* Plants and animals obtain energy by oxidizing carbohydrates in their cells to carbon dioxide and water (page 80). These products are excreted and the carbon dioxide returns once again to the environment.

(*b*) *Decay.* The organic matter of dead animals and plants is used by saprophytes, especially bacteria and fungi, as a source of energy. The micro-organisms decompose the plant and animal material, converting the carbon compounds to carbon dioxide.

(*c*) *Combustion.* In the process of burning carbon-containing fuels such as wood, coal, petroleum and natural gas, the carbon is oxidized to carbon dioxide. The hydrocarbon fuels originate from communities of plants such as prehistoric forests which have only partly decomposed over the millions of years since they were buried.

Thus, an atom of carbon which today is in a molecule of carbon dioxide in the air, may tomorrow be in a molecule of cellulose in the cell

Figure 26.3

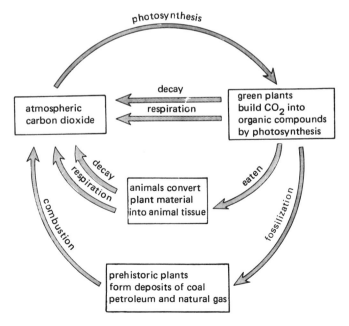

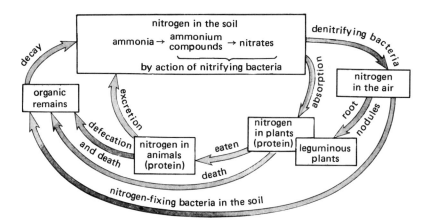

Figure 26.4

wall of a blade of grass. When the grass is eaten by a cow, the carbon atom may become one of many in a protein molecule in the cow's muscle. When the protein molecule is used for respiration, the carbon atom will enter the air once again as carbon dioxide. The same kind of cycling applies to nearly all the elements of the earth. No new matter is created but it is repeatedly rearranged. A great proportion of the atoms of which you are composed will, at one time, have been part of other organisms.

Today, man's activities affect these cycles. For example, the nitrogen present in his excretory products may not be recycled to the land which produces his food. The carbon fuels are being burned in great quantities, depleting their sources and adding more carbon dioxide to the atmosphere.

THE NITROGEN CYCLE

When a plant or animal dies its tissues decompose, largely as a result of the action of bacteria. One of the important products of this decomposition is ammonia, which is washed into the soil where it forms ammonium compounds.

Processes which add nitrates to the soil

Nitrifying bacteria. These are bacteria living in the soil which use the ammonia from decaying organisms as a source of energy. In the process of releasing energy from ammonia, the bacteria produce nitrates. Although plant roots can take up ammonium salts they take up nitrates more readily, so the nitrifying bacteria increase the fertility of the soil by making the nitrates available to the plants.

Nitrogen-fixing bacteria. This is a special group of nitrifying bacteria which can absorb gaseous nitrogen from the air in spaces in the soil and build it into compounds of nitrogen. Gaseous nitrogen is an unreactive element and cannot be used by plants. When it has been made into a compound of nitrogen, however, it can easily be changed to ammonia and nitrates by other nitrifying bacteria. The process of building gaseous nitrogen into a compound is called *nitrogen fixation*. Some of the nitrogen-fixing bacteria live freely in the soil. Others live in the roots of plants of the pea family (leguminous plants), where they cause swellings called *root nodules*. It follows that these leguminous plants are able to thrive in soils where nitrates are scarce, because the nitrogen-fixing bacteria in their nodules make compounds of nitrogen available to them. Leguminous plants are also included in crop rotations (see below) to increase the nitrogen content of the soil.

Processes which remove nitrates from the soil

Uptake by plants. Plant roots absorb nitrates from the soil and use them for making proteins.

Leaching. Nitrates are very soluble and as rain water passes through the soil it dissolves the nitrates and carries them away in the run-off or to deeper layers of the soil. This is called leaching.

Denitrifying bacteria. These are bacteria which derive their energy by breaking down nitrates to gaseous nitrogen which escapes to the atmosphere.

These processes are summarized in Figure 26.4. Although the diagram refers only to nitrogen, a similar cycle could be constructed for sulphur, phosphate, potassium and other minerals.

Manuring and crop rotation

Manuring. In a natural community of plants and animals, the processes which remove nitrates from, and add nitrates to, the soil are in balance. In agriculture, the crop is usually totally removed so that there would be no organic matter for the nitrifying bacteria to act on. In a mixed farm, i.e. one with animals as well as plant crops, the animal manure mixed with straw is ploughed back into the soil and thus replaces the nitrates and other minerals removed by the crop. It also maintains the soil structure and improves its water-holding properties.

When animal manure is not available in large enough quantities, artificial fertilizers are used. These are mineral salts made on an industrial scale (see page 273). Examples are ammonium nitrate (for nitrogen), ammonium sulphate (for nitrogen and sulphur), and compound NPK fertilizer for nitrogen, phosphorus and potassium. These are spread on the soil in carefully calculated amounts to provide the minerals, particularly nitrogen, phosphorus and potassium that the plants need. Figure 26.5 shows how these artificial fertilizers increase the yield of crops from agricultural land, though they do little to maintain the soil structure.

Crop rotation. Different crops make differing demands on the soil; potatoes and tomatoes use much potassium from the soil, for example. By changing the crop grown from year to year, the soil is not depleted of any particular group of minerals. Leguminous crops such as clover and beans may help to restore the nitrogen content of the soil because their root nodules contain nitrogen-fixing bacteria.

The use of artificial fertilizers has made crop rotation, at least for the reasons above, largely unnecessary. However, turning arable land over to grass for a year or two does improve the crumb structure of the soil and hence its drainage and other properties. Rotation also reduces the hazards from infectious diseases that can enter the crop through the soil. For example, successive crops of potatoes in the same field will increase the population of the fungus causing the disease 'potato blight'. A field freed for a few years from potatoes will show a reduced incidence of this disease when they are replanted.

Questions

1. Describe all the possible ways in which the following might be biologically interdependent: grass, earthworm, blackbird, oak tree, soil.

2. Explain how the following foodstuffs are produced, ultimately as a result of photosynthesis: wine, butter, eggs, beans.

3. An electric motor, a car engine and a racehorse can all produce energy. Show how this energy comes, ultimately from sunlight. What forms of energy on the earth are *not* derived from the sun?

4. Figures 26.3 and 26.4 (pages 97 and 98) show how carbon and nitrogen are recycled in nature. Construct a similar cycle for hydrogen, involving water (H_2O), carbohydrate ($C_6H_{12}O_6$), photosynthesis and respiration.

5. Outline the events that might happen to a carbon atom in a molecule of carbon dioxide which enters the stoma in a leaf of a potato plant, becomes part of a starch molecule in a potato tuber which is then eaten by a man. Finally, the carbon molecule is exhaled once again as carbon dioxide.

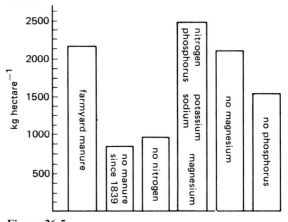

Figure 26.5

6. Figure 26.5 shows the average yearly wheat yield from a field at Rothamstead Experimental Station. From this chart judge which mineral element seems to have the most pronounced effect on the yield of wheat? Explain your answer.

27 DIFFUSION AND OSMOSIS

DIFFUSION

Many substances are made up of minute particles called *molecules* (page 14). A molecule of carbon dioxide, for example, consists of an atom of carbon joined to two atoms of oxygen. In a gas like carbon dioxide, the molecules are spaced far apart and are moving about rapidly in all directions (Figure 27.1), colliding with each other and with the walls of the vessel which contains them.

If some carbon dioxide is released from a cylinder in one corner of a room, the carbon dioxide molecules will move about at random. Sooner or later, however, because of this random movement the molecules will become evenly distributed throughout the room. This random movement of molecules which results in their even distribution is called *diffusion* (see also page 14). Diffusion is a slow process but is speeded up if there is a large difference in the concentration of molecules between two points. For example, in Figure 27.2 the molecules will diffuse from left to right. If there were even more molecules on the left and even fewer on the right, diffusion would be more rapid.

Diffusion of gases in plants and animals

Plants depend almost entirely on diffusion for the exchange of oxygen and carbon dioxide with the air. In darkness, the cells in a leaf are respiring (page 80) and using up oxygen. As a result, the concentration of oxygen in the air spaces in the leaf (Figure 25.6, page 93) falls to a level below that in the air outside the leaf. Oxygen, therefore, diffuses through the stomata (page 93) and into the leaf. The respiring cells are also producing carbon dioxide. The concentration of carbon dioxide, therefore, increases inside the leaf and causes it to diffuse out, through the stomata and into the air. In daylight, when rapid photosynthesis is taking place, carbon dioxide is being used up and oxygen produced inside the leaf. In these circumstances, the oxygen will diffuse out and carbon dioxide will diffuse in.

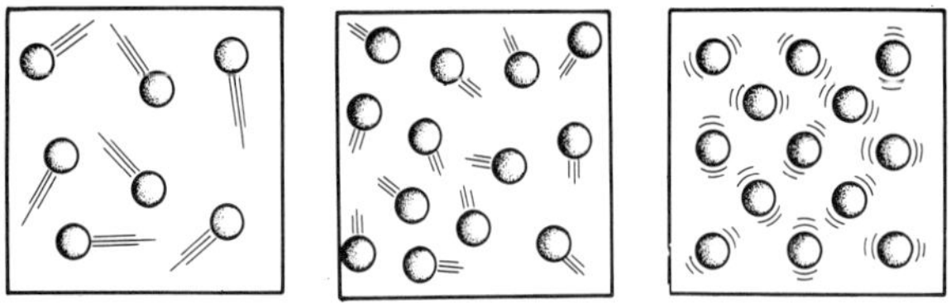

Diffusion is rapid enough to meet the needs of the plants because of the very small distances over which it has to take place, e.g. most leaves are less than 1 mm thick.

Many animals, such as insects and spiders, depend on diffusion for obtaining their oxygen and getting rid of carbon dioxide. For very small animals, the distances involved are small and diffusion is rapid enough. Larger animals have special organs, like lungs, which bring fresh supplies of air into the body. However, even in the lungs, the final stage in the exchange of gases depends on diffusion (see page 308).

Figure 27.1

gas	liquid	solid
molecules far apart moving rapidly	molecules closer together moving freely	molecules in fixed position, able to vibrate only

Figure 27.2

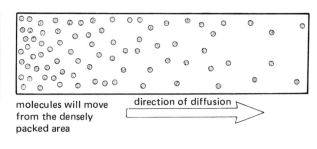

molecules will move from the densely packed area → direction of diffusion

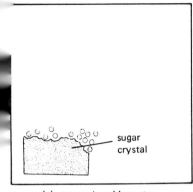
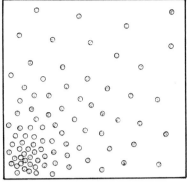
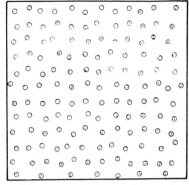

(a) sugar placed in water (b) sugar dissolves; molecules diffuse (c) sugar molecules evenly spaced

Figure 27.3

Diffusion in solutions

When a solid dissolves in a liquid, the molecules of the solid (or *solute*) move randomly through the liquid and so become evenly spread out (Figure 27.3). As in the case of gases, the dissolved molecules move from a region of high concentration to a region of low concentration.

Small animals living in water obtain oxygen and get rid of carbon dioxide by diffusion. This is similar to diffusion in air but the two gases are dissolved in the water.

Plants may take up some salts from the soil by diffusion. If the concentration of nitrates, for example, is higher in the soil water than it is in the root, then the nitrates will pass into the root by diffusion. The process by which plants take up mineral salts is, in fact, not known. It is possible that diffusion plays some part, but there are also thought to be special processes going on in the cytoplasm which take selected salts into the cells, even if the concentration outside is lower than inside. This process, called *active transport*, needs energy from respiration to drive it. Active transport and diffusion, to varying extents, account for the movement of dissolved substances into and out of cells and from one cell to another.

OSMOSIS

Osmosis is the diffusion of water across a membrane. The membrane in Figure 27.4 prevents the solutions mixing freely but does allow individual molecules to diffuse through. Solution A is a strong solution of sugar in water. The sugar molecules are shown grey and the water molecules white. The sugar molecules attract water molecules and thus reduce the number of water molecules which are able to diffuse freely. The dilute solution, B, has more free water molecules than solution A, and water molecules will therefore diffuse through the membrane from B to A. The sugar molecules will diffuse from A to B but more slowly than the water molecules. Since water molecules diffuse from B to A faster than sugar molecules diffuse from A to B, there will be a net flow of water from B to A. Thus the level of liquid in A will rise, while the level in B will fall. This flow of water across the membrane is called *osmosis* and can be demonstrated as shown in Experiment 1.

The type of membrane which permits osmosis is sometimes called *partially permeable* because it appears to allow water molecules to pass through it more easily than larger molecules.

Osmosis is the movement of water across a partially permeable membrane from a dilute to a concentrated solution.

Figure 27.4

Experiment 1 Demonstration of osmosis

Fill a piece of cellophane tubing, knotted at one end, with a strong sugar solution and fit it over the end of a glass capillary tube. Use an elastic band to hold the tubing in place and clamp the capillary tube upright as shown in Figure 27.5, with the cellophane tubing in a beaker of water.

Result. After a few minutes, note how the sugar solution will have risen up the capillary tubing.

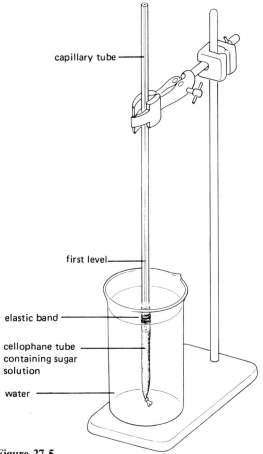

Figure 27.5

Interpretation. Pure water contains the greatest possible concentration of water molecules. The sugar solution contains a much lower concentration of free water molecules. Water molecules, therefore, pass through the cellophane tubing into the sugar solution, increasing its volume and forcing it up the capillary. The cellophane tube acts as a partially permeable membrane. Although sugar molecules diffuse out of the tube, they do so more slowly than the water molecules diffuse in.

Osmosis in plants

Turgor. The cellulose wall of a plant cell is freely permeable to water and dissolved substances. The cytoplasm lining the cell, however, is partially permeable. The cell sap in the vacuole is a solution of salts and sugars. If the cell were surrounded by water, osmosis would take place and water would enter the vacuole. The extra water in the vacuole would cause it to swell and push the cytoplasm against the cell wall (Figure 27.6). This has a similar effect to inflating a bicycle tyre. The outer cover, the tyre, is flexible but cannot be stretched. The inner tube is floppy and elastic but is impermeable to air. Air is pumped in, like water entering the cell by osmosis, until the pressure forces the inner tube outwards against the tyre and produces a rigid structure. When a cell is in this condition, it is said to be *turgid*. When all the cells in a leaf are turgid, the whole structure is firm. If the cells lose water faster than they can take it in, however, the pressure in the vacuoles falls, the cells lose their turgor and the leaf becomes limp and flabby. The plant is then said to be *wilting*. In young plants, it is the turgor of their cells which keeps the stems upright.

Growth. In a growing root, rapid cell division is taking place at the root tip producing many small cells with no vacuoles. Just behind the root tip, the cells develop vacuoles and take in water by osmosis. At this stage, the cell walls are plastic and can be extended lengthways. As the vacuole expands, it makes the cell larger (Figure 22.5, page

Figure 27.6 Turgor in a plant cell. (1) Cell sap contains dissolved salts and sugars; (2) water enters by osmosis, passing through the cell wall and the partially permeable cytoplasm; (3) the cell sap volume increases, creating an outward pressure on the cell wall and making the cell turgid.

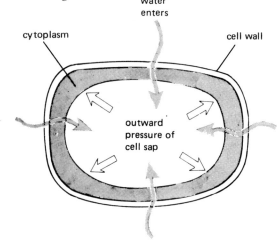

77). Hundreds of cells extending like this at the same time give rise to the rapid elongation of the root just behind the root tip (Figure 27.7)

Uptake of water by roots. In some regions of the root (Figure 27.7), the outermost cells produce hair-like outgrowths called *root hairs*. These root hairs grow between the soil particles and make very close contact with the film of water which surrounds them (Figure 27.8). Because the cell sap of the root hair cell is a stronger solution than the soil water, osmosis occurs and water enters the root hair. This extra water increases the pressure in the vacuole and so some water is forced out of the root hair cell into the cell next to it. In this way, water may be passed from cell to cell into the middle of the root. In fact, most of the water travels through the root in or between the cell walls without entering the vacuoles. From this flow of water, the cells can absorb water by osmosis if their turgor pressure falls for any reason.

Movement of water in leaves. Figure 27.9 shows some of the cells in the mesophyll of a leaf (see also Figure 28.1, page 106). The palisade cell (*a*) is

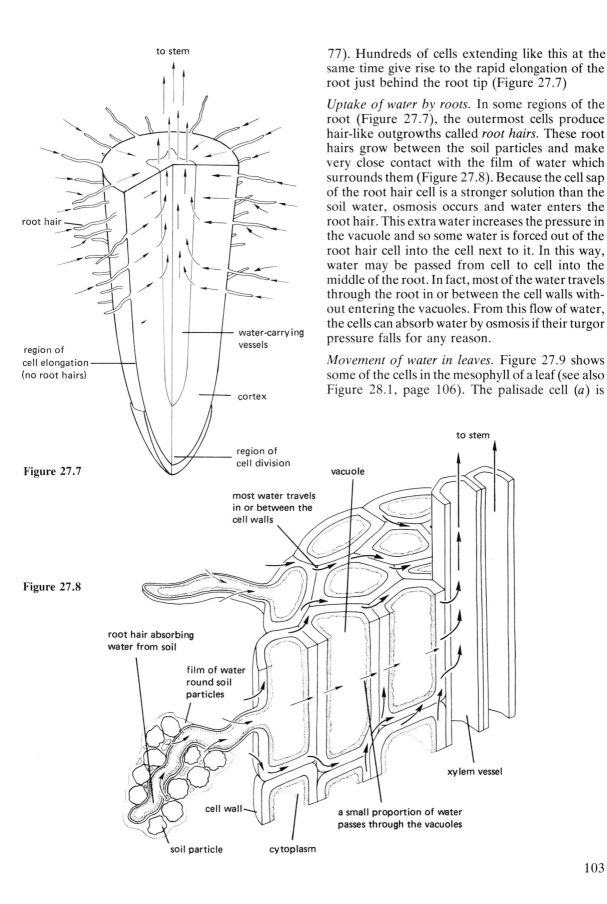

Figure 27.7

Figure 27.8

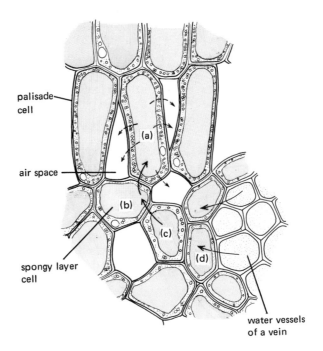

Figure 27.9

losing water by evaporation into the air spaces between the cells. This loss of water will reduce the cell's turgor and, at the same time, make its cell sap more concentrated. Its neighbouring cell (*b*), has greater turgor and a weaker solution in its cell sap, so water will pass by osmosis from cell (*b*) to cell (*a*). The turgor of cell (*b*) is thus reduced and allows water to enter from cell (*c*) which consequently absorbs water from (*d*). Cell (*d*) is next to a water vessel in a vein of the leaf and will absorb water from it, again by osmosis. The proportion of this water which travels through the leaf by osmosis is fairly small. Most of the water passes along or between the cell walls without entering the vacuoles, (Figure 28.2, page 107).

Experiment 2 Demonstrating osmosis in living tissue

Peel a very thin strip of epidermis from a red part of a rhubarb stalk, Figure 27.10a. Place a small piece of this tissue in a drop of water on a microscope slide and cover it with a cover slip, Figure 27.10b. Examine the slide with a microscope to find an area of epidermis where the cells are distinct and the cell sap is red. With a pipette, place two drops of sugar solution on the slide at the left-hand edge of the cover slip and draw the solution under the cover slip with a small piece of blotting paper as shown in Figure 27.11. Watch the cells for 2 or 3 minutes.

Figure 27.10

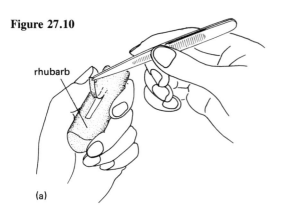

(a)

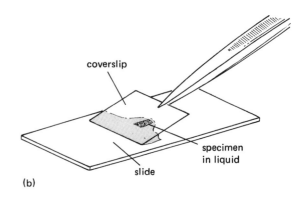

(b)

Figure 27.11

Result. The red vacuoles will shrink and pull the lining of cytoplasm away from the cell walls.

Interpretation. The sugar solution is more concentrated than the cell sap and so water passes by osmosis from the vacuole to the sugar solution causing the vacuole to shrink. It can be seen that the cell sap also becomes darker as it gets more concentrated. The procedure can be reversed by drawing water under the cover slip. The vacuoles should expand to fill the cells again.

Questions

1. Oxygen molecules can diffuse through water but can water molecules diffuse through air? Explain your answer with examples.

2. Study Figures 73.2a and b, page 306. Suggest why it is necessary to have a means of exchanging the air in the lungs rather than relying entirely on diffusion.

3. If the concentration of carbon dioxide outside a leaf is artificially increased, what effect might this have on the rate of photosynthesis in the leaf cells? Explain your answer.

4. An animal like a sea anemone depends on diffusion for its supply of oxygen. How might water movements in the rock pools help this process?

5. How could you speed up the process of osmosis in Experiment 1 (page 102).

6. If the cellophane tube in Figure 27.5 (page 102) is filled with sugar solution, knotted at both ends and placed in water, explain what might happen.

7. In Figure 27.6 (page 102) the cell is surrounded by water. Explain what would happen if the cell is surrounded with a sugar solution more concentrated than the cell sap.

8. If too much nitrate fertilizer is put in the soil, the plants wilt. Explain why this should happen.

28 TRANSPORT OF MATERIALS IN PLANTS

The flowering plant needs to transport water and dissolved salts from the soil to the leaves. In addition, sugars and other substances made in the leaves are transported to growing regions and storage organs. Water and salts move up through the plant, mainly in the xylem (page 92), in the transpiration stream. Compounds made by the plant travel in the phloem (page 92) which also carries some of the salts.

THE TRANSPIRATION STREAM

This is the upward flow of water and dissolved salts from the roots, through the stem, to the leaves. It is caused by a process called transpiration.

Transpiration is the evaporation of water, as water vapour, from the leaves of a plant into the atmosphere.

On pages 102–4 it was explained how the process of osmosis (i) causes turgor in plant cells and (ii) accounts for the movement of water from one cell to the next. Figure 28.1 shows a section through a small part of a leaf (see also Figure 27.9, page 104). The pressure of the cell sap in the leaf cells forces water out through the cell wall. From here, the water evaporates into the intercellular spaces. The high concentration of water vapour in these intercellular spaces of the leaf mesophyll, leads to an outward diffusion of water molecules. These escape into the air through the stomata (page 93).

The loss of water from the leaf cells reduces their turgor and allows water to pass into them by osmosis from their neighbours. Water also passes along the permeable cell walls (Figure 28.2). In either case the cells will eventually take water from the veins of the leaf and, in effect, will 'pull' more water up through the stem, causing the transpiration stream (Figure 28.3). Most of the water travelling through the plant in the transpiration stream is simply lost to the atmosphere when it evaporates from the leaves. Only a tiny fraction is retained for photosynthesis (page 86) and for maintaining the

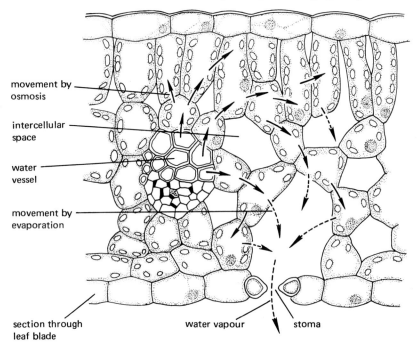

Figure 28.1 section through leaf blade

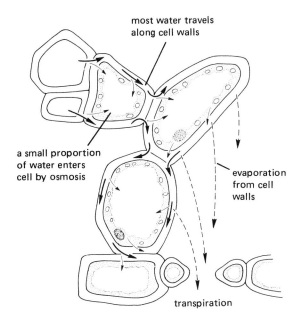

Figure 28.2

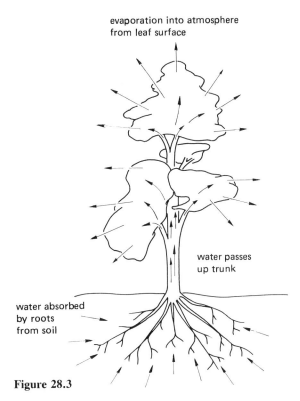

Figure 28.3

turgor of the cells. Nevertheless, transpiration causes the transpiration stream, and the transpiration stream carries dissolved salts from the roots to the leaves. The salts are used in the leaf to build up proteins, chlorophyll and other essential substances.

One other result of transpiration is that the evaporation of water from the leaves has a cooling effect. A leaf exposed to prolonged sunlight absorbs heat and could become hot enough to kill the living cells in it, were it not for the cooling effect of transpiration.

Rate of transpiration

Since transpiration is the evaporation of water from the leaves, any change in conditions which increases or reduces evaporation will have the same effect on transpiration.

(a) Light intensity. Light intensity itself does not affect evaporation but in daylight the stomata (page 93) of the leaves are open and therefore allow the water vapour in the leaves to diffuse out into the atmosphere. At night, when the stomata close, transpiration is greatly reduced.

(b) Humidity. If the atmosphere is very humid, i.e. contains a great deal of water vapour, it can accept very little more from the plants and so transpiration slows down. In dry air, the diffusion of water vapour from the leaf to the atmosphere will be rapid.

(c) Air movements. In still air, the region round a transpiring leaf will become saturated with water vapour so that no more can escape from the leaf. In these conditions, transpiration would slow down. In moving air, the water vapour will be swept away from the leaf as fast as it diffuses out. This would speed up transpiration.

(d) Temperature. (*i*) Warm air can hold more water vapour than cold air. Thus, evaporation or transpiration will take place more rapidly into warm air. (*ii*) When the sun shines on the leaves they will absorb heat as well as light. This warms them up and increases the rate of evaporation of water.

Experiment 1 Comparing rates of transpiration

For this experiment a *potometer* is used (Figure 28.4). Fill the apparatus with water and fit a freshly cut shoot into the rubber tubing at one end as shown. The shoot transpires and takes up water from the apparatus, causing the water column in the capillary tubing to move to the left. The rate of uptake is usually rapid enough for this movement

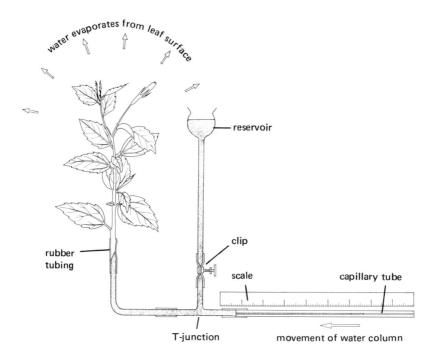

Figure 28.4

to be seen easily, and the distance it travels in, say, 2 minutes is recorded. When the liquid reaches the end of the scale, it can be returned to the start by opening the clip. Water then runs out of the reservoir and refills the capillary.

Repeat the experiment with the apparatus in different situations corresponding to the conditions (a) to (d) listed above.

(a) Compare the rates of water uptake with the apparatus in the shade and in bright sunlight. Sunlight dramatically increases the rate of uptake.
(b) Enclose the shoot in a transparent plastic bag so that the air round the leaves becomes saturated with water vapour. The plastic bag thus reduces the rate of water uptake.
(c) Compare the rates with the apparatus in still air and in front of a fan. The moving air from the fan will increase the rate of uptake.
(d) Compare the rates of uptake in a warm room and in a cold room (though this may be difficult to arrange).

In all these experiments, great care must be taken to change only *one* condition. For example, if the apparatus is taken from the laboratory to outside, there will be changes in light, humidity, temperature and air movement, and it will be impossible to say which of these is responsible for any change in uptake.

Limitations of the potometer

Not all the water taken up from the potometer will be transpired. Some will be retained for photosynthesis or to restore the leaf cells to full turgor. Alternatively, if the plant is wilting, more water may be lost by transpiration than is taken up from the potometer. In practice, these differences are small, and even though the rate of uptake of water is not exactly the same as the rate of transpiration, the two rates are proportional and very close to each other. Strictly speaking, however, the potometer should be used only to *compare* rates of *uptake* and not to *measure* actual rates of *transpiration*.

Experiment 2 To measure the rate of transpiration

Transpiration is the *loss* of water vapour, so the rate of transpiration must be measured by *weighing* a plant at intervals to see how much mass it loses. For example, water a potted plant and enclose the pot and soil in a plastic bag tied round the plant stem (Figure 28.5). This makes sure that any losses are due to evaporation from the shoot and *not* from the soil. Weigh the plant at intervals. If it loses, say 56 g in 4 hours, the rate of transpiration is 14 g water transpired per hour. This result assumes that any change in weight is entirely due to

Figure 28.5

transpiration. In fact, there may be small gains in weight due to absorption of carbon dioxide in photosynthesis (page 86), or small losses due to the escape of carbon dioxide from respiration (page 80). In practice, over a few hours, these changes are very small compared with the losses due to transpiration.

Experiment 3 To find which surface of a leaf loses more water vapour

Soak some filter paper in a five per cent solution of cobalt chloride. Then dry and cut it into 5 mm squares. When dry, the paper is blue but changes to pink when damp. Dry two squares of cobalt chloride paper, held in forceps, over a bench lamp or small Bunsen flame till they are blue. Then stick

Figure 28.6

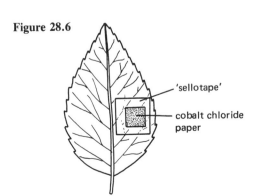

them to a leaf, using transparent adhesive tape, one square on the upper surface and one on the lower surface (Figure 28.6). The cobalt chloride paper will turn pink as water vapour reaches it from the leaf. The side of the leaf with more stomata will release water vapour more rapidly. On this side therefore, the cobalt chloride will turn pink first. In the leaves of most trees and shrubs, the stomata are on the underneath surface only. The leaves of grasses and related plants have stomata on both sides.

Experiment 4 Transport in the vascular bundles

Place the shoots of several leafy plants in a solution of one per cent methylene blue. Then leave the shoots in light for 30 minutes or more. In some cases, after this time, the blue dye will appear in the leaf veins. If some of the stems are cut across, the dye will be seen confined to the vascular bundles (see Figure 25.2, page 91). These results show that the dye, and therefore probably the water, travels up the plant in the vascular bundles. Closer study would show that it travels in the xylem vessels.

TRANSPORT IN THE PHLOEM

Amino acids (see page 292), made in the roots, are transported up the plant in the xylem vessels by the transpiration stream. However, most of the organic substances made in the plant are carried in the cells of the phloem. The leaf, in particular, makes carbohydrates during photosynthesis. These appear in the leaf as sugar and starch, but those which are not used in the leaf will be carried away in the form of sucrose (see page 292), in the phloem. From here, the sucrose may travel down the stem to be converted back into starch and stored in root tubers, such as carrots, or in stem tubers, such as potatoes. Sucrose and amino acids may travel up the stem to supply energy and raw materials to actively growing buds and shoots. These food materials may also travel up or down the stem to developing seeds and fruits.

The mechanism by which the dissolved substances are carried in the phloem is not understood. There are several theories but not enough evidence to support any of them convincingly. It is known, however, that the movement depends on living processes in the cells of the phloem. Chemicals or high temperatures which affect the living cytoplasm of the cells also reduce the transport of food materials.

Questions

1. What combination of climatic conditions is likely to lead to the maximum possible rate of transpiration?

2. A leafy shoot plus the beaker of water in which it is placed weigh 275 g. Two hours later it weighs 260 g. An identical beaker of water with no plant loses 3 g over the same period of time. What is the rate of transpiration of the shoot?

3. If you wanted to use a potometer to compare rates of transpiration of two different shoots, e.g. oak and holly, what precautions would you take to make sure the comparison was a fair one?

4. In Experiment 3 (page 109), (*a*) why do you think forceps are used to handle the cobalt chloride paper squares and (*b*) why can the colour change not be due to water vapour in the air rather than from the leaf?

5. Transpiration has been described in this chapter as if it took place only in leaves. What other parts of a plant might transpire? How could you test this experimentally?

6. A potometer is set up with a leafy shoot. After obtaining three similar readings for water uptake, the leaves are coated with petroleum jelly. Explain the effect you think this will have on water uptake.

7. The hydrogen atom of a water molecule taken up by a root might return to the same root some hours later as part of a sucrose molecule. Describe what has happened to it during these hours.

8. The bark of a tree contains the phloem. The wood consists entirely of xylem. When a ring of bark is cut away from a tree trunk, the xylem can still carry water and salts to the leaves and the leaves can still make food by photosynthesis. Nevertheless, the tree will die. Suggest an explanation for this.

29 FLOWERS, FERTILIZATION AND FRUITS

A flower is the reproductive structure of a plant. It contains sexual organs which produce male or female sex cells. The male cell is in the pollen grain and the female cell is in the ovary. After fertilization, the female sex cell forms the embryo of the seed and the ovary forms a fruit.

FLOWER STRUCTURE

The basic structure of a flower is shown in Figures 29.1 and 29.2.

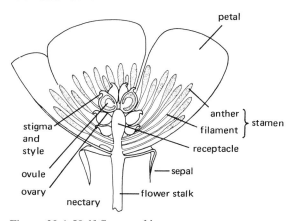

Figure 29.1 Half flower of buttercup.

Petals

These are usually brightly coloured, sometimes scented structures, often arranged in a circle (buttercup) or a cylinder (tulip). Most flowers have from five to ten petals and sometimes these are joined together to form a tube (foxglove), so the individual petals cannot be seen. The colour and scent of the petals attracts insects to the flowers for pollination.

Sepals

Outside the petals is a ring of sepals. They are often green and much smaller than the petals. They may protect the flower when it is in the bud.

Stamens

The stamens are the male reproductive organs of a flower. Each stamen consists of a stalk, the *filament*, with an *anther* on the end. Flowers such as the buttercup and wild rose have many stamens, others such as the lupin have a small number, often corresponding to the number of petals or sepals. The anther consists of four *pollen sacs* in which the pollen grains are produced by cell division. When the anthers are ripe, the pollen sacs split open and release their pollen.

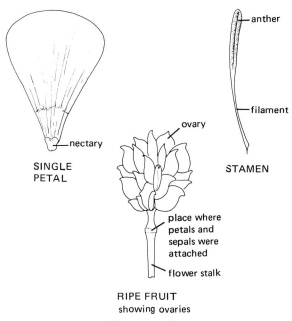

Figure 29.2 Parts of buttercup flower.

Ovary

This is the female reproductive organ. Flowers such as the buttercup and blackberry have a large number of ovaries while others, such as the lupin (Figure 29.3, page 112) have a single ovary. Inside the ovary there are one or more *ovules*. Each buttercup ovary contains one ovule but the lupin ovary contains several. The ovule, after fertilization, becomes the seed and the whole ovary becomes the fruit. Projecting from the ovary are the *style* and *stigma*. The stigma has a sticky surface and pollen grains stick to it during pollination. The style may be quite short (buttercup) or very long (lupin).

Receptacle

All the flower structures mentioned above are attached to the expanded end of a flower stalk. This is called the receptacle. In a few cases, after fertilization it becomes fleshy and edible, e.g. strawberry, apple and pear.

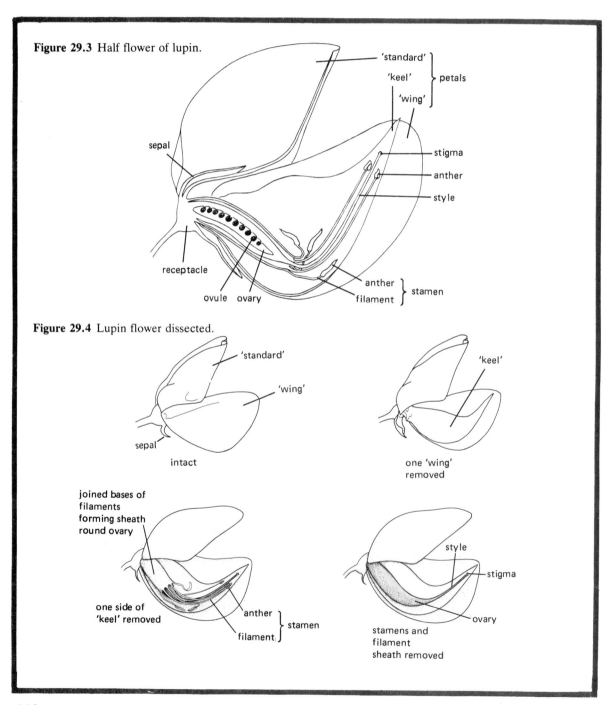

Figure 29.3 Half flower of lupin.

Figure 29.4 Lupin flower dissected.

TWO FLOWERS

Buttercup

The buttercup (Figure 29.1) is an example of a flower whose structure is fairly easy to see. In diagrams, a half-flower is drawn which shows the flower as it would appear if cut vertically down the middle. Thus, in the buttercup with five petals, two and a half would appear in the drawing. At the base of the petals are swellings called *nectaries*. These produce a sugary solution called *nectar*, which is collected by insects. There are five sepals and neither the sepals nor petals are joined together. There are about sixty stamens and thirty to forty ovaries, each containing a single ovule.

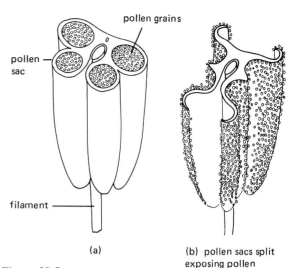

Figure 29.5

Lupin

The lupin (Figures 29.3 and 29.4) has five sepals but these are joined together forming a short tube. The five petals are of different shapes and sizes. The uppermost, called the 'standard', is held vertically. Two petals at the side are called 'wings' and are partly joined together. Inside the 'wings' are two more petals joined together to form a boat-shaped 'keel'.

The ovary is long, narrow and pod-shaped with about ten ovules in it. The long style ends in a stigma just inside the pointed end of the 'keel'. There are ten stamens, five long ones and five short ones. Their filaments are joined together at the base to form a sheath round the ovary.

POLLINATION

The process of transferring pollen from the anthers to the stigma is called pollination. The anthers split open exposing the microscopic pollen grains, (Figure 29.5). These are then carried away on the bodies of insects or simply blown by the wind, and reach the stigma of another flower. In *self-pollinating* flowers, the pollen which reaches the stigma comes from the same plant. In *cross-pollination*, the pollen is carried from the anthers of a flower on one plant to the stigma of a flower on another plant (of the same species). Flowers pollinated by insects usually have brightly coloured petals and produce nectar. In flowers pollinated by the wind, e.g. hazel catkins and grasses, the 'petals' are small and green and no nectar is produced.

Pollination of the lupin

Lupin flowers have no nectar. The bees which visit them come to collect pollen which they take back to the hive for food. Other members of the lupin family do produce nectar, e.g. clover.

The weight of the bee when it lands on the 'wings' pushes down these two petals and the petals of the 'keel'. The pollen from the anthers collects in the tip of the keel and as the petals are pressed down, the stigma and long stamens push the pollen out from the keel on to the underside of the bee (Figure 29.6). The bee, with pollen grains sticking to its body, then flies to another flower. If this flower is older than the first one, it will already have lost its pollen and when the bee's weight pushes down the 'keel', only the stigma comes out and touches the insect's body, picking up pollen grains on its sticky surface.

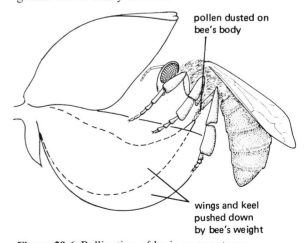

Figure 29.6 Pollination of lupin or sweet pea.

FERTILIZATION AND FRUIT FORMATION

Fertilization

The pollen grain absorbs fluid from the stigma and grows a microscopic pollen tube (Figure 29.7). This tube grows down the style and into the ovary

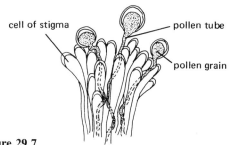

Figure 29.7

where it enters a small hole, the *micropyle*, in an ovule (Figure 29.8). The nucleus of the pollen grain travels down the pollen tube and enters the ovule. Here it combines with the female nucleus. This combination of male and female nuclei is called *fertilization* and must take place if the ovule is to develop into a seed. Each ovule needs to be fertilized by a separate pollen grain.

Although pollination must take place if the ovule is to be fertilized, pollination does not necessarily result in fertilization. For example, if the pollen of a buttercup arrived on the stigma of a lupin, it would not produce a pollen tube. The pollen is said to be *incompatible*. In some flowers there is an interval of 12 months between pollination and fertilization; in others, fertilization follows 16 hours after pollination.

From the description of pollination in the lupin, it will be evident that the pollen grains in the flower come into close contact with the stigma of the same flower but they do not form pollen tubes and therefore self-fertilization of the flower does not occur.

Fruit and seed formation

After fertilization, rapid cell division takes place in the ovule and a miniature plant, the *embryo*, is formed. Food is transferred to the ovule and stored in structures called *cotyledons*. The outer wall of the ovule becomes thicker and harder, forming the seed coat or *testa* (see page 117).

As the seeds grow, the ovary also becomes much larger and the petals and stamens wither and fall off (Figure 29.9). The ovary is now called a *fruit*.

A fruit is a fertilized ovary; it is not necessarily edible.

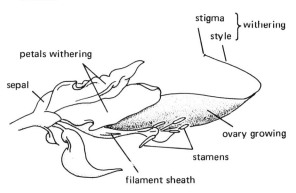

Figure 29.9 Lupin flower after fertilization.

In the lupin, the fertilized ovary forms a dry, hard pod, but in a related plant, the runner bean, the ovary wall becomes fleshy and edible. Tomatoes, cucumbers and plums are examples of edible fruits.

DISPERSAL OF FRUITS AND SEEDS

When flowering is over and the seeds are mature, the whole fruit or the individual seeds fall from the parent plant to the ground where germination may then take place. In many plants, the fruits or seeds

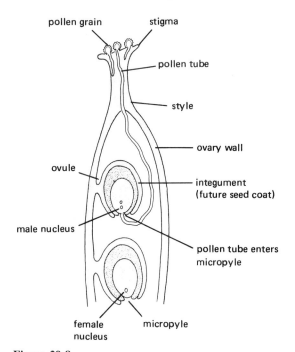

Figure 29.8

are adapted in such a way that they are carried considerable distances from the parent plant. This helps to reduce the competition for light and water between members of the same species. It may also result in new areas being colonized by the plant.

The principal adaptations are those which favour dispersal by wind and animals. In addition, some plants have 'explosive' pods or capsules that scatter the seeds, and others have fruits that are adapted to dispersal by water.

Wind dispersal

(a) *Censer mechanism.* Examples are the white campion, poppy (Figure 29.10) and antirrhinum. The flower stalk is usually long and the ovary becomes a dry, hollow capsule with one or more openings. The wind shakes the flower stalk and the seeds are scattered on all sides through the openings in the capsule.

(b) *'Parachute' fruits and seeds.* Clematis, thistles, willow-herb and dandelion (Figures 29.11) have seeds or fruits of this kind. Feathery hairs projecting from the fruit or seed increase its surface area so that air resistance to its movements is very great. Thus, it sinks to the ground very slowly and is likely to be carried great distances from the parent plant by slight air currents.

(c) *Winged fruits.* Fruits of the lime (Figure 29.12), sycamore (Figure 29.13) and ash trees have extensions from the ovary wall or leaf-like structures on the flower stalk which make 'wings'. These cause the fruit to spin as it falls from the tree and increase the time of fall, so its chances of being carried away in air currents are better.

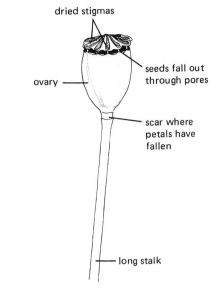

Figure 29.10 Poppy.

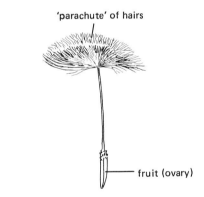

Figure 29.11 Dandelion.

Figure 29.12 Lime.

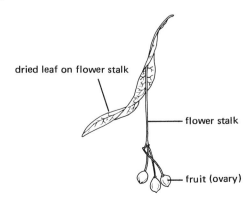

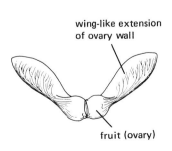

Figure 29.13 Sycamore.

Animal dispersal

(*a*) *Mammals—hooked fruits.* In herb bennet, agrimony (Figure 29.14) and goose-grass, hooks develop from the style, the receptacle or on the ovary wall. These hooks catch in the fur of passing mammals or in the clothing of people, so the fruit is carried away.

Later, at some distance from the parent, they fall off, or are brushed or scratched off.

(*b*) *Birds—succulent fruits.* Fruits like the blackberry and elderberry are eaten by birds. The hard pips containing the seed inside are undigested and pass out with the faeces of the bird, away from the parent plant. Even if the seeds are not swallowed, the fruit is often carried away before the seeds are dropped, e.g. rose-hip. The soft texture and bright colour of these fruits may be regarded as an adaptation to this method of dispersal.

Explosive fruits

The pods of flowers in the pea family, e.g. gorse, broom, lupin and vetches, dry in the sun and shrivel. The tough fibres in the fruit wall shrink and set up a tension. When the fruit splits in half down two lines of weakness, the two halves curl back suddenly and flick out the seeds (Figure 29.15).

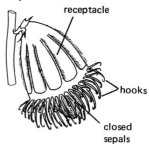

Figure 29.14 Agrimony.

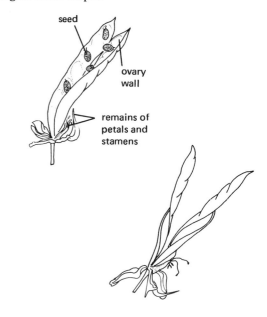

Figure 29.15 Lupin.

Questions

1. Working from outside to inside, list the parts of a flower.

2. What features of flowers attract insects?

3. How does a lupin flower differ from a buttercup flower?

4. If plants are growing in a greenhouse where insects cannot enter, how could you make sure that the flowers were pollinated?

5. Why do you think a large insect such as a bee can pollinate a lupin flower whereas a small insect like a fly cannot?

6. At what stage would you say that pollination ends and fertilization begins? Justify your answer.

7. Give a short definition of fertilization.

8. Which parts of a tomato flower (*a*) grow to form the fruit, (*b*) fall off after fertilization and (*c*) remain attached to the fruit.

9. What are the advantages of a plant dispersing its seeds over a wide area? What disadvantages might there be?

10. Which methods of dispersal are likely to result in seeds travelling the greatest distances? Explain your answer.

30 GERMINATION OF SEEDS

The last chapter described how a seed is formed from the ovule of a flower as a result of fertilization, and is then dispersed from the parent plant. If the seed lands in a suitable place it will *germinate*, i.e. grow into a mature plant. To understand the process of germination, the structure of two seeds, the maize and the french bean, will first be studied. The french bean is similar to the lupin seed in its structure and germination but is much larger and, therefore, easier to study.

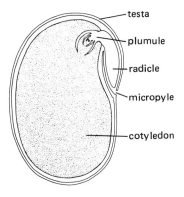

(a) LONGITUDINAL SECTION

SEED STRUCTURE

The french bean

The seed of the french bean (Figures 30.1 and 30.2) contains a miniature plant, the *embryo*, which consists of a root or *radicle*, and a shoot or *plumule*. The embryo is attached to two leaves called *cotyledons*, which are swollen with stored food. This stored food, mainly starch, is used by the embryo when it starts to grow. The embryo and cotyledons are enclosed in a tough seed coat or *testa*. The *micropyle*, through which the pollen tube entered (page 114), remains as a small hole in the testa and allows the seed to take up water before germinating.

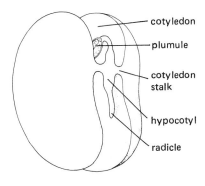

(b) EMBRYO PLANT

Figure 30.2

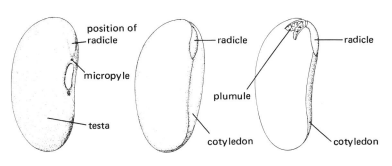

Figure 30.1

(a) EXTERNAL APPEARANCE   (b) TESTA REMOVED   (c) ONE COTYLEDON REMOVED

The maize grain

The embryo of the maize grain (Figures 30.3 and 30.4), like that of the french bean, has a plumule and radicle but each of these is enclosed in a protective sheath. The sheath over the plumule is called the *coleoptile*. The radicle is enclosed in the *coleorhiza*. Unlike the bean, there is only one

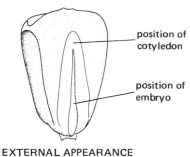

EXTERNAL APPEARANCE

Figure 30.3

cotyledon attached to the embryo, and it contains very little stored food. The bulk of the food reserve, starch, is contained in the *endosperm* on the outside of the cotyledon. The function of the cotyledon is to digest this starch and pass it to the growing embryo when germination starts.

The embryo, cotyledon and endosperm are enclosed in a testa, which itself is still covered by a thin ovary wall. Thus the maize grain is, strictly speaking, a fruit which contains a single large seed.

Experiment 1 The role of the cotyledons in the bean seed

Soak four runner bean seeds in water for 24 hours. Peel off the testas and separate the cotyledons. Discard the cotyledons which do not have embryos attached to them. Pin one of the cotyledons to a piece of expanded polystyrene ceiling tile covered with blotting paper, with the embryo just touching the blotting paper (Figure 30.5). Cut away and discard three quarters of the second cotyledon and pin the remaining piece, with the embryo, to the polystyrene. Cut away as much as possible of the third cotyledon, leaving a small piece attached to the embryo. Pin this to the polystyrene. Finally, dissect off the embryo from the fourth cotyledon and pin it to the polystyrene.

Place the polystyrene strip in a tall, screw-top jar, with a little water, as shown in Figure 30.5 and leave it for a week.

Result. The embryo on its own will grow a little but not as much as the others which will probably grow

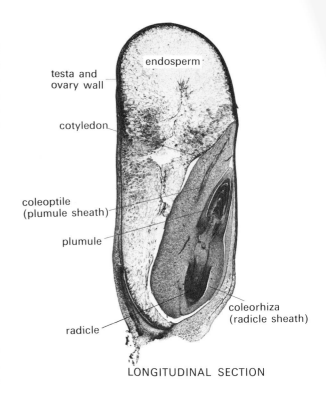

LONGITUDINAL SECTION

Figure 30.4

in proportion to amount of cotyledon attached to them. It may take another week, however, to see a difference between the whole and the quarter cotyledon.

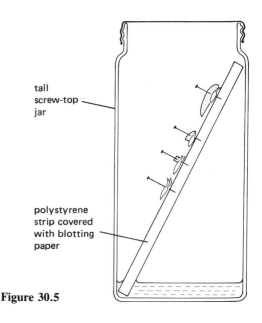

Figure 30.5

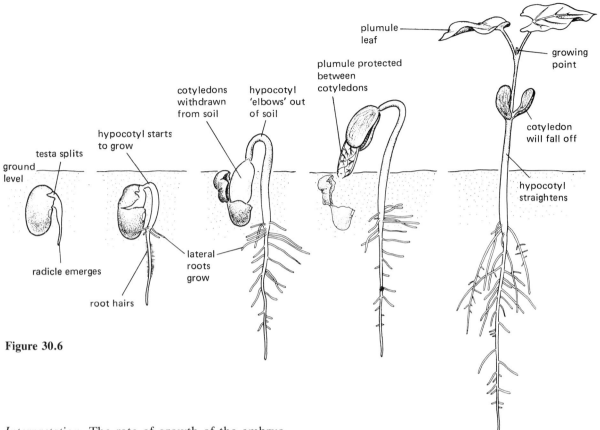

Figure 30.6

Interpretation. The rate of growth of the embryo seems to depend on the amount of cotyledon. This could be because the cotyledon supplies the embryo with food or because it contains a growth-promoting substance.

GERMINATION

The french bean

The seed of the french bean (Figure 30.6) absorbs water and swells. After 3 or 4 days, the radicle grows, bursts through the testa and grows down into the soil, pushing its way between soil particles and small stones. Branches, called *lateral roots*, grow out from the side of the main root and help to anchor it firmly in the soil. On both the main roots and lateral roots, microscopic *root hairs* grow. These are fine projections from some of the outermost cells which make close contact with the soil particles and absorb water from the spaces between them (page 103).

A region of the embryo's stem, the *hypocotyl*, just above the radicle, now starts to get longer. The radicle, by now, is firmly anchored in the soil so the rapidly growing hypocotyl arches upwards through the soil, pulling the cotyledons with it. Sometimes the cotyledons are pulled out of the testa, leaving it below the soil and sometimes the cotyledons remain enclosed in the testa for a time. In either case, the plumule is well protected from damage while being pulled through the soil, because it is enclosed between the cotyledons.

Once the cotyledons are above the soil, the hypocotyl straightens up and the two plumule leaves open out. Up to this point, all the food needed for making new cells and producing energy has come from the cotyledons which now shrivel and fall off. The plumule leaves turn green and expand and absorb sunlight to start making their own food by photosynthesis (see page 86). Between the plumule leaves is a growing point which continues the upward growth of the stem and the production of new leaves. The embryo has now become an independent plant, absorbing water and mineral salts from the soil, carbon dioxide from the air and making food in its leaves.

119

Figure 30.7

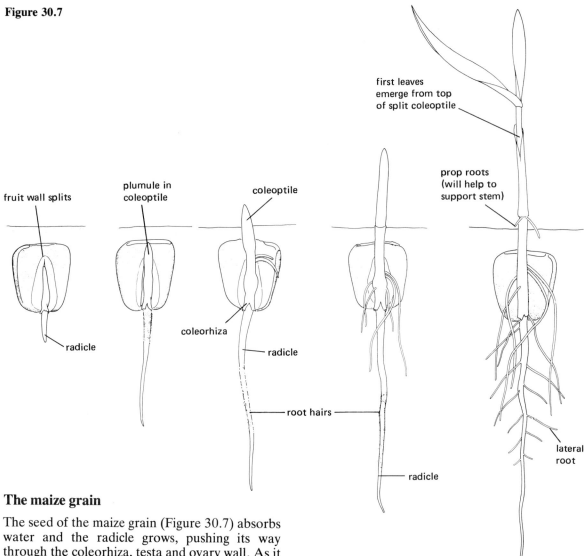

The maize grain

The seed of the maize grain (Figure 30.7) absorbs water and the radicle grows, pushing its way through the coleorhiza, testa and ovary wall. As it grows down into the soil, root hairs develop on that part of the root which has just ceased elongating. The plumule grows but does not yet push through the coleoptile. Instead, the coleoptile grows rapidly, forcing its way through the soil and so protecting the delicate leaves inside it from damage. Lateral roots develop and, from the base of the shoot, additional roots grow.

Once the coleoptile is above the soil, the plumule leaves push through its tip and continue their growth, turning green and making food by photosynthesis (see page 86). The cotyledon has been digesting the starch in the endosperm and passing it to the growing embryo to provide it with energy and the raw materials for making new cells. When this food store is exhausted the fruit wall and cotyledons shrivel up.

Conditions for germination

Water. All seeds need a supply of water to start them germinating and to continue their growth as seedlings and mature plants.

Oxygen. Most seeds require oxygen for germination. This normally comes from the air spaces in the surface layers of the soil. If the soil is waterlogged or heavily compacted, the lack of oxygen may prevent germination taking place.

Temperature. Germination is slowed down or even stopped by low temperatures. You can easily demonstrate this by placing equal numbers of soaked peas on moist cotton wool in a refrigerator (4 °C) and in a room (18 °C). After 2 or 3 days, most of the peas at room temperature will have radicles about 10 mm long. Those in the refrigerator will not have started to germinate. They will do so, however, if taken from the refrigerator and left for 2 days at room temperature.

Conditions for continued growth

Once the seedling has used up the food stores in the cotyledons or endosperm, it starts to make food in its leaves. For this it needs energy from sunlight, carbon dioxide from the air, water and mineral salts from the soil (see 'Photosynthesis and Nutrition in Plants', page 86).

REGIONS OF GROWTH

Experiment 1 Region of growth in a root

Immerse some peas in water for a day and then wrap them in a roll of blotting paper as shown in Figure 30.8. After 3 days, the radicles will have grown about 10 mm. Select the seedlings with straight radicles and mark them with ink lines about 2 mm apart. Figure 30.9 shows one way of doing this.

Place two or three marked seedlings between two strips of moist cotton wool in a Petri dish so that the seeds are held firmly but the radicles are exposed and clearly visible (Figures 30.10 and 11). Hold the Petri dish lid in place with an elastic band and leave the whole dish on its edge, with the radicles pointing downwards, for 2 days.

Result. The ink marks will have become most widely spaced in the region just behind the root tip (Figure 30.12).

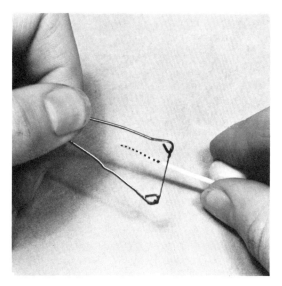

Figure 30.9

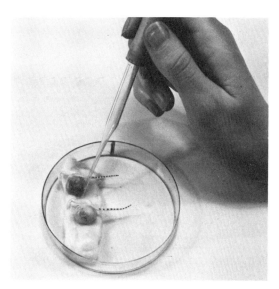

Figure 30.10

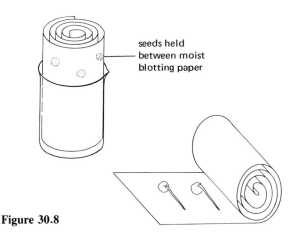

Figure 30.8

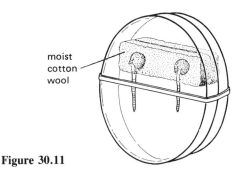

Figure 30.11

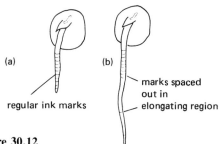

(a) regular ink marks
(b) marks spaced out in elongating region

Figure 30.12

Interpretation. The wide spacing of the marks indicates the region of most rapid elongation, i.e. just behind the tip. In fact, it is known that, although cell division takes place in the tip of the root, cell elongation occurs a short distance behind the tip (see Figure 27.7, page 103).

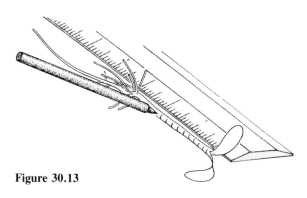

Figure 30.13

Experiment 2 Region of growth in a shoot

Soak some french bean seeds in water for one day and then roll them in blotting paper as shown in Figure 30.8, and keep moist for 12 days. After this time, the shoots will be growing straight up, clear of the blotting paper. Use a felt-tip pen to mark lines on the stem at 2 mm intervals (Figure 30.13).

Return the seedlings to the beaker and roll them in fresh blotting paper with the shoots exposed. Place in a cupboard for 2 or 3 days, keeping the blotting paper moist.

Results and Interpretation. The ink marks will be most widely spaced in the region of most rapid growth. In young seedlings, there may be some growth in the region just below the cotyledons but most growth will take place in the stem above the cotyledons. In older plants, it will be the stem just below the tip of the shoot which shows the greatest elongation.

Questions

1. How does the maize grain differ from the french bean seed in the structure of its (*a*) plumule, (*b*) radicle and (*c*) cotyledon?

2. How does a root hair differ from a lateral root?

3. During germination, how is the plumule protected from damage during its passage through the soil in (*a*) the french bean, (*b*) maize?

4. At what stage of development is a seedling able to stop depending on its cotyledons or endosperm for food?

5. What uses does the germinating seedling make of the food stored in its endosperm or cotyledons?

6. What is the likely advantage to a germinating seed of its radicle emerging some time before the shoot starts to grow?

7. (*a*) In principle, how would you try to design an experiment to test whether oxygen is necessary for the germination of certain seeds?
   (*b*) Why does it seem unlikely that light is an essential requirement for germination?

8. Figure 30.12*b* shows the result of an experiment to find the region of maximum growth in a root. Draw a diagram to show how the result would have appeared if (*a*) the root grew simply by adding new cells at the tip, (*b*) if the root grew mainly at the point just below its attachment to the cotyledons.

Forces and Energy

31 MEASUREMENT

Before any measurement can be made, we must choose a standard, or *unit* of the quantity. An *instrument* is then needed which has a marked scale using that unit. Figure 31.1 shows an aircraft and the many measuring instruments used on the flight deck.

In science, the SI (Système International) family of units is used. It is a decimal family in which units are divided or multiplied by 10 to give smaller or larger units.

Three important basic quantities we have to measure are *length, mass* and *time*.

Length

Figure 31.1*a*

The unit of length is the *metre* (m) which is divided into centimetres (cm) and millimetres (mm).

1 cm = 1/100 m = 0·01 m ∴ 100 cm = 1 m
1 mm = 1/1000 m = 0·001 m ∴ 1000 mm = 1 m
Also
1 mm = 1/10 cm = 0·1 cm ∴ 10 mm = 1 cm

For large lengths, the kilometre (km) is used.

1 km = 1000 m (about 5/8 mile)

Many length measurements are made with a ruler. The correct way to read one is shown in Figure 31.2. The reading is 7·6 cm (or 76 mm).

Experiment 1 Measuring lengths

Use a rule marked in centimetres and millimetres.

Figure 31.1*b*

a Measure the lengths of different things. Record your results in a table, giving your answers as shown below.

Measurement	Result
Length of pencil	12 cm 4 mm = 12·4 cm
Width of book	
Your hand span	
Length of your foot	
Your height	

b Measure the thickness of a coin.

• How could you get a more accurate result?

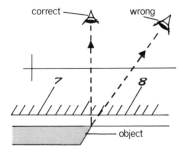

Figure 31.2

c Wind a piece of wire round a pencil ten times so that there are no spaces between the turns (Figure 31.3). Make a measurement from which you can *work out* the thickness (diameter) of the wire.

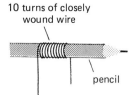

Figure 31.3

Vernier scales

When a reading falls between two marks on a scale of an instrument, it is possible to estimate the position and often give a reasonable guess for the measurement. The reading can be obtained more accurately if the instrument also has a *vernier* scale. Such scales are used on sliding calipers for finding the inside or outside diameter of a tube (Figure 31.4a).

A vernier scale for use with a millimetre scale is 9 mm long and has 10 divisions (Figure 31.4b).

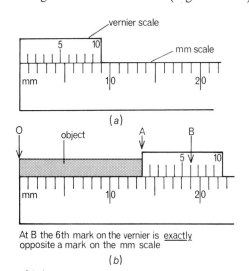

Figure 31.4

1 vernier division = 9/10 mm = 0·9 mm

The object in Figure 31.4b is between 13 mm and 14 mm long, i.e. it is 13·x mm long. x is found by noting which mark on the vernier scale is exactly opposite (or nearest to) the mark on the mm scale. Here it is the sixth mark so the reading is 13·6 mm since:

$$OA = OB - AB = 19·0 - 6 \times 0·9$$
$$= 19·0 - 5·4 = 13·6 \text{ mm}$$

Micrometer screw gauge

This measures very small lengths such as the diameter of a wire (Figure 31.5). One complete turn of the drum, which is attached to the screw, opens the jaws by 1 division on the scale of the shaft. This is usually 0·5 mm. The drum has 50 divisions round it, so turning it by 1 division opens the jaws by 0·5/50 = 0·01 mm. The thickness t of the object shown

= 2·5 on shaft scale + 33 divisions on drum scale
= 2·5 + 33 × 0·01 = 2·5 + 0·33 = 2·83 mm.

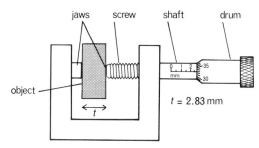

Figure 31.5

Area

The sides of the square in Figure 31.6a are each 1 cm long. The *area* of the square is 1 square centimetre (1 cm²). In Figure 31.6b the rectangle measures 2 cm by 1 cm. Its area is 2 cm² since it contains two squares each of area 1 cm². The rectangle in Figure 31.6c has length 4 cm and breadth 3 cm and contains 4 × 3 = 12 squares each of area 1 cm². The area is therefore 12 cm².

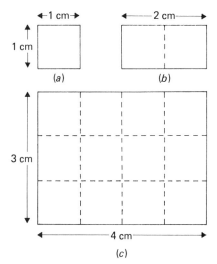

Figure 31.6

To find the area of a square or rectangle is simple: multiply length by the breadth. For a rectangle:

AREA = LENGTH × BREADTH

Area is measured in square centimetres (cm²), square metres (m²), square kilometres (km²), etc.

Volume

Volume is the amount of space occupied. The SI unit of volume is the *cubic metre* (m³). The cubic metre is rather large for many purposes, so we will use the cubic centimetre (cm³). The volume of a cube with edges each 1 cm long is 1 cm³ (Figure 31.7).

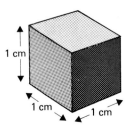

Figure 31.7

(*a*) *Regular solid objects.* The volume of a regular-shaped object such as a box can be found from measurements of length, breadth and height. The volume of the box in Figure 31.8a is 4 cm × 3 cm × 2 cm = 24 cm³. You can see that this is so from Figure 31.8b in which the same box is divided up into cubes each of volume 1 cm³. The volume equals the number of unit cubes contained.

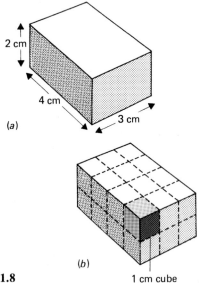

Figure 31.8

For a cube, or a rectangular object:

VOLUME = LENGTH × BREADTH × HEIGHT

(*b*) *Liquids.* The volume of a liquid can be found by pouring it into a measuring cylinder (Figure 31.9). When making a reading, the base of the measuring cylinder should be on a horizontal surface and your eye level with the *bottom* of the meniscus (Figure 31.9a). The meniscus formed by mercury is curved the opposite way to that of other liquids. In the case of mercury, read the top (Figure 31.9b).

Figure 31.9

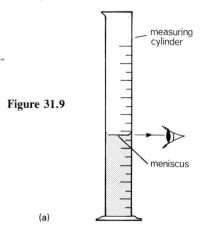

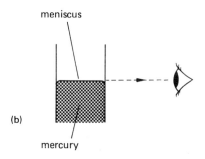

Volumes of liquids are measured in litres (l) or cubic decimetres (dm³).

1 litre = 1 cubic decimetre = 1000 cm³

(about 1¾ pints)

Mass

We can think of mass as a measure of the amount of matter it contains. The mass of an object is *not* the same as its weight. Weight is the downward pull (force) of the earth on an object. For some people, the two are confused because mass is usually measured on a balance by a process called weighing!

The unit of mass is the kilogram (kg). The gram (g) is one-thousandth of a kilogram.

1 g = 1/1000 kg   1 kg = 1000 g (about $2\frac{1}{4}$ lb)

Weight is a *force* that acts on masses. A 1 kg mass experiences a pull from the earth of about 10 newtons ((N): see page 132). One kilogram therefore weighs about 10 N.

There are several types of balance. The quickest and easiest to use is the lever balance (Figure 31.10a). It has two ranges (e.g. 0–250 g and 0–1000 g) and a simple adjustment is needed to change one range to the other. Figure 31.10b shows a very accurate modern top pan electric balance.

Figure 31.10a

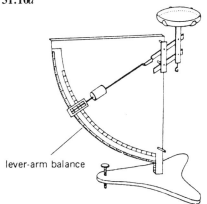

lever-arm balance

Figure 31.10b

Density

In everyday language, steel is said to be 'heavier' than wood. This doesn't mean that a steel needle has more mass than a tree-trunk. What it does mean is that a piece of steel has more mass than a piece of wood when *both have the same volume*. In science, we have to be exact and 'heaviness' is compared by considering the *mass per unit volume* or *density* of each substance.

The density of steel is 8 grams per cubic centimetre (shortened to 8 g per cm³ or 8 g/cm³). This means that a piece of steel of volume 1 cm³ has a mass of 8 g; 2 cm³ of steel would have a mass of 16 g and so on.

If we know the mass and the volume of a substance, its density can be calculated from:

$$\text{DENSITY } (d) = \frac{\text{MASS } (m)}{\text{VOLUME } (V)}$$

For example, if a piece of lead has a mass of 550 g and a volume of 50 cm³, then:

$$\text{density of lead} = \frac{\text{mass}}{\text{volume}} = \frac{550}{50} = 11 \text{ g/cm}^3$$

The SI unit of density is the kilogram per cubic metre (kg/m³). For solids and liquids, the densities measured in kg/m³ are inconveniently large (e.g. the density values for steel and water are 8000 kg/m³ and 1000 kg/m³). To change from kg/m³ to g/cm³, we divide by 1000. For water, the density is 1000 kg/m³ = 1 g/cm³.

Experiment 2 Measuring densities

a *Regular solid* (e.g. a glass block). Find the mass of the solid on a balance. Measure the length, breadth and height with a ruler.

Mass of solid = g
Volume = length × breadth × height
       = cm × cm × cm
       = cm³

$\frac{\text{Density}}{\text{of solid}} = \frac{\text{mass}}{\text{volume}} = \frac{\text{g}}{\text{cm}^3}$
       = g/cm³

b *Irregular solid* (e.g. a pebble). Find the mass of the solid. Measure its volume by one of the methods shown in Figure 31.11a and b. In a the volume is the difference between the first and second readings. In b it is the volume of water collected in the measuring cylinder. Work out the density.

c *Liquid* (e.g. water). Weigh the beaker then pour in a known volume of the liquid from a measuring cylinder. Reweigh the beaker to get the mass of the liquid.

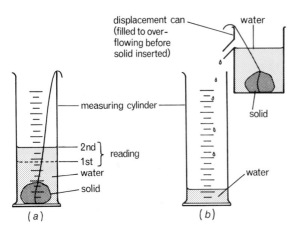

Figure 31.11

Mass of beaker empty = ___ g
Mass of beaker + liquid = ___ g
∴ Mass of liquid = ___ g
Volume of liquid = ___ cm³
Density of liquid = $\dfrac{\text{___ g}}{\text{___ cm}^3}$ = ___ g/cm³

Density of air

We have already seen (page 16) that the molecules which make up a gas are much farther apart than those in solids and liquids. It follows that the density of air (a mixture of gases, page 20) is very small. An approximate value for density of air can be found as described in Experiment 3.

Experiment 3 Measuring the density of air

Weigh a 500 cm³ round-bottomed flask (full of air) on an accurate balance. Remove the air using a good vacuum pump and reweigh the evacuated flask. The difference between the two readings equals the *mass of air in the flask*.

Fill the flask with water to the level of the rubber stopper. Then pour the water into a measuring cylinder and measure the volume of the water. This is the *volume of air*.

From these two measurements the density can be found.

Relative density

The methods described for finding density all require separate measurements of mass and volume. Usually, the volume measurement is not as accurate as the mass one. Using the idea of *relative density*, it is possible to measure density more reliably by methods requiring only weighings.

Relative density of a substance is defined by:

$$\text{relative density} = \frac{\text{density of substance}}{\text{density of water}}$$

The density of aluminium is 2·7 g/cm³. The density of water is 1·0 g/cm³. So the relative density of aluminium is 2·7/1·0 = 2·7. Note that relative density has no units, but is numerically equal to the density of the substance in g/cm³. A relative density measurement is, in effect, a density measurement. Working from the last equation, we can obtain a new equation which is useful for relative density measurements. We have:

$$\text{relative density} = \frac{\text{density of substance}}{\text{density of water}} =$$

$$\frac{\text{mass of substance}}{\text{volume of substance}} \div \frac{\text{mass of water}}{\text{volume of water}} =$$

$$\frac{\text{mass of substance}}{\text{volume of substance}} \times \frac{\text{volume of water}}{\text{mass of water}}$$

If the substance and water have equal volumes, then:

$$\text{RELATIVE DENSITY} = \frac{\text{MASS OF SUBSTANCE}}{\text{MASS OF SAME VOLUME OF WATER}}$$

Experiment 4 Relative density of a liquid by relative density bottle

A density bottle has a well-fitting glass stopper with a narrow hole through it (Figure 31.12). When the bottle is full and the stopper inserted, the excess liquid rises up the hole and spills over. The

Figure 31.12

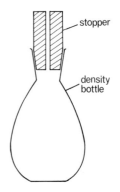

bottle always contains exactly the same volume when the liquid level is at the top of the hole.

Weigh a dry density bottle with stopper when empty and then when full of liquid, e.g. methylated spirit. Return the liquid to its stock bottle, rinse out the density bottle with water, fill it with water and weigh again. Always ensure that the outside of the bottle is dry before weighing.

The measurements may be set out as shown below and the result worked out using the equation for relative density.

Mass of empty bottle	=	g
Mass of bottle full of liquid	=	g
∴ Mass of liquid	=	g
Mass of bottle full of water	=	g
∴ Mass of same volume of water	=	g

Floating and sinking

An object sinks in a liquid of smaller density than its own; otherwise it floats, partly or wholly submerged. For example, a piece of glass of density 2.5 g/cm^3 sinks in water (density 1.0 g/cm^3) but floats in mercury (density 13.6 g/cm^3). An iron nail sinks in water but an iron ship floats because it contains air so its average density, which includes the air in it, is less than that of water.

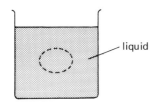

Figure 31.13

Figure 31.13 shows a container of liquid. The dotted line is drawn round a 'piece' of the liquid. This piece of liquid stays in the same position. But if it was taken out of the rest of the liquid it would fall. It follows that when it is in the liquid, this piece of liquid is supported by an upward (buoyancy) force equal to its own weight.

Suppose that a stone is weighed (Figure 31.14a) and then immersed in the liquid. It displaces liquid and experiences an upward force equal to the weight of liquid displaced. This upward force is less than the weight of the stone, so the stone sinks, although it weighs less (Figure 31.14b). An immersed piece of cork displaces liquid which exerts a bigger buoyancy than the weight of the cork. It therefore bobs to the surface and floats (Figure 31.14c).

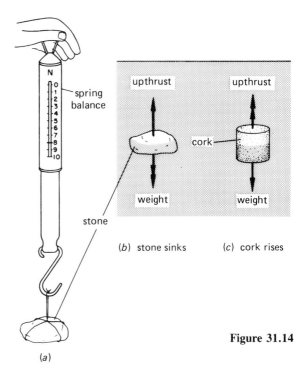

Figure 31.14

Formulas for density calculations

If we write d for density, m for mass and V for volume, we get the *formula*:

$$d = \frac{m}{V} \qquad (1)$$

It can also be written in two other ways:

$$m = d \times V \qquad (2)$$

and

$$V = \frac{m}{d} \qquad (3)$$

Time

The unit of time is the *second* (s). In clocks and watches time is measured using some kind of action which is constantly repeated. Many have a small balance wheel which swings to and fro. In some modern ones the movement is controlled by vibrations of a quartz crystal. A pendulum clock is controlled by a swinging pendulum.

Experiment 5 Simple pendulum

A simple pendulum consists of a piece of thread which is fixed at one end and has a metal ball (called the bob) on the other end.

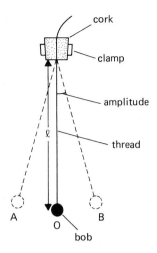

Figure 31.15

The periodic time, or *period* (T) of a pendulum is the time for it to make a complete swing (oscillation) from O to A to O to B and back to O again (or A to B and back to A) (Figure 31.15).

The length l of the pendulum is the distance from the bottom of the support to the centre of the bob.

The *amplitude* α (alpha) of the swing is the angle between the rest position of the thread and the furthest position it reaches on one side of the swing.

Set up the pendulum as shown, making the length l about 25 cm.

a *Effect of amplitude.* Find the time in seconds for 100 complete swings when $\alpha = 10°$. To do this, pull the bob sideways so the thread makes an angle of 10° with the rest (vertical) position, as shown by a protractor. Let the bob go and count 3, 2, 1, 0, 1, ... 100 as it passes the zero in the *same direction*. Start the clock on 0 and stop it on 100. The time it shows is 100 T.

Repeat for $\alpha = 20°$ and 30° and draw a table as shown below for the results.

α	100 T	T
10°		
20°		
30°		

• Does T depend on α?

b *Effect of length.* Increase l to 50 cm and measure the time for 100 complete swings with $\alpha = 10°$. Repeat with $l = 100$ cm. Copy and complete the following table.

l	100 T	T
25 cm		
50 cm		
100 cm		

• What happens to T when l is increased?

c *Effect of the bob.* Replace the bob by a heavier or a lighter one and find the time for 100 complete swings when $l = 100$ cm and $\alpha = 10°$.

Time for 100 swings (100 T) =   s
∴ T =   s

• Does the mass of the bob affect T?

You should find that T:

a does not depend on α (unless α is made very large)
b increases when l increases
c does not depend on the mass of the bob.

Questions

1. How many millimetres are there in (*a*) 1 cm, (*b*) 4 cm, (*c*) 0.5 cm, (*d*) 6.7 cm, (*e*) 1 m?

2. Write the following lengths in metres (m): (*a*) 300 cm, (*b*) 550 cm, (*c*) 870 cm, (*d*) 43 cm, (*e*) 100 mm.

3. Write down the lengths of the objects shown in Figure 31.16.

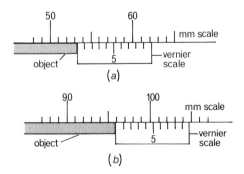

Figure 31.16

4. Write down the screw gauge readings in Figure 31.17.

Figure 31.17

5. A rectangular metal block measures 10 cm × 2 cm × 2 cm. What is its volume? How many blocks each 2 cm × 2 cm × 2 cm could be made from the block?

6. How many cases each 1 m × 0·5 m × 0·2 m can be stored in a space measuring 4 m × 2 m × 1 m?

7. A perspex box has a 6 cm square base and contains water to a height of 7 cm (Figure 31.18).

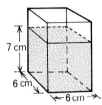

Figure 31.18

(a) What is the volume of the water?
(b) A stone is lowered into the water so that it is completely covered and the water rises to a height of 9 cm. What is the volume of water displaced?

8. (a) If the density of wood is 0·5 g/cm³ what is the mass of (i) 1 cm³, (ii) 2 cm³, (iii) 10 cm³?
(b) What is the density of a substance of (i) mass 100 g and volume 10 cm³, (ii) volume 3 m³ and mass 9 kg?
(c) The density of gold is 19 g/cm³. Find the volume of (i) 38 g, (ii) 95 g of gold?

9. A piece of steel has a volume of 12 cm³ and a mass of 96 g. What is its density in (a) g/cm³, (b) kg/m³?

10. What is the mass of 5 m³ of cement of density 3000 kg/m³?

11. What is the mass of air in a room measuring 10 m × 5 m × 2 m if the density of air is 1·3 kg/m³?

12. A perspex box has a 10 cm square base and contains water to a height of 10 cm. A piece of rock of mass 500 g is lowered into the water and the level rises to 12 cm.

(a) What is the volume of water displaced by the rock?
(b) What is the volume of the rock?
(c) Calculate the density of the rock.

13. A density bottle has mass 70 g when empty, 90 g when full of water and 94 g when full of a liquid. What is the relative density of the liquid?

32 INTRODUCTION TO FORCES

A force is a push or a pull that acts on an object. A force can:

a cause a body which is at rest to move
b change the speed or direction of motion of a moving body
c change the shape or size of a body.

The weight-lifter shown in Figure 32.1 exerts a pull then a push.

With some forces, like those exerted by our muscles, there has to be *contact* with the objects on which they act. Other kinds of forces can act through space and do not need to touch objects. We call these *action-at-a-distance* forces. Magnetic and electrical forces are of this type. Weight (gravitational) forces are also of the action-at-a-distance type.

Weight

We all constantly experience the force of gravity, i.e. the pull of the earth. It causes an unsupported body to fall from rest to the ground.

The weight of a body is the force of gravity on it.

The nearer a body is to the centre of the earth the more does the earth attract it. Since the earth is not a perfect sphere the weight of a body varies over the earth's surface. It is greater at the poles than at the equator.

A unit of force: the newton

The SI unit of force is the *newton* (N). It will be defined later (page 156) but the definition is based on the change of speed a force can produce on a body. Weight is a force and therefore should be measured in newtons. The weight of an average-sized apple is about 1 N.

The weight of a body can be measured by hanging it on a *spring balance* marked in newtons (Figure 32.2) and letting the pull of gravity stretch the spring in the balance. The greater the pull the more does the spring stretch. On most of the earth's surface *the weight of a body of mass 1 kg is 9·8 N*.

Often this is taken as 10 N; a mass of 2 kg has a weight of 20 N, and so on. The mass of a body is the same wherever it is and does not depend on the presence of the earth as does weight.

Mass is measured by a lever, beam or top pan balance; weight is measured by a spring balance. Many spring balances are marked in kg or g; a reading of 1 kg would correspond to 10 N and 100 g to 1 N (on the earth).

Experiment 1 Stretching a spring

Set up a steel spring as in Figure 32.3. Read the scale opposite the bottom of the hanger. Add a 100 g load and take the reading. Repeat with loads of 200 g, 300 g, etc. Enter the readings in a table, as shown below.

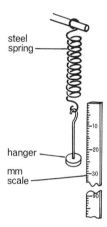

Figure 32.3

Load g	Scale reading mm	Total stretch (extension) mm
0		
100		
200		
etc.		

- Do the results suggest any rule about how the spring behaves when it is stretched?

Sometimes it is easier to find the laws (rules) of science by showing results on a graph. Do this with the results obtained above by plotting *load* readings along the *y*-axis and *total stretch* readings along the *x*-axis. Every pair of readings will give a point. Mark them with small crosses and draw a smooth line through them.

- What is the shape of the line?

Hooke's law

A man called Robert Hooke investigated springs about 300 years ago. He discovered the following:

Total stretch is proportional to the stretching force (load) so long as the spring is not permanently stretched.

This means that doubling the force doubles the stretch; trebling the force trebles the stretch, and so on. Using the mathematical sign for proportionality, we can write *Hooke's law* as:

TOTAL STRETCH ∝ STRETCHING FORCE

It is true only if the *elastic limit* of the spring is not exceeded, i.e. only if the spring goes back to its original length when the force is removed.

The graph shown in Figure 32.4 is for a spring that has been stretched beyond the elastic limit E.

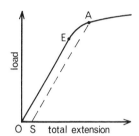

Figure 32.4

OE is a straight line passing through the origin O. This straight line is graphical proof of Hooke's law. If the force for point A is applied, the elastic limit is passed, so when the force is removed, some of the stretch (OS) remains.

- Over which part of Figure 32.4 does a spring balance work?

Friction

Friction is a *contact force* that opposes one surface when it moves, or tries to move over another. Friction is a very useful force. For example, walking is only possible because when your foot pushes against the ground, a friction force opposes it. If the frictional force is greatly reduced, e.g. by a patch of oil or a banana skin, the foot can slide straight back—with the usual result!

There are many other examples of friction being useful—road wheels on bicycles and other vehicles, objects staying still on slightly sloping surfaces, brakes, and even the ability to pick up and hold an object.

Friction can also be a nuisance, particularly between moving parts of machinery. Here, engineers reduce friction with the aid of lubricants, such as oils.

The block shown in Figure 32.5 is placed on a table. A gradually increasing force P is applied using a spring balance. At first, the block does not move. This is because the force applied is opposed by an equal frictional force (F) which acts where the block touches the table. As P increases, F increases, so at any instant $P = F$.

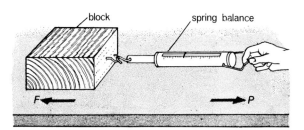

Figure 32.5

If P is increased more, the block eventually moves. Just before motion takes place, F has its greatest value, called the *starting* or *limiting* friction. When the block is moving at a steady speed, the balance reading is slightly less. *Sliding* or *dynamic* friction is less than starting friction.

If a weight is added to the block, the force pressing the surfaces together also increases. This makes both starting friction and sliding friction greater.

Surface tension forces

Although a needle made of steel is denser than water, you can float it on a *clean* water surface. A film formed by dipping an inverted funnel in a detergent solution, rises *up* the funnel (Figure 32.6a). When the film inside the cotton loop in Figure 32.6b is broken, the loop forms a *circle* (Figure 32.6c).

These observations suggest that the surface of a liquid (in a vessel or on a film) behaves as if it was covered with an elastic skin that is trying to shrink. The effect is called *surface tension*. It can be reduced if the liquid is 'contaminated' by adding, for example, detergent: a needle floating in water then sinks.

Surface tension is due to the molecules in a liquid surface being slightly farther apart than normal, like those in a stretched wire.

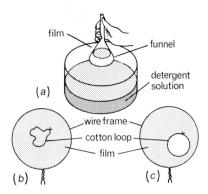

Figure 32.6

Adhesion and cohesion

The force of attraction (page 16) between molecules of the same substance is known as *cohesion*. The force of attraction between molecules of different substances is *adhesion*. The adhesion of water to glass is greater than the cohesion of water, and water spilt on clean glass wets it by spreading to a thin film. By contrast, mercury on glass forms small spherical drops or large flattened ones because cohesion of mercury is stronger than its adhesion to glass.

The cleaning action of detergents depends on their ability to weaken the cohesion of water. Instead of forming drops as it normally does on greasy clothes, the water penetrates the fabric and releases dirt.

Capillarity

If you dip a glass tube of very narrow bore (a capillary tube) into water, the water rises up inside the tube to a height of a few centimetres, (Figure 32.7a). The narrower the tube the greater the rise. Adhesion between water and glass exceeds cohesion between water molecules, the meniscus curves up and the water rises.

The action of blotting paper is due to capillary rise in the narrow spaces between the fibres it contains. The rise of oil up a lamp wick occurs in

Figure 32.7

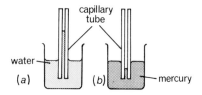

the same way. The damp course in a house, i.e. the layer of non-porous material in the walls just above ground level but below the floor, prevents water rising up the pores in the bricks by capillary action from the ground and causing dampness.

In mercury, the force of cohesion is very much bigger than the force of adhesion. This is why mercury is depressed in a capillary tube (Figure 32.7b).

Questions

1. Why are gases more easily squeezed than liquids or solids?

2. Why is water vapour less dense than water?

3. Explain why (a) diffusion occurs more quickly in a gas than in a liquid, (b) diffusion is still quite slow even in a gas (whose molecules move very fast).

4. A small needle is floated on some clean water. Explain why (a) the needle can float, even though steel is denser than water, (b) adding detergent to the water surface causes the needle to sink.

5. Some salt is placed into a standard flask (Figure 32.8). Water is added until the 250 cm^3 mark is reached.

Figure 32.8

The flask is shaken until the salt dissolves. Explain why the level of the water goes down.

33 FORCES AND TURNING EFFECTS

When a force acts on an object which is hinged or pivoted, e.g. a door, it may cause turning motion. The turning effect of a force depends not only on the size of the force but also on how far it is from the pivot. For example, the handle of a door is at the outside edge so that it opens and closes easily. You would need a much larger force if the handle were near the hinge. In a similar way it is easier to loosen a nut with a long spanner than with a short one.

We can find out more about the turning effect of a force using a lever, that is, a small 'see-saw' balanced at its centre.

Experiment 1 Law of the lever

Set the lever with its central groove resting on a wedge (Figure 33.1). You may have to stick some

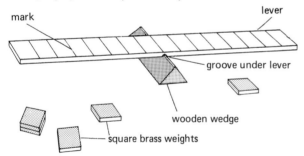

Figure 33.1

Plasticine under the lever at some point to get it to stay level (or nearly so). Place a square brass weight on the *first mark* out from the centre on the left and balance this with another similar weight on the right-hand side (Figure 33.2a). It is best to place the weights on the marks *diagonally*.

Now place 2, 3, 4, etc. weights in turn on the *first mark on the left* and balance each pile by moving the one weight on the right (Figure 33.2b). Record your results in a table, as shown opposite. Try other arrangements of the weights with two piles on one side.

- When the lever is balanced what can you say about the *number* of weights and their *distance* on each side from the pivot?

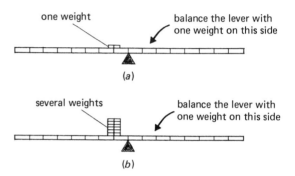

Figure 33.2

Number of weights on left-hand side								Number of weights on right-hand side						
7	6	5	4	3	2	1	Marks from centre	1	2	3	4	5	6	7
							1							
							2							
							3							

Moment of a force

The turning effect or *moment* of a force depends both on the size of the force and the *distance* of its point of action from the pivot. In view of this we measure the moment of a force by multiplying these two quantities together, but the distance taken is the *perpendicular* one from the force to the pivot (or *fulcrum*).

MOMENT OF FORCE = FORCE × PERPENDICULAR DISTANCE FROM PIVOT

The unit of moment is the newton metre (N m). In Figure 33.3a AB is a see-saw, the moment of the force acting at A trying to cause anticlockwise turning about O is $400 \times 3 = 1200$ N m.

If another force F also acts, as in Figure 33.3b, so that the lever is balanced, the moment of F trying to rotate the lever in a clockwise direction

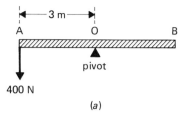
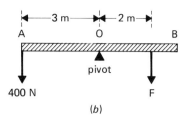

Figure 33.3

about O must equal the moment of the 400 N force acting anticlockwise. The moment of F about O is $F \times 2$ N m, therefore:

$$F \times 2 = 400 \times 3$$
$$\therefore F = \frac{400 \times 3}{2} = \frac{1200}{2} = 600 \text{ N}$$

In general, if a lever is balanced (or as we say is in *equilibrium*) then the sum of the clockwise moments about any point equals the sum of the anticlockwise moments. This is known as the *Law of the Lever* or the *Law of Moments* and is a useful rule.

Worked examples.
1. *A girl weighing 300 N (30 kg) sits at a distance of 2 m from the central pivot of a plank being used as a see-saw. Where must a boy weighing 400 N (40 kg) sit to balance the see-saw (Figure 33.4).*

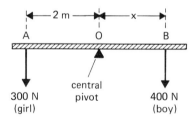

Figure 33.4

Taking moments about the pivot O:

anticlockwise moment = $300 \times OA$
$= 300 \times 2 = 600$ N m
clockwise moment = $400 \times OB = 400 \times x$ N m

By the Law of Moments, when the see-saw is balanced:

clockwise moment = anticlockwise moment
$$\therefore 400 \times x = 600$$
$$\therefore x = \frac{600}{400} = \frac{6}{4} = 1 \cdot 5 \text{ m}$$

2. *The see-saw in Figure 33.5 balances when Sue of weight 320 N is at A, Tom of weight 540 N is at B and Harry of weight W is at C. Find W.*

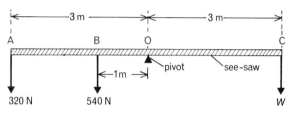

Figure 33.5

Taking moments about the pivot O:

anticlockwise moment = $320 \times 3 + 540 \times 1$
$= 1500$ N m
clockwise moment = $(W \times 3)$ N m

By the Law of Moments:

clockwise moments = anticlockwise moments
$$3W = 1500$$
$$W = 500 \text{ N}$$

Levers

A lever is a device which can turn about a pivot. In a lever a force called the *effort* is used to overcome a resisting force called the *load* acting at some other point on the lever.

When we use a crowbar to move a heavy boulder (Figure 33.6a), our hands apply the effort at one end of the bar and the load is the force exerted by the boulder on the other end. If distances from the fulcrum O are as shown and the load is 1000 N, the effort needed can be calculated from the Law of Moments. As the boulder *just begins* to move we can say, taking moments about O, that:

clockwise moment = anticlockwise moment
effort $\times 200 = 1000 \times 10$
$\therefore$ effort = 50 N

The crowbar in effect magnifies the effort twenty

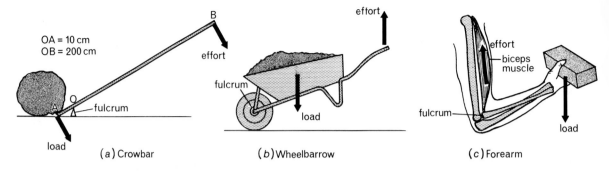

Figure 33.6

times but the effort must move farther than the load.

Other examples of levers are shown in Figure 33.6b and 33.6c. In b the load is between the effort and the fulcrum; in this case as in a the effort is less than the load. In c the effort (applied by the biceps muscle) is between the load and the fulcrum and is greater than the load but the load moves farther than the effort.

Questions

1. Metre rulers marked off at intervals of 10 cm are shown in Figure 33.7. Identical coins are placed on the rulers as shown. State in each case whether the rulers will turn clockwise, anticlockwise or are balanced?

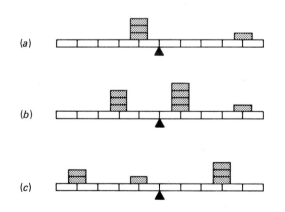

Figure 33.7

2. (a) Is the lever in Figure 33.8a balanced?
(b) If the lever in Figure 33.8b is balanced what is the value of F?

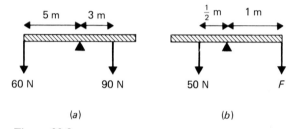

Figure 33.8

3. In an experiment to weigh a boy a 6 m long plank is pivoted at its centre. The boy stands on the plank 0·4 m from the pivot and is balanced by a load of 80 N (8 kg) at the other end of the plank. What does he weigh?

4. In Figure 33.9 the distance XY = YZ. Find in each case the force F which will keep the lever balanced.

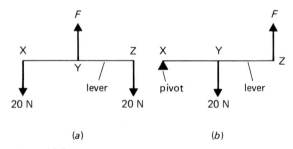

Figure 33.9

34 CENTRE OF GRAVITY

A body behaves as if its whole weight were concentrated at one point, called its *centre of gravity* (c.g.), even though the earth attracts every part of it. The c.g. of a ruler is at its centre and when supported there it balances (Figure 34.1a). If it is supported at any other point it topples because the moment of its weight W about the point of support is not zero (Figure 34.1b).

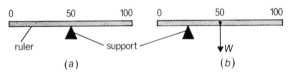

Figure 34.1

Your c.g. is near the centre of your body and the vertical line from it to the floor must be within the area enclosed by your feet or you will fall over. You can test this by standing with one arm and the side of one foot pressed against a wall (Figure 34.2). Now try to raise the other leg sideways.

Figure 34.2

- Can you do it without falling over?

A tight-rope walker has to keep his c.g. exactly above the rope. Some carry a long pole to help them to balance (Figure 34.3). The combined weight of the walker and pole is then spread out more and if the walker begins to topple to one side he moves the pole to the other side.

Figure 34.3

The c.g. of a regular-shaped body of the same density all over is at its centre. In other cases it can be found by experiment.

Experiment 1 Finding c.g. using a plumb line

Suppose we have to find the c.g. of an irregular-shaped lamina (a thin sheet) of cardboard.

Make a hole A in the lamina and hang it so that it can swing freely on a nail clamped in a stand. It will come to rest with its c.g. vertically below A. To locate the vertical line through A tie a plumb line (a thread and a weight) to the nail (Figure 34.4) and mark its position AB on the lamina. The c.g. lies on AB.

Hang the lamina from another position C and

139

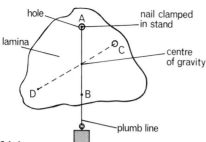

Figure 34.4

mark the plumb line position CD. The c.g. lies on CD and must be at the point of intersection of AB and CD. Check this by hanging the lamina from a third hole. Also try balancing it at its c.g. on the tip of your forefinger.

Toppling

The position of the c.g. of a body affects whether or not it topples over easily. This is important in the design of such things as tall vehicles (which tend to overturn when rounding a corner), racing cars, reading lamps and even teacups.

A body topples when the vertical line through its c.g. falls outside its base (Figure 34.5a). Otherwise it remains stable (Figure 34.5b).

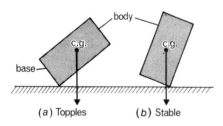

Figure 34.5

Toppling can be investigated by placing an empty can on a plank (with rough surface to prevent slipping) which is slowly tilted. The angle of tilt is noted when the can falls over. This is repeated with 1 kg in the can.

- How does this affect the position of the c.g.?

The same procedure is followed with a second can of the same height as the first but of greater width. It will be found that the can with the weight in it can be tilted through the greater angle.

The stability of a body is therefore increased by:

Figure 34.6a

Figure 34.6b

(i) lowering its c.g.
(ii) increasing the area of its base.

In Figure 34.6a the c.g. of a tractor is being found. It is necessary to do this when testing a new design since tractors are often driven over sloping surfaces and any tendency to overturn must be discovered. The stability of a double-decker bus is being tested in Figure 34.6b. When the top deck only is fully laden with passengers (represented by sand bags in the test), it must not topple if tilted through an angle of 28°. Racing cars have a low c.g. and a wide wheel base.

Stability

Three terms are used in connection with stability.

a A body is in *stable equilibrium* if when slightly displaced and then released it returns to its previous position. The ball at the bottom of the dish in Figure 34.7a is an example. Its c.g. rises when it is displaced. It rolls back because its weight has a moment about the point of contact.

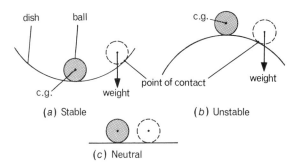

Figure 34.7

b A body is in *unstable equilibrium* if it moves farther away from its previous position when slightly displaced. The object in Figure 34.7b behaves in this way. Its c.g. falls when it is displaced slightly.

c A body is in *neutral equilibrium* if it stays in its new position when displaced (Figure 34.7c). Its c.g. does not rise or fall.

Questions

1. An irregular-shaped sheet of card is hung by a thread at two points A and B in turn. Copy Figure 34.8b and mark the c.g. of the sheet. Draw a line on your diagram which would be vertical if the sheet were hung from point C.

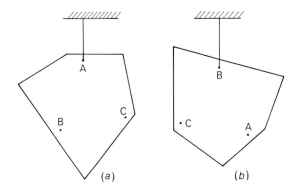

Figure 34.8

2. A Bunsen burner is shown in three different positions in Figure 34.9. State in which position it is in (*a*) stable equilibrium, (*b*) unstable equilibrium, (*c*) neutral equilibrium.

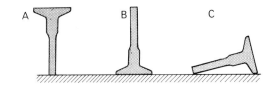

Figure 34.9

3. The weight of the uniform (equally thick all along its length) bar in Figure 34.10 is 10 N. Does it balance, tip to the right or tip to the left? Why?

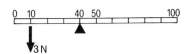

Figure 34.10

35 FORCES AND PRESSURE

Pressure in liquids

To explain some effects in which a force acts on a body we have to consider not only the force but also the area on which it acts. For example, wearing skis prevents a skier from sinking into soft snow because his weight is spread over a greater area. We say the *pressure* is less.

The greater the area over which a force acts the less is the pressure. This is why a tractor with wide wheels can move over soft ground. The pressure is large when the area is small and accounts for a nail having a sharp point.

Pressure is the force (or thrust) *acting on unit area* and is calculated from:

$$\text{PRESSURE} = \frac{\text{FORCE}}{\text{AREA}}$$

The unit of pressure is the *pascal* (Pa); it equals 1 newton per square metre (N/m²) and is quite a small pressure. An apple in your hand exerts about 1000 Pa.

The pressure exerted on the floor by the same box (*a*) standing on end, (*b*) lying flat, is shown in Figure 35.1.

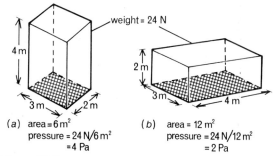

(a) area = 6 m²
pressure = 24 N/6 m²
= 4 Pa

(b) area = 12 m²
pressure = 24 N/12 m²
= 2 Pa

Figure 35.1

Liquid pressure

(*a*) *Pressure in a liquid increases with depth.* In Figure 35.2 water spurts out fastest and furthest from the lowest hole because the farther down you go the greater the weight of liquid above.

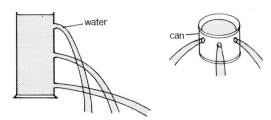

Figure 35.2   **Figure 35.3**

(*b*) *Pressure at one depth acts equally in all directions.* The can of water in Figure 35.3 has similar holes all round it at the same level. Water comes out as fast and as far from each hole. Hence, the pressure exerted by the water at this depth is the same in all directions.

(*c*) *A liquid finds its own level.* In the U-tube of Figure 35.4a the liquid pressure at the foot of P is greater than at the foot of Q because the left-hand

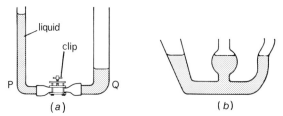

Figure 35.4

column is higher than the right-hand one. When the clip is opened the liquid flows from P to Q until the pressure and the levels are the same, i.e. the liquid finds its own level. Although the weight of liquid in Q is now greater than in P, it acts over a greater area since tube Q is wider.

In Figure 35.4b the liquid is at the same level in each tube and confirms that the pressure at the foot of a liquid column depends only on the *vertical* depth of the liquid and not on the width or shape of the tube.

Water-supply system

A town's water-supply often comes from a reservoir on high ground. Water flows from it through

Figure 35.5

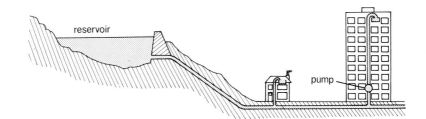

pipes to any tap or storage tank that is below the level of water in the reservoir (Figure 35.5). The lower the level of the place supplied the greater is the water pressure at it.

- Why may it be necessary in very tall buildings to first pump the water to a large tank in the roof?

Reservoirs for water-supply or for hydroelectric power stations are often made in mountainous regions by building a dam at one end of a valley. The dam must be thicker at the bottom than at the top due to the large water pressure at the bottom (Figure 35.6).

Hydraulic machines

Liquids are almost incompressible (i.e. their volume cannot be reduced by squeezing) and they pass on any pressure applied to them. Use is made of these facts in hydraulic machines. Figure 35.7 shows the principle on which they work. Suppose a downward force of 1 N acts on a piston A of area $1/100$ m². The pressure transmitted through the liquid is:

$$\text{pressure} = \frac{\text{force}}{\text{area}} = \frac{1}{1/100} = 100 \, \text{Pa}$$

This pressure acts on piston B of area $\frac{1}{2}$ m². The total upwards force or thrust on B is given by:

$$\text{force} = \text{pressure} \times \text{area} = 100 \times \tfrac{1}{2} = 50 \, \text{N}$$

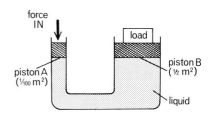

Figure 35.7

A force of 1 N thus produces a force of 50 N.

A *hydraulic jack* (Figure 35.8) has a platform on top of piston B and is used in garages to lift cars. Both valves open only to the right and they allow B to be raised a long way when A moves up and down

Figure 35.6

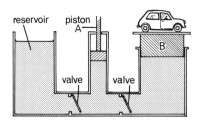

Figure 35.8

repeatedly. A *hydraulic fork lift truck* works similarly. In a *hydraulic press* there is a fixed plate above B, and one like that in Figure 35.9 shapes car bodies from sheets of steel placed between B and the plate. *Hydraulic car brakes* are shown in Figure 35.10. When the brake pedal is pushed the piston in the master cylinder exerts a force on the brake fluid and the resulting pressure is transmitted equally to eight other pistons (four are shown). These force the brake shoes or pads against the wheels and stop the car.

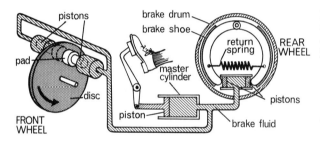

Figure 35.10

PUMPS

Syringe

The syringe is used by doctors to give injections and by gardeners to spray plants. It consists of a tight-fitting piston in a barrel (Figure 35.11) and

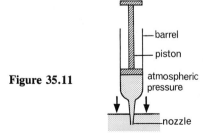

Figure 35.11

Figure 35.9

may be filled by putting the nozzle under the liquid and drawing back the piston. This reduces the air pressure in the barrel and atmospheric pressure forces the liquid up into it. Pushing down the piston drives liquid out of the nozzle.

Lift pump

From early times lift pumps were used to raise water from wells and many are preserved in villages today. They have two valves, one in the piston and one at the bottom of the barrel. The valves, which allow water to flow upwards only, are pieces of leather weighted by metal discs that normally keep them closed.

The pump is primed by first pouring water on top of the piston to prevent air leaking past it. After a few strokes the pump fills with water as a result of the following action.

On the upstroke (Figure 35.12a), valve A is closed and the pressure falls in the barrel below the piston. Atmospheric pressure acting on the surface of the water in the well then pushes water up the pipe through valve B. Also, water on top of the piston is lifted upwards and flows out of the spout.

On the downstroke (Figure 35.12b), valve B closes by the pressure of water on it and valve A opens, allowing water to pass upwards into the barrel above the piston.

Atmospheric pressure can only support a column of water 10 m high and this is the maximum height through which water can be raised by a lift pump. In practice, it is less because of leaks at the valves and piston.

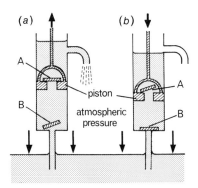

Figure 35.12

Force pump

Force pumps can raise water more than 10 m if the pump itself is within 10 m of the water-supply.

On the upstroke (Figure 35.13), valve A closes and atmospheric pressure forces water up into the barrel of the pump through valve B.

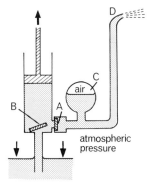

Figure 35.13

On the downstroke, valve B closes and water is forced through valve A into the reservoir C and also out of the spout D. The air in C is compressed and on the next upstroke it expands so keeping up the supply of water from D.

Bicycle pump

When the piston is pushed in, the air between it and the tyre valve is compressed. This pushes the rim of the plastic washer against the wall of the barrel to form an airtight seal (Figure 35.14). Air is forced past the tyre valve into the tyre when the pressure of the air between the plastic washer and the valve exceeds the pressure of the air in the tyre.

When the piston is withdrawn, the tyre valve is

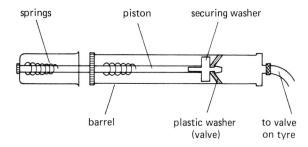

Figure 35.14

closed by the greater pressure in the tyre. Atmospheric pressure then forces air past the plastic washer, no longer pressed hard against the wall, into the barrel.

Vacuum pump

A modern vacuum pump is shown in Figure 35.15. Figure 35.16 shows how it works. The drum and vanes in Figure 35.16 are rotated by an electric motor. The volume of space A increases and the

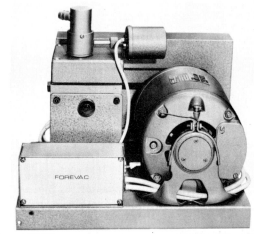

Figure 35.15

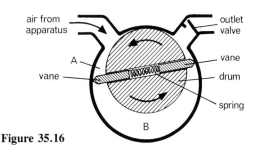

Figure 35.16

145

pressure decreases so drawing air in from the apparatus to be evacuated. The air in space B is then compressed and forced out of the outlet valve. This happens every half rotation until hardly any air is left.

Siphon

Using a siphon is often the only way to empty a tank. To start the siphon (Figure 35.17), it must be full of liquid and end A must be below the liquid level in the tank.

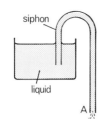

Figure 35.17

ATMOSPHERIC PRESSURE

The air forming the earth's atmosphere stretches upwards a long way. Air has weight. The air in an ordinary sized room weighs about the same as you do, e.g. 500 N.

The air molecules are in constant motion in all directions (page 16). They constantly bombard all objects, exerting pressure on them. Since there are so many molecules present, the pressure at sea-level is great. This pressure acts equally in all directions.

We do not normally feel atmospheric pressure because the pressure inside our bodies is almost the same as that outside. This is also the case with other objects but if air is removed from one place, the effect of atmospheric pressure is noticed.

A space from which all (or nearly all) the air has been taken away is called a *vacuum*.

Air pressure demonstrations

(a) *Collapsing can.* If air is removed from a can (Figure 35.18a), by a vacuum pump, the can collapses because the air pressure inside becomes less than that outside (Figure 35.18b).

(b) *Magdeburg hemispheres.* The vacuum pump was invented by von Guericke, the Mayor of Magdeburg. About 1650 he used it to remove the air from two large hollow metal hemispheres, fitted together to give an airtight sphere. So good was his

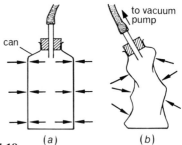

Figure 35.18

pump that it took two teams, each of eight horses, to separate the hemispheres. A similar experiment can be done by two people with small brass hemispheres (Figure 35.19).

Figure 35.19

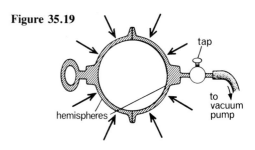

Using air pressure

(a) *Drinking straw* (Figure 35.20). When you suck, your lungs expand and air passes into them from the straw. Atmospheric pressure pushing

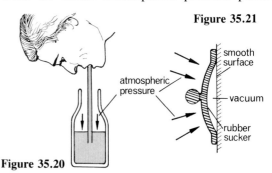

Figure 35.21

Figure 35.20

down on the surface of the liquid in the bottle is now greater than the pressure of the air in the straw and so forces the liquid up to your mouth.

(b) *Rubber sucker* (Figure 35.21). When the sucker is moistened and pressed on a smooth flat surface, the air inside is pushed out. Atmospheric pressure then holds it firmly against the surface. Suckers are used as holders for towels in the home, for attaching car licences to windscreens and in industry for lifting metal sheets.

Experiment 1 Measuring air pressure

Attach a spring balance marked 0–100 N to a small rubber sucker which has been moistened and stuck firmly to a smooth, flat surface. Pull the balance slowly but steadily and when the sucker comes off the surface, note the balance reading (say F newtons). Repeat several times and take the average value.

Find the area of the sucker (say A cm^2) *either* by drawing round it on a piece of centimetre graph paper and estimating the number of 1 cm^2 squares it encloses *or* by measuring its radius r and working out πr^2 (the area of a circle). Then, air pressure $= F/A$ newtons per square centimetre. The result is only very rough.

- What are the causes of error?

At sea-level, air pressure is about 10 N/cm^2 or 100 000 N/cm^2 (since 1 m$^2 = 100 \times 100$ cm^2) or 100 000 Pa (100 kPa).

Pressure gauges

These measure the pressure exerted by a fluid, i.e. by a liquid or a gas.

(*a*) *Bourdon gauge.* It works like the toy in Figure 35.22*a*. The harder you blow into the paper tube, the more it uncurls. In a Bourdon gauge, Figure 35.22*b*, when a fluid pressure is applied, the curved

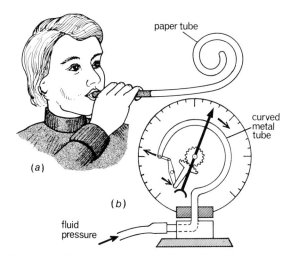

Figure 35.22

metal tube tries to straighten out and rotates a pointer over a scale. Car oil pressure gauges and the gauges on gas cylinders are of this type.

(*b*) *U-tube manometer.* In Figure 35.23*a* each surface of the liquid is acted on equally by atmospheric pressure and the levels are the same. If one side is connected to, for example, the gas supply, Figure 35.23*b*, the gas exerts a pressure on surface A and level B rises until:

pressure = atmospheric + pressure due
of gas   pressure   to liquid in
                             column BC

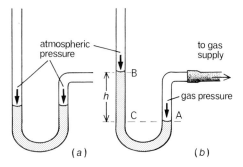

Figure 35.23

The pressure of the liquid column BC therefore equals the amount by which the gas pressure *exceeds* atmospheric pressure. It equals $10\,hd$ (in Pa) where h is the vertical height of BC (in metres) and d is the density of the liquid (in kg/m^3). The height h is called the *head of liquid* and sometimes, instead of stating a pressure in Pa, we say that it is so many centimetres of water (or mercury for higher pressures).

Mercury barometer

A barometer is a manometer which measures atmospheric pressure. A simple barometer can be made by using a small funnel to nearly fill a thick-walled glass tube about 1 m long with mercury. If the tube is slowly inverted several times, with a finger over the open end, the large air bubble runs up and down collecting any small air bubbles trapped in the mercury.

The tube is then filled with mercury, closed with the finger and inverted into a bowl of mercury. When the finger is removed the mercury falls until it is about 760 mm above the level in the bowl, Figure 35.24*a*. The pressure at X due to the weight of the column of mercury XY equals the atmospheric pressure on the surface of the mercury in the bowl. XY measures the atmospheric pressure in millimetres of mercury (mm Hg).

The *vertical* height of the column is unchanged if the tube is tilted.

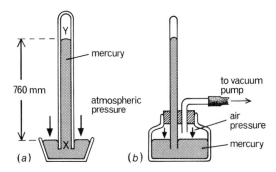

Figure 35.24

Questions

1. A girl wearing stiletto heeled shoes is more likely to damage a wooden floor than is an elephant. Explain why. (Stiletto heels come almost to a point.)
2. (a) What is the pressure on a surface when a force of 50 N acts on an area of (i) 2 m², (ii) 100 m², (iii) 0·5 m²?
   (b) A pressure of 10 Pa acts on an area of 3 m². What is the force acting on the area?
3. A block of concrete weighs 800 N and its base is a square of side 2 m. What pressure does the block exert on the ground?
4. (a) What is the volume of the block in Figure 35.26?
   (b) What is the mass of the block if its density is 2000 kg/m³?

Figure 35.26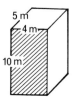

- Would it be different with a wider tube? Why?

The space above the mercury in the tube is a vacuum (except for a little mercury vapour).

- How could you test this?

The apparatus in Figure 35.24b may be used to show that it is atmospheric pressure which holds up the column. When the air above the mercury in the bottle is pumped out, the column falls.

(c) What is the weight of the block? (Assume a mass of 1 kg has weight 10 N.)
(d) What pressure is exerted on the ground by the block?
(e) If the shaded side of the block is on the ground, what effect, if any, will this have on (i) the force exerted by the block on the ground, (ii) the pressure exerted by the block on the ground?

5. In a hydraulic press a force of 20 N is applied to a piston of area 0·2 m². The area of the other piston is 2 m². What is (a) the pressure transmitted through the liquid, (b) the force on the other piston?

Aneroid barometer

An aneroid (no liquid) barometer consists of a partially evacuated, thin metal box with corrugated sides to increase its strength, Figure 35.25.

6. (a) Why must a liquid and not a gas be used as the 'fluid' in a hydraulic machine?
   (b) On what other important property of a liquid do hydraulic machines depend?
7. What is the pressure 100 m below the surface of sea water of density 1150 kg/m³?
8. A siphon tube full of water is shown in Figure 35.27 dipping into two cans of water. Copy and complete the

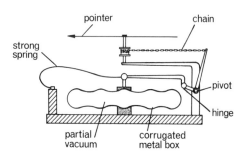

Figure 35.25

The box is prevented from collapsing by a strong spring. If the atmospheric pressure increases, the box caves in slightly, if it decreases the spring pulls it out. A system of levers magnifies this movement and causes a chain to move a pointer over a scale.

Aneroid barometers are used as weather glasses, high pressure being associated with fine weather. They are also used as altimeters to measure the height of an aircraft since the pressure decreases the higher it goes.

Figure 35.27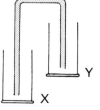

diagram showing the water-level in each can, so that water will flow from X to Y.

9. A bicycle pump can be used as a vacuum pump if its end is sawn off and the plastic washer reversed. With the help of a diagram explain how it works.

10. A simple mercury barometer is shown in Figure 35.28.

Figure 35.28

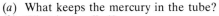

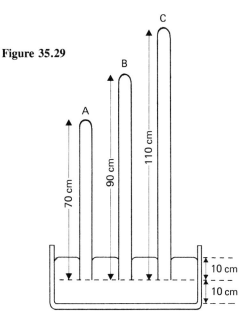

Figure 35.29

(a) What keeps the mercury in the tube?
(b) What is A called?
(c) What is the value of the atmospheric pressure shown on the barometer?
(d) Would the value be different if the tube was (i) wider, (ii) tilted?
(e) How would the value be affected if the barometer was taken up a mountain?

11. Three thick glass tubes A, B and C of lengths 70 cm, 90 cm and 110 cm, all closed at one end, are completely filled with mercury and then turned upside down in a bowl of mercury as in Figure 35.29. The atmospheric pressure is 75 cm of mercury.

(a) Copy the diagram and show what happens to the mercury in each tube. Mark distances.
(b) Tube B is pushed 5 cm farther down into the vessel. What happens to the mercury level in the tube?

(c) What happens if tube C is raised 5 cm out of the vessel?

12. In one method of moulding plastics, air pressure is used to form a sheet of plastic, previously softened by heating, into a complicated shape such as a bowl. Explain from Figure 35.30 how this happens.

13. A metal block is weighed (a) in air, (b) half-submerged in water, (c) fully submerged in water, (d) fully submerged in a strong salt solution.

The readings obtained, though not necessarily in the correct order, were 5 N, 8 N, 10 N and 6 N. Which reading was obtained for each weighing?

14. A block of wood of volume 50 cm³ and density 0·6 g/cm³ floats on water. What is (a) the mass of the block, (b) the mass of water displaced, (c) the volume immersed in the water? (Density of water = 1 g/cm³.)

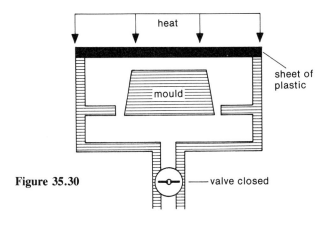

Figure 35.30

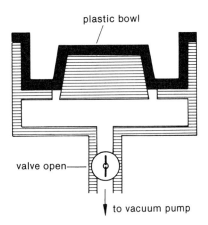

36 FORCES AND MOTION

Speed

If a car travels 300 km in 5 hours, its *average speed* is 300/5 = 60 km/h. The speedometer would certainly not read 60 km/h for the whole journey but might vary considerably from this value. This is why we state the average speed. If a car could travel at a constant speed of 60 km/h for 5 hours, the distance covered would still be 300 km. It is *always* true that:

$$\text{AVERAGE SPEED} = \frac{\text{DISTANCE MOVED}}{\text{TIME TAKEN}}$$

$$v = \frac{s}{t}$$

To find the *actual speed* at any instant we would need to know the distance moved in a very short interval of time.

Velocity

Speed is the distance travelled in unit time; *velocity is the distance travelled in unit time in a stated direction*. If two cars travel due north at 60 km/h, they have the same speed of 60 km/h and the same velocity of 60 km/h *due north*. If one travels north and the other south, their speeds are the same but *not* their velocities since their directions of motion are different.

$$\text{VELOCITY} = \frac{\text{DISTANCE MOVED IN A STATED DIRECTION}}{\text{TIME TAKEN}}$$

The velocity of a body is *uniform* (or *constant*) if it moves with a steady speed in a straight line. If it moves in a curved path, the direction is changing all the time. The speed may be the same, but the velocity is changing.

The units of speed and velocity are the same, e.g. km/h, m/s, and

60 km/h = 60 000 m/3600 s = 17 m/s

Acceleration

When the velocity of a body increases we say it accelerates. If a car starting from rest and moving due north has a velocity 2 m/s after 1 s, its velocity has increased by 2 m/s in 1 s and its acceleration is 2 m/s per second due north. We write this as 2 m/s^2.

Acceleration is the change of velocity in unit time.

$$\text{ACCELERATION} = \frac{\text{CHANGE OF VELOCITY}}{\text{TIME TAKEN FOR CHANGE}}$$

For a steady increase of velocity from 40 km/h to 70 km/h in 5 s

$$\text{acceleration} = \frac{(70-40)}{5} = 6 \text{ km/h per second}$$

Acceleration, like velocity, is a vector quantity and both its magnitude and direction should be stated. However, at present we will consider only motion in a straight line and so the magnitude of the velocity will equal the speed, and the magnitude of the acceleration will equal the change of speed in unit time.

The speedometer readings of a car accelerating on a straight road are shown below.

Time (s)	0	1	2	3	4	5	6
Speed (km/h)	0	5	10	15	20	25	30

The speed increases by 5 km/h every second and the acceleration is said to be *uniform*.

An acceleration is positive if the velocity increases and negative if it decreases. A negative acceleration is also called a *deceleration* or *retardation*.

Worked example.

A car has a speed of 50 km/h. Its speed 5 s later is 60 km/h.
(a) What is its acceleration in km/h *per second*?
(b) What will its speed be after another 2 s if the acceleration stays the same?
(c) How long will it take to reach 80 km/h *at this acceleration*?

(a) We have:

$$\text{acceleration} = \frac{\text{change of velocity}}{\text{time taken for change}}$$

$$= \frac{(60-50)}{5}$$

$$= \frac{10}{5} = 2 \text{ km/h per second}$$

(b) The car's speed is increasing by 2 km/h every second, therefore after another 2 s the increase of speed will be 4 km/h.

$$\therefore \text{new speed} = 60 + 4 = 64 \text{ km/h}$$

(c) To reach 80 km/h its speed will have to increase by $(80 - 64) = 16$ km/h. If it carries on increasing by 2 km/h every second then:

$$\text{time required} = \frac{16}{2} = 8 \text{ s}$$

Tickertape timer: tape charts

A tickertape timer (Figure 36.1) enables us to measure speeds and hence accelerations. It has a steel strip which vibrates fifty times a second and makes dots at 1/50 s intervals on the paper tape being pulled through it. 1/50 s is called a 'tick'.

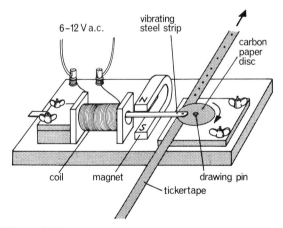

Figure 36.1

The distance between successive dots equals the average speed of whatever is pulling the tape in, say, centimetres per 1/50 s, i.e. centimetres per tick. The 'ten-tick' (1/5 s) is also used as a unit of time. Since ticks and tenticks are small we drop the 'average' and just refer to the 'speed'.

Tape charts are made by sticking successive strips of tape, usually tentick lengths, side by side.

That in Figure 36.2 represents a body moving with *uniform velocity* since equal distances have been moved in each tentick interval.

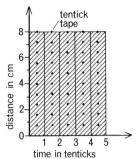

Figure 36.2

The chart in Figure 36.3 is for *uniform acceleration*: the 'steps' are of equal size showing that the speed increases by the same amount in every tentick. The acceleration (average) can be found from the chart as follows.

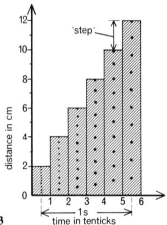

Figure 36.3

The speed during the *first* tentick is $2/\frac{1}{5}$ or 10 cm/s. During the *sixth* tentick it is $12/\frac{1}{5}$ or 60 cm/s. And so during this interval of 5 tenticks, i.e. 1 second, the change of speed is $(60 - 10)$ cm/s = 50 cm/s.

$$\text{acceleration} = \frac{\text{change of speed}}{\text{time taken}}$$

$$= \frac{50}{1}$$

$$= 50 \text{ cm per s per s}$$

$$= 50 \text{ cm/s}^2$$

Experiment 1 Investigating motion

a *Your own motion.* Pull a 2 m length of tape through a tickertape timer as you walk away from it quickly, then slowly, then speeding up again and finally stopping (Figure 36.4a).

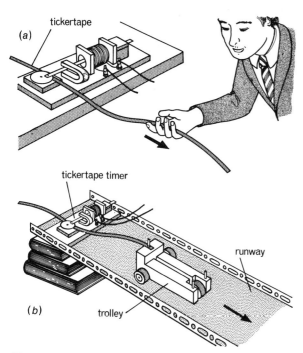

Figure 36.4

Cut the tape into tentick lengths and make a tape chart. Write labels on it to show where you speeded up, slowed down, etc.

b *Trolley on a sloping runway.* Attach a length of tape to a trolley and release it at the top of a runway (Figure 36.4b). The dots will be very crowded at the start—ignore them; but beyond cut the tape into tentick lengths. Make a tape chart.

• Is the acceleration uniform?

Velocity–time graphs

If the velocity of a body is plotted against the time, the graph obtained is a velocity–time graph. It provides a way of solving motion problems. Tape charts are crude velocity–time graphs which show the velocity changing in jumps rather than smoothly, as occurs in practice.

The area under a velocity–time graph measures the distance travelled.

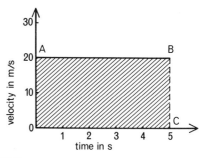

Figure 36.5

In Figure 36.5 AB is the velocity–time graph for a body moving with a *uniform velocity* of 20 m/s. Since distance = average velocity × time, after 5 seconds it will have moved 20 × 5 = 100 m. This is the shaded area under the graph, i.e. rectangle OABC.

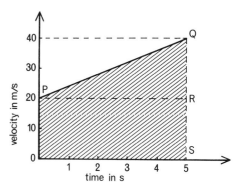

Figure 36.6

In Figure 36.6 PQ is the velocity–time graph for a body moving with *uniform acceleration*. At the start of the timing the velocity is 20 m/s but it increases steadily to 40 m/s after 5 s. If the distance covered equals the area under PQ, i.e. the shaded area OPQS, then:

distance = area of rectangle OPRS
          + area of triangle PQR
    = OP × OS + ½ × PR × QR
(area of triangle = ½ base × height)
    = 20 × 5 + ½ × 5 × 20
    = 100 + 50 = 150 m

Notes.

(*a*) When calculating the area from the graph the unit of time must be the same on both axes.
(*b*) This rule for finding distance travelled is true even if the acceleration is not uniform.

The slope or gradient of a velocity–time graph represents the acceleration of the body.

In Figure 36.5 the slope of AB is zero, as is the acceleration, In Figure 36.6 the slope of PQ is QR/PR = 20/5 = 4: the acceleration is 4 m/s².

Distance–time graphs

A body travelling with uniform velocity covers equal distances in equal times. Its distance–time graph is a straight line, like OL in Figure 36.7 for a

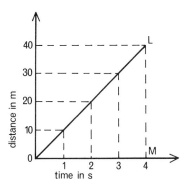

Figure 36.7

velocity of 10 m/s. The slope of the graph = LM/OM = 40/4 = 10, which is the value of the velocity. The following statement is true in general.

The slope or gradient of a distance–time graph represents the velocity of the body.

When the velocity of the body is changing, the slope of the distance–time graph varies (Figure 36.8) and at any point equals the slope of the tangent. For example, the slope of the tangent at T is AB/BC = 40/2 = 20. The velocity at the instant corresponding to T is therefore 20 m/s.

Figure 36.8

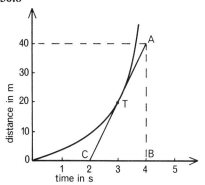

Falling bodies

In air, a coin falls faster than a bit of paper. In a vacuum they fall at the same rate as may be shown with the apparatus of Figure 36.9. The difference is due to *air resistance* being greater to light bodies than to heavy ones—even more so if the light bodies have a larger surface area.

Figure 36.9

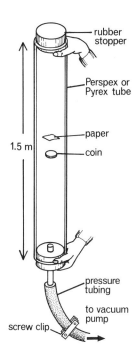

However, heavy bodies, whatever their size, are only slightly affected by air resistance. There is a story, untrue we now think, that in the sixteenth century the Italian scientist Galileo dropped a small iron ball and a large cannon ball ten times heavier, from the top of the Leaning Tower of Pisa in Italy (Figure 36.10). And we are told, to the surprise of onlookers who expected the cannon ball to arrive first, they reached the ground almost at the same time.

Experiment 2 Motion of a falling body

Set up the apparatus as shown in Figure 36.11 and investigate the motion of a 100 g mass falling from a height of about 2 m.

Construct a tape chart using *one* tick lengths. The dots at the start will be too close together, so choose as dot 'O' the first one you can see clearly.

- What does the tape chart tell you about the motion of the falling mass?

Figure 36.10

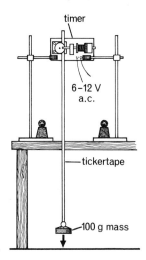

Figure 36.11

Acceleration due to gravity

All bodies falling freely under the force of gravity do so with *uniform acceleration* if air resistance is negligible (i.e. the 'steps' in the previous tape chart should all be equal).

This acceleration, called the *acceleration due to gravity*, is denoted by the italic letter g. Its value varies over the earth, but it is about 9.8 m/s^2 or near enough 10 m/s^2. The velocity of a free falling body therefore increases by 10 m/s every second. A ball thrown straight upwards with a velocity of 30 m/s decelerates by 10 m/s every second and reaches its highest point after 3 s.

Worked example.

A stone is dropped from the top of a high tower.

(a) What is its velocity after 2 s?
(b) How far has it fallen after 2 s?

Ignore air resistance and take $g = 10$ m/s^2.

(a) Since $g = 10$ m/s^2, the velocity of the stone increases by 10 m/s every second. Therefore since it falls from rest, i.e. initial velocity is zero, then:

after 1 s its velocity $= 10$ m/s
after 2 s its velocity $= 10 + 10 = 20$ m/s

(b) Average velocity during first 2 s

$$= \frac{\text{initial velocity} + \text{final velocity}}{2}$$

$$= \frac{0 + 20}{2} = 10 \text{ m/s}$$

But,
$$\text{average velocity} = \frac{\text{distance fallen}}{\text{time taken}}$$

$$\therefore 10 = \frac{\text{distance fallen}}{2}$$

$\therefore$ distance fallen $= 10 \times 2 = 20$ m

Newton's first law of motion

Friction and air resistance cause a car to come to rest when the engine is switched off. If these forces were absent we believe that a body, once set in motion, would go on moving forever with a constant speed in a straight line. That is, force is not needed to keep a body moving with uniform velocity so long as no opposing forces act on it.

This idea was proposed by Galileo and is summed up in Newton's first law.

A body stays at rest, or if moving it continues to move with uniform velocity, unless an external force makes it behave differently.

It seems that the question we should ask about a moving body is not 'what keeps it moving' but 'what changes or stops its motion'. The smaller the external forces opposing a moving body, the smaller is the force needed to keep it moving with uniform velocity. Friction is much reduced for a hovercraft floating on a cushion of air.

Mass and inertia

The first law is another way of saying that all matter has a built-in opposition to being moved if it is at rest or, if it is moving, to having its motion changed. This property of matter is called *inertia* (from the Latin word for laziness).

Its effect is evident on the occupants of a car which stops suddenly; they lurch forward in an attempt to continue moving. The reluctance of a stationary object to move can be shown by placing a large coin on a piece of card on your finger (Figure 36.12). If the card is flicked *sharply* the coin stays where it is while the card flies off.

Figure 36.12

The larger the mass of a body the greater is its inertia, i.e. the more difficult is it to move when at rest and to stop when in motion. Because of this we consider that *the mass of a body measures its inertia*. This is a better definition of mass than the one given earlier (page 126) in which it was stated to be the 'amount of matter' in a body.

Experiment 3 Force, mass and acceleration

The apparatus consists of a trolley to which a force is applied by a stretched length of elastic (Figure 36.13). The velocity of the trolley is found from a timer.

Figure 36.13

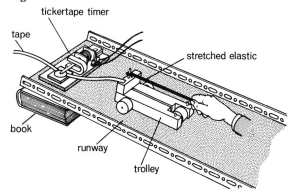

First compensate the runway for friction by raising one end until the trolley runs down with uniform velocity when given a push. The dots on the tape should be equally spaced. There is now no resultant force on the trolley and any acceleration produced later will be due only to the force caused by the stretched elastic.

a *Force and acceleration (mass constant)*. Fix one end of an elastic to the rod at the back of the trolley and stretch it until the other end is level with the front of the trolley. Before attaching a tape, practice pulling the trolley down the runway, keeping the same stretch on the elastic. After a few trials you should be able to produce a steady accelerating force. Now do it with a tape.

Repeat with fresh tapes using first two and then three *identical* elastics, stretched side by side by the same amount, to give two and three units of force.

Make a tape chart for each force and use it to find the acceleration produced in cm/tentick^2 (see page 151). Ignore the start of the tape (where the dots are too close) and the end (where the force may not be steady).

- Does a steady force cause a steady acceleration?

  Put the results in a table, as shown below.

- Do they suggest any relationship between a and F?

Force (F) (no. of elastics)	1	2	3
Acceleration (a) (cm/tentick^2)			

b *Mass and acceleration (force constant)*. Do the experiment as in **a** using two elastics (i.e. constant F) to accelerate first one trolley, then two (stacked one above the other) and finally three. Check the friction compensation of the runway each time.

Find the accelerations from the tape charts and tabulate the results.

- Do they suggest any relationship between a and m?

Mass (m) (no. of trolleys)	1	2	3
Acceleration (a) (cm/tentick^2)			

Newton's second law of motion

The previous experiment should show roughly that the acceleration a is:

(i) directly proportional to the applied force F for a fixed mass, i.e. $a \propto F$
(ii) inversely proportional to the mass m for a fixed force, i.e. $a \propto 1/m$.

Combining the results into one equation, we get
$$a \propto F/m \text{ or } F \propto ma$$
Therefore, $\quad F = kma$

where k is the constant of proportionality.

One newton is defined as the force which gives a mass of 1 kg an acceleration of 1 m/s².

Hence if $m = 1$ kg and $a = 1$ m/s² then $F = 1$ N. Substituting in $F = kma$ we get $k = 1$ and so we can write:

$$F = ma$$

This is Newton's second law of motion. When using it two points should be noted. First, F is the resultant (or unbalanced) force causing the acceleration a. Second, F must be in newtons, m in kilograms and a in metres per second squared, otherwise k is not 1.

Worked example.

A block of mass 2 kg is pushed along a table with a constant velocity by a force of 5 N. When the push is increased to 9 N what is (a) the resultant force, (b) the acceleration?

When the block moves with constant velocity the forces acting on it are balanced. The force of friction opposing its motion must therefore be 5 N.
(a) When the push is increased to 9 N the resultant (unbalanced) force F on the block is $(9 - 5)$ N $= 4$ N (since the frictional force is still 5 N).
(b) The acceleration a is obtained from $F = ma$ where $F = 4$ N and $m = 2$ kg.

$$\therefore a = F/m = 4/2 = 2 \text{ m/s}^2$$

Newton's third law of motion

If a body A exerts a force on body B, then body B exerts an equal but opposite force on body A.

The law states that forces never occur singly but always in pairs as a result of the action between two bodies. For example, when you step forward from rest your foot pushes backwards on the earth and the earth exerts an equal and opposite force forward on you. Two bodies and two forces are involved. The small force you exert on the large mass of the earth gives no noticeable acceleration to the earth but the equal force it exerts on your very much smaller mass causes you to accelerate.

Note that the equal and opposite forces *do not act on the same body*; if they did, there could never be any resultant forces and acceleration would be impossible.

An appreciation of the third law and the effect of friction is desirable when stepping from a rowing boat (Figure 36.14). You push backwards on the boat and, although the boat pushes you forwards

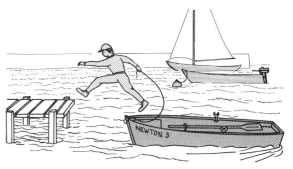

Figure 36.14

with an equal force, it is itself now moving backwards (because friction with the water is slight), and this reduces your forward motion by the same amount—and you might fall in!

Air resistance

When small dense objects, e.g. ball-bearings, fall in air, the air resistance is negligible. And as we have seen, these fall with a constant acceleration of 9·8 m/s². It is appreciable, however, for objects that are light, e.g. raindrops, or have a large surface area, e.g. parachutes or the 'Red Devils' sky divers in free fall (Figure 36.15). In these cases, as the

Figure 36.15

falling object speeds up, air resistance increases and reduces its acceleration. Eventually air resistance acting upwards equals the weight of the object acting downwards. The resultant force is then zero and the object falls with a constant velocity called its *terminal* velocity.

Figure 36.16

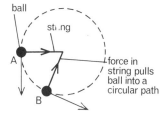

Circular motion

In Figure 36.16 a ball on a string is being whirled round in a horizontal circle. Its direction of motion is changing all the time. Its velocity is also changing because an object only moves with constant velocity if *both* its speed and its direction of motion do not change. The ball is accelerating even although its speed may be steady.

The force causing this acceleration is the *inwards* pull of the string on the ball and is called the *centripetal* (centre-seeking) *force*. When anything moves in a circle (or curved path) there must be a centripetal force acting on it. A car rounding a bend has a frictional force exerted inwards on its tyres by the road.

Questions

1. What is the average speed in km/h of (*a*) someone who walks 15 km in 3 hours, (*b*) an athlete who runs 40 km in 5 hours, (*c*) a cyclist who travels 30 km in 2 hours?

2. A car travels on a straight road at a steady speed of 80 km/h. What distance does it cover in 90 minutes? How long will it take to travel 240 km at this speed?

3. A motor cyclist starts from rest and reaches a speed of 6 m/s after travelling with uniform acceleration for 3 s. What is his acceleration?

4. A train has a speed of 60 km/h. Its speed 5 s later is 70 km/h.
(*a*) What is its acceleration in km/h per second?
(*b*) What will its speed be after another 2 s if the acceleration remains the same?
(*c*) How long will it take to reach 105 km/h at this acceleration?

5. An aircraft travelling at 600 km/h accelerates steadily at 10 km/h per second. Taking the speed of sound as 1100 km/h at the aircraft's altitude, how long will it take to reach the 'sound barrier'?

6. Figure 36.17*a* shows the distance–time graph for a moving object and Figure 36.17*b* shows the speed–time graph for another moving object.

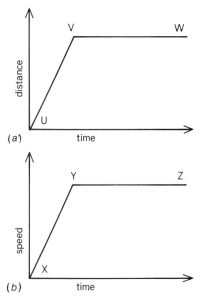

Figure 36.17

Describe the motion, if any, of the objects in the regions (*a*) UV, (*b*) VW, (*c*) XY, (*d*) YZ.

7. The graph in Figure 36.18 represents the distance travelled by a car plotted against time.

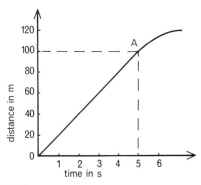

Figure 36.18

(*a*) How far has the car travelled at the end of 5 s?
(*b*) What is the speed of the car during the first 5 s?
(*c*) What has happened to the car after A?

157

8. Figure 36.19 shows an uncompleted velocity–time graph for a boy running a distance of 100 m.

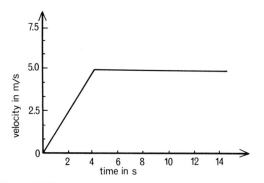

Figure 36.19

(a) What is his acceleration during the first 4 s?
(b) How far does the boy travel during (i) the first 4 s, (ii) the next 9 s?
(c) Copy and complete the graph showing clearly at what time he has covered the distance of 100 m. Assume his speed remains constant at the value shown by the horizontal portion of the graph.

9. An object falls from rest over a high cliff with an acceleration of $10\,\text{m/s}^2$.

(a) What is its velocity after (i) 1 s, (ii) 2 s, (iii) 3 s, (iv) 4 s, (v) 5 s?
(b) Using the answers from (a) plot a velocity–time graph for the object for its first 5 s of fall.
(c) From the velocity–time graph work out the distance the object has fallen after (i) 1 s, (ii) 2 s, (iii) 3 s, (iv) 4 s, (v) 5 s.

10. A ball is thrown vertically upwards with a velocity of 20 m/s. If the acceleration due to gravity is $10\,\text{m/s}^2$ what is its velocity after (a) 1 s, (b) 2 s?

11. Which one of the diagrams in Figure 36.20 shows the arrangement of forces which gives the block M the greatest acceleration?

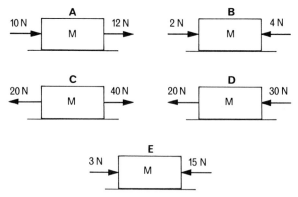

Figure 36.20

12. In Figure 36.21 if P is a force of 20 N and the object moves with *constant velocity* what is the value of the opposing force F?

Figure 36.21

13. (a) What resultant force produces an acceleration of $5\,\text{m/s}^2$ in a car of mass 1000 kg?
(b) What acceleration is produced in a mass of 2 kg by a resultant force of 30 N?

14. A car of mass 500 kg accelerates steadily from rest to 40 m/s in 20 s.

(a) What is its acceleration?
(b) What resultant force produces this acceleration?
(c) The actual force will be greater. Why?

15. What provides the centripetal force for
(a) a *coin* placed on a rotating turntable,
(b) a *space capsule* circling the earth?

37 ENERGY

Animals and plants live and grow because they have something called *energy*. Machines can do many useful jobs but only if they have a supply of energy. It is usually easy to recognize when a job is being done and energy is being 'used'. Energy does, however, take a number of apparently very different forms and it is not always easy to see that something possesses energy. For example, there is nothing apparently very 'energetic' about a gallon of petrol, although when burned it will release enough energy to enable a 1 tonne motor car to travel up to 60 km at quite high speeds.

Most of our energy comes from the sun in the form of heat and light radiations (page 217). The heat energy keeps this planet at temperatures which allow life to exist. Light enables us to see. Light is also used by plant life to manufacture materials we use as foods (see *'Photosynthesis'*, page 86). Modern civilization is very dependent on 'fossil fuels' like petroleum and coal, which were formed by the absorption of the sun's energy many millions of years ago by animal and plant life. These reserves will be used up in the foreseeable future and much modern research is concerned with finding suitable alternative sources of energy.

Forms of energy

(*a*) *Kinetic energy (K.E.)*. Any moving body has K.E. The faster it moves, the more K.E. it has. When a moving body is stopped, e.g. when a hammer hits a nail, or brakes stop a car, most of the K.E. becomes heat energy.

(*b*) *Mechanical potential energy (P.E.)*. This comes in two forms.

Elastic P.E. is energy stored by a stretched or squeezed body. For example, a stretched bow or catapult has elastic P.E. which can be used to give K.E. to a missile. Similarly, a wound-up spring can drive a watch.

Gravitational P.E. is possessed by a raised body. A stone raised to the top of a building has more energy than one at the bottom of the building, since if it is released it will fall and its P.E. will become K.E.. If water is stored behind a dam, it has more P.E. than water at a lower level. When released, the water loses P.E. and gains K.E. This K.E. can be used to drive machinery which generates electrical energy. Note that the energy of the water originally came from the sun. Heat from the sun evaporated water which rose, formed clouds, then fell as rain in high places.

Mechanical P.E. and K.E. are usually called *mechanical energy*.

(*c*) *Chemical P.E.* Food and fuels are stores of chemical energy. Foods release energy as a result of chemical reactions in the body. In a typical human being, about 25 per cent of the chemical energy is available as mechanical energy. The rest is released as heat needed to maintain a steady body temperature.

When fuels are burnt they release energy mainly as heat. The heat is then used to drive an engine, e.g. a petrol engine (in cars), gas turbine (in aircraft) or steam turbine (in ships and power stations).

(*d*) *Other forms of energy. Heat energy* causes a change in the internal energy of the particles of which matter is made. Light energy and infra-red radiation (radiant heat) are forms of electromagnetic energy (page 217).

Electrical energy can be in the form of both P.E. and K.E. A charged cloud stores electrical P.E. When the charge is released through a lightning discharge, the moving charged particles (a current) are a form of K.E. Since electrical energy can be transmitted easily through wires and converted into other energy forms very simply, much of our energy is first converted into electrical energy before transmission and use.

Sound energy exists when materials (e.g. air) are set in vibration (page 220). Human ears are very sensitive, so even for a very loud sound, the amount of energy is very small.

Nuclear energy is the source of an appreciable amount of the electrical energy used in the western world. In a *nuclear reactor*, heavy atoms of either uranium 235 or plutonium are broken down

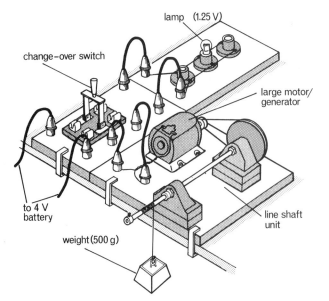

Figure 37.1

(fissioned) into smaller atoms by bombardment with neutrons (page 384). This breaking down causes some of the potential energy of atomic nuclei to drive the steam turbines which power electrical generators.

Energy changes

There are many devices which can cause one energy form to change into another.

The apparatus in Figure 37.1 may be used to show a battery changing *chemical* energy to *electrical* energy, which in the electric motor becomes K.E. The motor raises the weight, giving it *gravitational P.E.* If a change-over switch is joined to the lamp, and the weight is allowed to fall, the P.E. of the weight drives the motor which now acts as an electrical generator (or dynamo). This drives current through the lamp where *heat* and *light* energies are produced.

Some other examples of energy changes are shown in Figure 37.2.

Figure 37.2

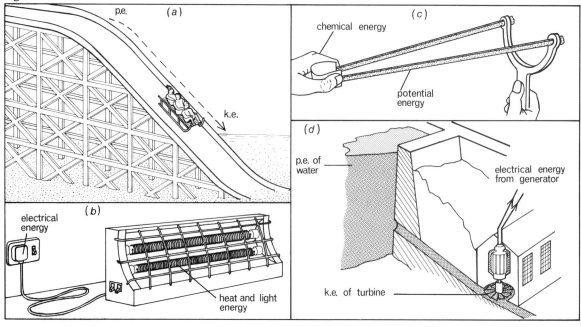

Experience shows that in a great many energy changes the final form of energy is heat, generally at a low temperature. In the case of a car travelling at a steady speed the chemical energy of the petrol used becomes heat and warms slightly the air through which the car passes. When the car comes to a halt the brakes, the tyres, the road and the air are all heated.

It is unfortunate that most energy becomes heat because heat, especially at low temperatures, is not easily changed into other forms of energy. There is enough low temperature heat, sometimes called low quality heat, in the sea to meet all our energy needs if only we could use it.

Work

In science, the term *work* means the amount of energy changed or transferred.

To measure scientific work we use the fact that frequently an energy change causes a force to push something through a distance. For example, if you lift a 3 kg load 2 m higher (Figure 37.3), chemical

Figure 37.3

energy is transferred *from* your muscles *to* the load. Since the mass of the load is 3 kg its *weight* will be near enough 30 N and so in lifting it 2 m you will have to apply a force of 30 N. We say the work done is 30×2, i.e. 60 newton metres (Nm). By this we mean that 60 Nm of chemical energy has been changed to mechanical energy. *Work* or energy change from one form to another is therefore given by:

WORK = FORCE × DISTANCE MOVED IN DIRECTION OF FORCE

Energy is rather like money, and work done like a bill that has been paid. The bill states how much money has been transferred from one person to another.

The unit of work is the *joule* (J) and is the *work done when a force of 1 newton* (N) *moves through 1 metre* (m). For example, if you have to pull with a force of 50 N to move a box steadily 3 m in the direction of the force, the work done is $50 \times 3 = 150 \, \text{Nm} = 150 \, \text{J}$. That is:

JOULES = NEWTONS × METRES

All forms of energy, as well as work, are measured in joules.

Conservation of energy

The principle of conservation of energy states that energy cannot be destroyed, it only changes from one form to another.

If it seems that some energy has 'disappeared' during a change, the 'lost' energy has often been converted into heat. For example, when a moving motor car is stopped by the brakes, the K.E. 'lost' has been converted into heat at the brakes and tyres due to friction. When a brick falls, its P.E. becomes K.E. At the ground, the K.E. becomes mainly heat, with some sound.

Although the total energy remains constant, we are continually using useful sources of energy like fossil fuels and getting low temperature (low grade) heat in return. Figure 37.4 shows one of the alternative energy sources—solar, wind and tidal—which are now being investigated.

Figure 37.4

Power

The *power* of something which uses energy (e.g. a motor) is the *rate at which it changes energy from one form to another*. This is the same as the rate at which it does work, i.e. the work it does each second. The more powerful a car is, the greater is the rate at which it changes useful forms of energy.

$$\text{POWER} = \frac{\text{ENERGY CHANGE}}{\text{TIME TAKEN}} = \frac{\text{WORK DONE}}{\text{TIME TAKEN}}$$

The unit of power is the *watt* (W). One watt is a rate of working of 1 joule per second. 1 W = 1 J/s. If a machine changes 500 J of energy in 10 s, then its power is:

$$P = \frac{500}{10} = 50 \text{ J/s} = 50 \text{ W}$$

Larger units of power are:

1 kilowatt (kW) = 1000 W
1 megawatt (MW) = 1 000 000 W

An electric oven of 5 kW is converting electrical energy into heat at a rate of 5000 J each second.

Questions

1. Name one device which changes (*a*) electrical energy to light, (*b*) sound to electrical energy, (*c*) chemical energy to electrical energy, (*d*) P.E. to K.E., (*e*) electrical energy to K.E.

2. How much work is done when a mass of 3 kg (weighing 30 N) is lifted vertically through 6 m?

3. A boy of mass 50 kg climbs a wall 2 m high and then jumps to the ground.

(*a*) What is his weight in newtons?
(*b*) How much work is done in climbing the wall?
(*c*) What kind of energy and how much of it does he have *just before* he lands on the ground?
(*d*) What happens to this energy after he lands?

4. In each of the following say what energy changes occur and calculate the work in joules.

(*a*) A crane, driven by a Diesel engine, lifts a crate, on which the earth pulls with a force of 400 N, through a vertical height of 8 m.
(*b*) A man pushes a car *steadily* along a level road for 100 m with a force of 250 N.

5. A boy whose weight is 600 N runs up a flight of stairs 10 m high in 15 s. What is the average power he develops?

38 SIMPLE MACHINES

A machine is any device which enables a force (the *effort*) acting at one point to overcome another force (the *load*) acting at some other point. A lever (page 136) is a simple machine. So are pulleys, gears, etc. Simple machines are often combined to build more complicated machines like the crane in Figure 38.4.

Definitions

If a load of 50 N is moved by applying an effort of 10 N to a machine, the *mechanical advantage* of the machine is 50/10 = 5.

$$\text{MECHANICAL ADVANTAGE (M.A.)} = \frac{\text{LOAD}}{\text{EFFORT}}$$

Machines with M.A. greater than 1 allow heavy loads to be moved by smaller efforts. But the effort has to move farther than the load. We define

$$\text{VELOCITY RATIO (V.R.)} = \frac{\text{DISTANCE MOVED BY EFFORT}}{\text{DISTANCE MOVED BY LOAD}}$$

V.R. is fixed by the design of the machine. M.A. has to be found experimentally. The V.R. of the lever in Figure 38.1 is 10. The effort needed to

Figure 38.1

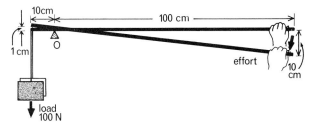

raise the load is found by taking moments about the fulcrum O. Do this (ignore the weight of the lever).

- What is the M.A.?

Efficiency of a machine

Machines make work easier. No machine is perfect and in practice more work is done by the effort on a machine than by the machine on the load. Work measures energy transfer (or change) and so we can also say that the energy input to a machine is greater than its energy output. Some energy is always wasted to overcome friction and move parts of the machine itself. In Figure 38.1 there is friction at the fulcrum and the lever has weight; the effort will therefore be more than 10 N. We define

$$\text{EFFICIENCY} = \frac{\text{ENERGY OUTPUT}}{\text{ENERGY INPUT}}$$

$$= \frac{\text{WORK DONE BY LOAD}}{\text{WORK DONE BY EFFORT}}$$

This is expressed as a percentage and is always less than 100 per cent.

There is a useful relation between M.A., V.R. and efficiency. From the above equation we can say, remembering that work = force × distance

$$\text{efficiency} = \frac{\text{load} \times \text{distance load moves}}{\text{effort} \times \text{distance effort moves}}$$

$$= \text{M.A.} \times \frac{1}{\text{V.R.}}$$

$$\therefore \text{PERCENTAGE EFFICIENCY} = \frac{\text{M.A.}}{\text{V.R.}} \times 100\%$$

Pulleys

(a) Single fixed pulley (Figure 38.2). This enables us to *lift* a load L more conveniently by applying a

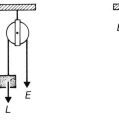

Figure 38.2   **Figure 38.3**

downwards effort E. E need be only slightly greater than L and if friction in the pulley bearings is negligible then $E = L$ and M.A. = 1.

- What is the V.R.?

(b) Single moving pulley (Figure 38.3). If the effort applied to the free end of the rope is E the total upward force on the pulley is $2E$ since two parts of the rope support it. A load $L = 2E$ can therefore be raised if the pulley and rope are frictionless and weightless. That is, M.A. = 2 (but less in practice).

To raise the load by 1 m requires each side of the rope to shorten by 1 m. The free end has therefore to take up 2 m of slack and so V.R. = 2.

(c) Block and tackle. This type of pulley system is used in cranes (Figure 38.4) and lifts. It consists of two blocks each with one or more pulleys. In the arrangement of Figure 38.5a the pulleys in the blocks are shown one above the other for clarity; in practice they are side by side on the same axle (Figure 38.5b). The rope passes round each pulley in turn.

Figure 38.4

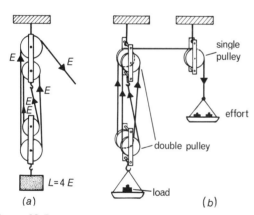

Figure 38.5

The total upward force on the lower block is $4E$ since it is supported by four parts of the rope and a load $L = 4E$ can be raised. Hence M.A. = 4 if the pulleys are frictionless and weightless. Using the same reasoning as in (b) we see that V.R. = 4, i.e. the number of times the rope passes from one block to the other.

Experiment 1 Efficiency of a pulley system

Set up the system of Figure 38.5b or a similar one. Starting with 50 g in the load pan, add weights to the effort pan until the load just rises steadily.

Record the load and effort in a table, as shown below (100 g has a weight of 1 N) and repeat for greater loads.

The V.R. can be obtained as explained before. Work out the M.A. and the efficiency for each pair of readings of load and effort.

Load N	Effort N	M.A.	Efficiency = $\frac{M.A.}{V.R.} \times 100\%$

Notes.

a The lower pulley block and the load pan are also raised by the effort but are not included as part of the load. They become less important as the load increases and the M.A. and the efficiency both increase for this reason. The V.R. is constant for a particular system.

b The efficiency is less than 100 per cent because the system is not frictionless and the moving parts not weightless.

Other simple machines

(a) Wheel and axle. A screwdriver and the steering wheel of a car (Figure 38.6) use the wheel and axle

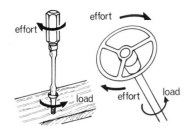

Figure 38.6

principle. This is shown in Figure 38.7a; the effort is applied to a rope wound round the wheel and the load is raised by another rope wound oppositely on

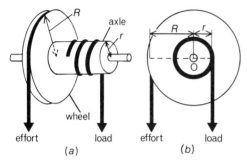

Figure 38.7

the axle. For one complete turn of the wheel the effort moves a distance $2\pi R$, i.e. the circumference of the wheel, and the load a distance $2\pi r$. Hence:

$$\text{V.R.} = \frac{2\pi R}{2\pi r} = \frac{R}{r}$$

The M.A. of a frictionless wheel and axle is also R/r as you can prove from Figure 38.7b by taking moments about O.

(b) *Inclined plane.* It is easier to push a barrel up a plank on to a lorry than to lift it vertically. In Figure 38.8 to raise the load L through a *vertical* height h, the smaller effort E moves a greater distance d equal to the length of the incline.

Figure 38.8

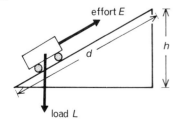

$$\text{V.R.} = \frac{\text{length of incline}}{\text{height of incline}} = \frac{d}{h}$$

The M.A. of a *perfect* incline may be calculated from the principle of Conservation of Energy (page 161) by assuming:

work done on load = work done by effort

i.e. $L \times h = E \times d$

$$\therefore \text{M.A.} = \frac{L}{E} = \frac{d}{h}$$

In fact, because of friction, it will be less.

(c) *Screw jack.* In a car jack a screw passes through a nut carrying an arm that fits into the car chassis (Figure 38.9a). When the effort applied to the

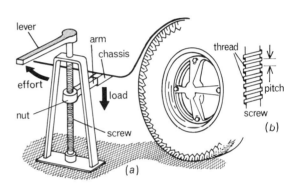

Figure 38.9

lever at the top of the screw makes one complete turn, the screw (and the load) rises a distance equal to its pitch, i.e. the distance between successive threads (Figure 38.9b).

$$\text{V.R.} = \frac{\text{circumference of circle made by effort}}{\text{pitch of screw}}$$

The M.A. of a perfect jack can also be calculated assuming conservation of energy.

(d) *Gears.* The V.R. and M.A. of a machine can be changed by gears. In Figure 38.10 the gear with 10 teeth makes two revolutions for each complete

Figure 38.10

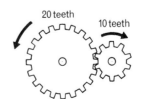

revolution of the one with 20 teeth. The V.R. is therefore 2 if the effort is applied to the small gear which drives the large gear. Hence:

$$\text{V.R.} = \frac{\text{no. of teeth on driven gear}}{\text{no. of teeth on driving gear}}$$

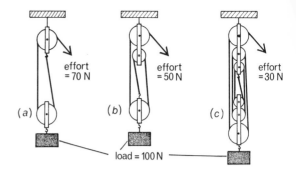

Figure 38.11

Questions

1. A machine of V.R. 6 is used to raise a load of weight 250 N. The effort needed is 50 N.
   (a) The M.A. of the machine is (i) 1/5, (ii) 4, (iii) 5, (iv) 10, (v) 50?
   (b) The efficiency of the machine is (i) 50%, (ii) 65%, (iii) 75%, (iv) 83%, (v) 110%?

2. An effort of 250 N raises a load of 1000 N through 5 m using a pulley system. If the effort moves through 30 m, what is (a) the work done in raising the load, (b) the work done by the effort, (c) the efficiency of the pulley system?

3. For each pulley system shown in Figure 38.11 what is (i) the M.A., (ii) the V.R., (iii) the efficiency?

4. A machine consisting of a wheel of radius 50 cm and an axle of radius 10 cm is used to lift a load of 400 N with an effort of 100 N. For this system calculate (a) the M.A., (b) the V.R., (c) the efficiency.

5. A trolley of weight 10 N is pulled from the bottom to the top of the inclined plane in Figure 38.8 (page 165) at a steady speed by a force of 2 N. If $h = 2$ m and $d = 20$ m what is (a) the M.A., (b) the V.R., (c) the efficiency?

6. If gear A has 30 teeth and drives gear B with 75 teeth, how many times does A rotate for each rotation of B?

39 HEAT ENGINES

A heat engine is a machine which changes heat energy, obtained by burning a fuel, to kinetic energy. In an *internal combustion* engine, e.g. petrol, diesel, jet engine, the fuel is burnt in the cylinder or chamber where the energy change occurs. This is not so in other engines, e.g. steam turbine.

SOME TYPES OF HEAT ENGINE

Petrol engine

Figure 39.1 shows the action of a petrol engine. On the *intake stroke*, the piston moves down (due to the starter motor in a car or the kickstart in a motor cycle turning the crankshaft) so reducing the pressure inside the cylinder. The inlet valve opens and the petrol–air mixture from the carburettor is forced into the cylinder by atmospheric pressure.

On the *compression stroke*, both valves are closed and the piston moves up, compressing the mixture.

On the *power stroke*, a spark jumps across the points of the sparking plug and explodes the mixture, forcing the piston down.

On the *exhaust stroke*, the outlet valve opens and the piston rises, pushing the exhaust gases out of the cylinder.

The crankshaft turns a flywheel (a heavy wheel) whose momentum keeps the piston moving between power strokes.

Most cars have at least four cylinders on the same crankshaft (Figure 39.2). Each cylinder 'fires' in turn in the order 1–3–4–2, giving a power stroke every half revolution of the crankshaft. Smoother running results.

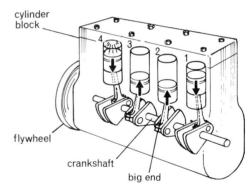

Figure 39.2

Figure 39.1

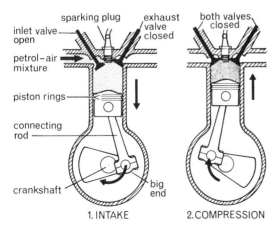
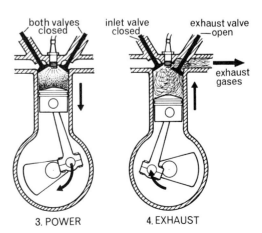

167

Jet engines (gas turbines)

If you release an inflated balloon with its neck open, it flies off in the opposite direction to that of the escaping air (Figure 39.3). This is the principle of both jet engines and rockets.

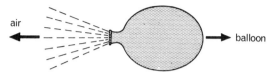

Figure 39.3

There are several kinds of jet engine; Figure 39.4 is a simplified diagram of a turbo-jet.

To start the engine, an electric motor sets the compressor rotating. The compressor is like a fan, its blades draw in and compress air at the front of

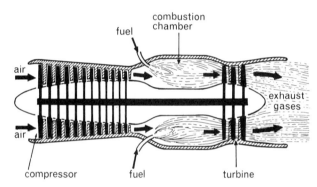

Figure 39.4

the engine. Compression raises the temperature of the air before it reaches the combustion chamber. Here, fuel (kerosene, i.e. paraffin) is injected and burns to produce a high-speed stream of hot gas which escapes from the rear of the engine, so thrusting it forward. The exhaust gas also drives a turbine (another fan) which, once the engine is started, turns the compressor, since both are on the same shaft.

Turbo-jet engines have a high power-to-weight ratio (i.e. produce large power for their weight) and are ideal for use in aircraft. They are also used in ships, trains and cars.

Rockets

Rockets, like jet engines, obtain their thrust from the hot gases they eject by burning a fuel. They

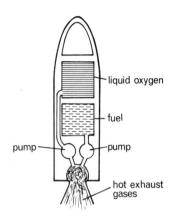

Figure 39.5

can, however, travel where there is no air since they carry the oxygen needed for burning, instead of taking it from the atmosphere as does a jet engine. Space rockets use liquid oxygen (at −183 °C). Common fuels are kerosene and liquid hydrogen (at −253 °C), but solid fuels are also used. Figure 39.5 is a simplified drawing of a rocket.

Figure 39.6

Figure 39.7

Steam turbine

Steam turbines are used in power stations and nuclear submarines (Figure 39.6). They have efficiencies of about 30 per cent.

The action of a steam turbine resembles that of a water-wheel but moving steam not moving water causes the motion. Steam produced in a separate boiler enters the turbine and is directed by the *stator* (sets of fixed blades) on to the *rotor* (sets of blades on a shaft that can rotate). The rotor revolves and drives whatever is coupled to it, e.g. an electrical generator or a ship's propeller. The steam expands as it passes through the turbine and the size of the blades increases along the turbine to allow for this. Figure 39.7 shows the rotor of a steam turbine.

Rotary engines like the steam turbine are smoother than piston (reciprocating) engines.

Questions

1. In a motor car engine chemical energy is converted into mechanical energy. What other forms of energy are produced? Some of the mechanical energy is not usefully employed. What happens to it? Describe the energy changes involved when the brakes are applied.

2. The cylinder of a four-stroke petrol engine is shown in Figure 39.8. What is wrong with the diagram if the sparking plug has just fired?

Figure 39.8

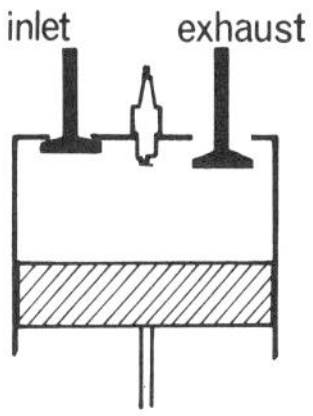

3. Why does a rocket work better outside the atmosphere?

Heat, Light and Sound Energy

40 THERMOMETERS

The temperature of anything tells us how hot it is and is measured by a thermometer, usually in *degrees Celsius* (°C). There are different kinds of thermometers, each kind being suitable for a certain job.

Liquid-in-glass thermometer

In this type the liquid in a glass bulb expands up a capillary tube when the bulb is heated. The liquid must be easily seen and expand (or contract) rapidly by a large amount over a wide range of temperature. It must not stick to the inside of the tube or the reading will be low when the temperature is falling.

Mercury and coloured alcohol are in common use. Mercury freezes at $-39\,°C$ and boils at $357\,°C$; alcohol freezes at $-115\,°C$ and boils at $78\,°C$ and is therefore more suitable for low temperatures.

Scale of temperature

A scale and unit of temperature are obtained by choosing two temperatures, called the *fixed points*, and dividing the range between them into a number of equal divisions or *degrees*.

On the Celsius scale (named after the Swedish scientist who suggested it), *the lower fixed point is the temperature of pure melting ice* and is taken as $0\,°C$. Impurities in the ice lower its melting point (page 185).

The upper fixed point is the temperature of the steam above water boiling at normal atmospheric pressure of 760 mmHg and is taken as $100\,°C$. The temperature of the boiling water itself is not used because any impurities in the water raise its boiling point; the temperature of the steam is not affected however (page 185).

Methods of finding the fixed points are shown in Figure 40.1*a*, *b*. When they have been marked on the thermometer, the distance between them is divided into 100 equal degrees (Figure 40.2). The thermometer now has a scale, i.e. it has been calibrated or graduated.

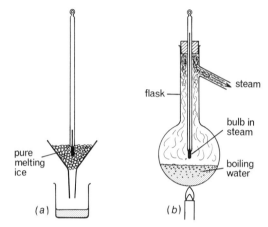

Figure 40.1

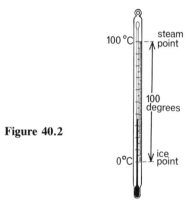

Figure 40.2

Clinical thermometer

This is a special type of mercury-in-glass thermometer used by doctors and nurses. Its scale only extends over a few degrees on either side of the normal body temperature of $37\,°C$ (Figure 40.3).

The tube has a constriction (i.e. a narrower part) just beyond the bulb. When the thermometer is placed under a patient's tongue the mercury expands, forcing its way past the constriction. When the thermometer is removed (after 1 minute) from the mouth, the mercury in the bulb cools and contracts, breaking the mercury thread at the constriction. The mercury beyond the con-

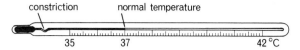
Figure 40.3

striction stays in the tube and shows the body temperature. After use, the mercury is returned to the bulb—by a flick of the wrist.

Heat and temperature

It is important not to confuse the temperature of a body with the heat energy that can be obtained from it. For example, a red-hot spark from a fire is at a higher temperature than the boiling water in a saucepan. If the spark landed in the water, heat would pass from it to the water even though much more heat energy could be supplied by the water.

Temperature decides the direction in which heat flows just as pressure does in liquid flow. Heat flows naturally from a body at a higher temperature to one at a lower temperature. A liquid flows of its own accord from a higher pressure to a lower one.

Questions

1. From the list of temperatures below choose the one which is most likely for each of the following: (*a*) temperature of a healthy person, (*b*) melting point of pure ice, (*c*) temperature of a warm room in a house, (*d*) temperature of a steel-making furnace.

1600 °C, 100 °C, 60 °C, 37 °C, 20 °C, 10 °C, 0 °C, −10 °C.

2. Why will a thermometer measure small changes of temperature more accurately if the diameter of the capillary tube is narrow?

3. (*a*) Why does a clinical thermometer have a very narrow part just above the bulb?
   (*b*) Why must a clinical thermometer never be put into hot water to wash it?

41 EXPANSION OF SOLIDS AND LIQUIDS

Solids, liquids and gases expand when they are heated and contract when they are cooled.

Experiment 1 Expansion of solids

a Arrange an iron rod as in Figure 41.1. The 1 kg weight keeps the rod pressed down on the needle.

Heat the rod strongly by moving a burner along it. The rod expands most easily at the end resting on the needle which rolls, rotating the straw and showing up the very small expansion.

b Arrange the 'bar-breaker' as in Figure 41.2 with the cast iron rod *inside* the pillar and the knurled nut screwed tight so that the rod is hard against the pillar. Heat the bar *strongly*; it will expand and exert so large a force that the rod breaks.

c Rearrange the 'bar-breaker' with a new cast iron rod through a hole in the bar which is *outside* the pillar. Also move the knurled nut outside the other pillar and again heat the bar strongly for a few minutes. Then tighten the nut and cool the bar by pouring cold water on it. The rod should snap.

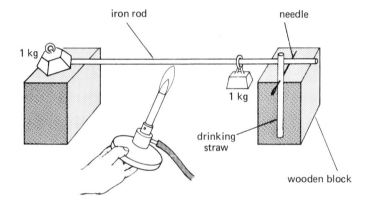

Figure 41.1

Precautions against expansion

The expansion of solids can sometimes be a nuisance and engineers have to take precautions to prevent damage occurring.

(*a*) *Railway lines.* Previously, gaps were left between the lengths of rail to allow for expansion in summer. They caused a 'clickety-click' sound as the train passed over them.

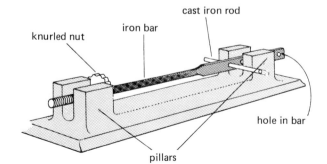

Figure 41.2

174

Figure 41.3*a*

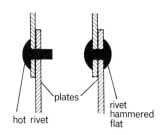

Figure 41.4

Figure 41.3*b*

Today rails are welded into lengths of about 1 km and are held by concrete sleepers that can withstand the large forces created, without buckling. Also, at the joints the ends are tapered and overlap (Figure 41.3*a*). This gives a smoother journey and allows some expansion near the ends of each length of rail.

(*b*) *Bridges*. One end is fixed and the other rests on rollers which permit movement (Figure 41.3*b*).

Uses of expansion

(*a*) *Riveting metal plates*. A white-hot rivet is placed in the hole drilled in the plates and its end is hammered flat. On cooling it contracts and pulls the plates together (Figure 41.4). Steel plates are riveted to make ships and boilers.

(*b*) *Bimetallic thermostat*. A bimetallic strip consists of equal lengths of two different metals, e.g. copper and iron, riveted together so that they cannot move separately (Figure 41.5*a*). Copper expands more than the same length of iron when they are heated through the same temperature. To allow this to happen the strip has to bend, with copper on the outside (Figure 41.5*b*).

Figure 41.5

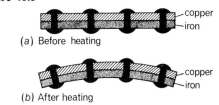

A thermostat keeps the temperature of a room or an appliance constant. The one in Figure 41.6

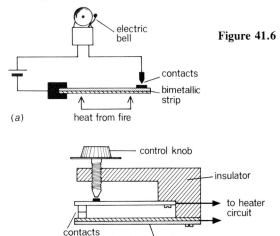

Figure 41.6

uses a bimetallic strip in the electrical heating circuit of, for example, an electric iron.

When the iron reaches the required temperature the strip bends down, breaks the circuit at the contacts and switches off the heater. After cooling a little the strip remakes contact and turns the heater on again. A near-steady temperature results.

If the control knob is screwed down, the strip has to bend more to break the heating circuit and this needs a higher temperature.

Expansion of liquids

The expansion of a liquid can be shown by observing the level of the liquid in the capillary tube in Figure 41.7 when the test-tube is placed in hot water. Different liquids expand by different amounts.

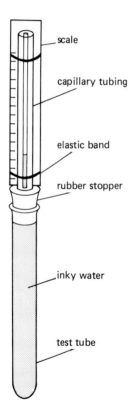

Figure 41.7

As water is cooled to 4 °C it contracts, as we would expect. However, between 4 °C and 0 °C it expands slightly. At 0 °C when it freezes, a large expansion occurs and every 100 cm³ of water becomes 109 cm³ of ice—which accounts for the bursting of water pipes in very cold weather.

Expansion of gases

Gases expand much more than liquids as may be shown by warming the air in the flask in Figure 41.8 with your hands. The plug of water can be forced out of the tube.

Figure 41.8

Questions

1. Explain why (*a*) the metal lid on a glass jam jar can be unscrewed easily if the jar is inverted for a few seconds with the lid in very hot water, (*b*) furniture may creak at night after a warm day, (*c*) concrete roads are laid in sections with pitch between them.

2. A bimetallic strip is made from aluminium and copper. When heated it bends in the direction shown in Figure 41.9. Which metal expands more for the same rise in temperature?

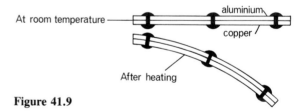

Figure 41.9

Draw a diagram to show how the bimetallic strip would appear if it were cooled to below room temperature.

42 GAS LAWS

The volume of a certain mass of a gas, e.g. air, depends on both its temperature and its pressure. To find how each of these quantities affects the volume we must keep one fixed while the other is changed.

BOYLE'S LAW (VOLUME AND PRESSURE)

Changes in the volume of a gas due to pressure changes can be studied using the apparatus in Figure 42.1. The volume V of air trapped in the glass

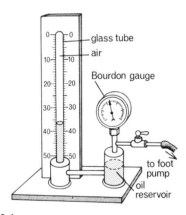

Figure 42.1

tube is read off on the scale behind. The pressure is altered by pumping air from a foot pump into the space above the oil reservoir. This forces more oil into the glass tube and increases the pressure p on the air in it; p is measured by the Bourdon gauge.

If p is doubled, V is halved. That is, V is *inversely proportional* to p.

$$V \propto \frac{1}{p} \quad \text{or} \quad V = \text{constant} \times \frac{1}{p}$$

$$\therefore pV = \text{constant}$$

If several pairs of readings p_1V_1, p_2V_2, etc. are taken, then $p_1V_1 = p_2V_2 = $ a constant.

Boyle's law states that the volume of a fixed mass of gas is inversely proportional to the pressure, if the temperature is unchanged.

Experiment 1 Charles' law (volume and temperature)

Arrange the apparatus as in Figure 42.2. The index of concentrated sulphuric acid traps the air column

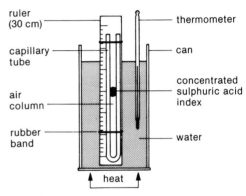

Figure 42.2

to be investigated and also dries it. Adjust the capillary tube so that the bottom of the air column is opposite a convenient mark on the ruler.

Note the length of the air column (to the *lower* end of the index) at different temperatures. Put the results in a table but before taking a reading, stop heating and stir well to make sure that the air has reached the temperature of the water.

Plot a graph of volume (in centimetres, since the length of the air column is a measure of it) on the y-axis and temperature (in °C) on the x-axis.

Absolute zero

The volume–temperature graph for a gas (obtained by an experiment like the above) is a straight line (Figure 42.3). It shows that a gas expands *uniformly* with temperature as measured on a mercury thermometer, i.e. equal temperature

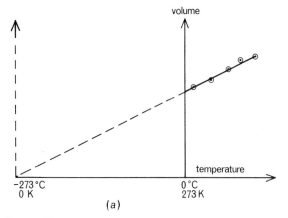

Figure 42.3

increases cause equal volume increases.

The graph does not pass through the Celsius temperature origin (0 °C). If it is produced backwards it cuts the temperature axis at about -273 °C. This temperature is called *absolute zero* because we believe it is the lowest temperature possible. It is the zero of the *absolute* or *Kelvin scale of temperature*.

Degrees in this scale are called *kelvins* and are denoted by K. They are exactly the same size as Celsius degrees, i.e. 1 °C = 1 K. Since -273 °C = 0 K, conversions from °C to K are made by adding 273. For example:

$$0\,°C = 273\,K$$
$$15\,°C = 273 + 15 = 288\,K$$
$$100\,°C = 273 + 100 = 373\,K$$

Kelvin or absolute temperatures are represented by the letter T and if θ stands for a Celsius scale temperature then, in general:

$$T = 273 + \theta$$

The gas laws

Using absolute temperatures the gas laws can be stated in a convenient form for calculations.

(a) *Charles' law*. In Figure 42.3 the volume–temperature graph does pass through the origin if temperatures are measured on the Kelvin scale. That is, if we take 0 K as the origin. We can then say that the volume V is directly proportional to the absolute temperature T, i.e. doubling T doubles V, etc. Therefore:

$$V \propto T \quad \text{or} \quad V = \text{constant} \times T$$
$$\text{or} \quad V/T = \text{constant} \quad (1)$$

Charles' law may be stated as follows:

The volume of a fixed mass of gas is directly proportional to its absolute temperature if the pressure is kept constant.

(b) *Boyle's law*. We have:

$$pV = \text{constant} \quad (2)$$

These two equations can be combined giving:

$$\frac{pV}{T} = \text{constant}$$

It is useful for cases in which p, V and T all change from say p_1, V_1 and T_1 to p_2, V_2 and T_2, then:

$$\frac{p_1 V_1}{T_1} = \frac{p_2 V_2}{T_2} \quad (3)$$

(c) *Pressure law*. Experiments can be done to see how the pressure of a gas changes with temperature when its volume is kept constant. The results enable us to state the Pressure law as follows.

The pressure of a fixed mass of gas is directly proportional to its absolute temperature if the volume is kept constant.

If p represents pressure and T absolute temperature then we can write:

$$p \propto T \quad \text{or} \quad p = \text{constant} \times T$$

that is,

$$\frac{p}{T} = \text{constant}$$

Worked example.

A cycle pump contains 50 cm³ of air at 17 °C and a pressure of 1·0 atmosphere. Find the pressure when the air is compressed to 10 cm³ and its temperature rises to 27 °C.

We have:

$p_1 = 1 \cdot 0$ atmosphere $p_2 = ?$
$V_1 = 50$ cm³ $V_2 = 10$ cm³
$T_1 = 273 + 17 = 290$ K $T_2 = 273 + 27 = 300$ K

From equation (3) above we get:

$$p_2 = p_1 \times \frac{V_1}{V_2} \times \frac{T_2}{T_1}$$

Replacing:

$$p_2 = 1 \times \frac{50}{10} \times \frac{300}{290} = 5 \cdot 2 \text{ atmosphere}$$

Notes.

a All temperatures must be in K.
b Any units can be used for p and V so long as they are the same on both sides of the equation.

c In some calculations the volume of the gas has to be found at s.t.p. (standard temperature and pressure). This is 0 °C and 760 mmHg pressure.

GASES AND THE KINETIC THEORY

The pressure exerted by a gas is due to the molecules bombarding the walls of the vessel. When the temperature of a gas rises the average speed of its molecules increases.

Questions

1. In Figure 42.4b the piston is pushed in so that the air enclosed occupies one-quarter of the length of the cylinder that it occupied in Figure 42.4a. What has happened to (a) the volume of air, (b) the pressure of the air, (c) the number of molecules of air?

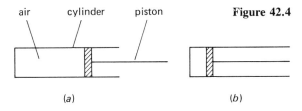

Figure 42.4

2. If a certain quantity of gas has a volume of 30 cm³ at a pressure of 1×10^5 Pa, what is its volume when the pressure is (a) 2×10^5 Pa, (b) 5×10^5 Pa?

3. A gas of volume 2 m³ at 27 °C is (a) heated to 327 °C, (b) cooled to −123 °C, at constant pressure. What are its new volumes?

4. A gas has a volume of 150 cm³ at 27 °C and a pressure of 1 atmosphere. Calculate the volume when (a) the absolute temperature is trebled at constant pressure, (b) the pressure is trebled at constant temperature, (c) the pressure is 2 atmospheres and the temperature 127 °C.

43 CONDUCTION, CONVECTION, RADIATION

To keep a building or a house at a comfortable temperature, in winter and summer, requires a knowledge of how heat travels, if it is to be done economically and efficiently.

CONDUCTION

The handle of a metal spoon held in a hot drink soon gets warm. Heat passes along the spoon by *conduction*.

Conduction is the flow of heat through matter from places of higher to places of lower temperature without movement of the matter as a whole.

A simple demonstration of the different conducting powers of various metals is shown in Figure 43.1. A matchstick is fixed to one end of each

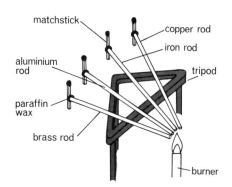

Figure 43.1

rod using a little melted wax. The other ends of the rods are heated by a burner. When the temperatures of the far ends reach the melting point of wax, the matches drop off. The match on copper falls first showing it is the best conductor, followed by aluminium, brass and perhaps iron.

Most metals are good conductors of heat; materials such as wood, glass, cork, plastics and fabrics are bad conductors. The apparatus in Figure 43.2 can be used to show the difference between brass and wood. If the rod is passed

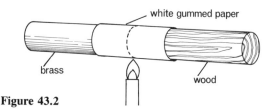

Figure 43.2

through a flame several times, the paper over the wood scorches but not over the brass. The brass conducts the heat away from the paper quickly and prevents it reaching the temperature at which it burns. The wood only conducts it away slowly.

Metal objects below body temperature *feel* colder than those made of bad conductors because they carry heat away faster from the hand—even though all the objects are at the same temperature.

Liquids and gases also conduct heat but only very slowly. Water is a very poor conductor as may be shown in Figure 43.3. The water at the top of the tube can be boiled before the ice at the bottom melts.

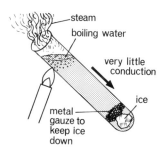

Figure 43.3

Uses of conductors

(a) *Good conductors*. These are used whenever heat is required to travel quickly through something. Kettles, saucepans, boilers and radiators are made of metals such as aluminium, iron and copper.

(b) *Bad conductors* (insulators). The handles of teapots, kettles and saucepans are made of wood or plastic. Cork is often used for tablemats.

Air is one of the worst conductors, i.e. best insulators. This is why houses with cavity walls (i.e. two walls separated by an air space), and double-glazed windows, keep warmer in winter and cooler in summer.

Materials which trap air, e.g. wool, felt, fur, feathers, polystyrene, fibre glass, are also very bad conductors. Some are used as 'lagging' to insulate water pipes, hot water cylinders, ovens, refrigerators and the roofs (and walls) of houses. Others make warm winter clothes.

CONVECTION

Convection in liquids

Convection is the usual method by which heat travels through fluids, i.e. liquids and gases. It can be shown in water by dropping a few crystals of potassium permanganate down a tube to the bottom of a flask of water. When the tube is removed and the flask heated just below the crystals by a *small* flame (Figure 43.4), purple streaks of water rise upwards and fan outwards.

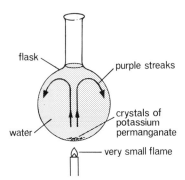

Figure 43.4

Streams of warm moving fluids are called *convection currents*. They arise when a fluid is heated because it expands, becomes less dense and is forced upwards by surrounding cooler, denser fluid which moves under it. We say 'hot water (or hot air) rises'. Warm fluid behaves like a cork released under water: being less dense it bobs up. In convection, however, a fluid floats in a fluid, not a solid in a fluid.

Convection is the flow of heat through a fluid from places of higher to places of lower temperature by movement of the fluid itself.

Convection in air

Black marks often appear on the wall or ceiling above a lamp or radiator. They are caused by dust being carried upwards in air convection currents produced by the hot lamp or radiator.

A laboratory demonstration of convection currents in air can be given using the apparatus of Figure 43.5. The direction of the convection current created by the candle is made visible by the

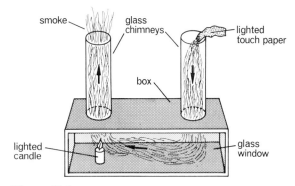

Figure 43.5

smoke from the touch paper (made by soaking brown paper in strong potassium nitrate solution and drying it).

Convection currents set up by electric, gas and oil heaters help to warm our homes. Many so-called 'radiators' are really convector heaters.

Natural convection currents

(a) *Coastal breezes.* During the day the temperature of the land increases more quickly than that of the sea (because the specific heat capacity (see page 188) of the land is much smaller). The hot air above the land rises and is replaced by colder air from the sea. A breeze from the sea results (Figure 43.6a).

At night the opposite happens. The sea has more heat to lose and cools more slowly. The air above the sea is warmer than that over the land and a breeze blows from the land (Figure 43.6b).

(b) *Gliding.* Gliders, including 'hang-gliders' (Figure 43.7), depend on hot air currents, called *thermals*. By flying from one thermal to another gliders can stay airborne for several hours.

RADIATION

Radiation is a third way in which heat can travel but whereas conduction and convection both need

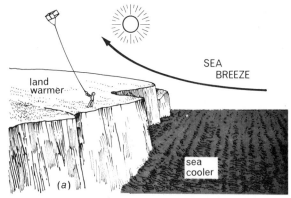
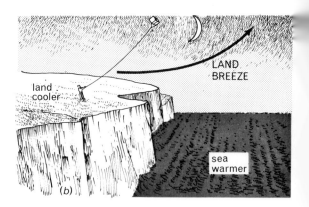

Figure 43.6

Figure 43.7

matter to be present, radiation can occur in a vacuum. It is the way heat reaches us from the sun.

Radiation has all the properties of electromagnetic waves (page 217), e.g. it travels at the speed of radio waves. When it falls on an object, it is partly reflected, partly transmitted and partly absorbed; the absorbed part raises the temperature of the object.

Radiation is the flow of heat from one place to another by means of electromagnetic waves.

Radiation consists mostly of infra-red radiation (page 217), but light and ultraviolet are also present if the body is very hot (e.g. the sun).

Good and bad absorbers

Dull black surfaces are better absorbers of radiation than white shiny surfaces—the latter are good reflectors of radiation. This is why buildings in hot countries are often painted white and why light-coloured clothes are cooler in summer. Also, reflectors on electric fires are made of polished metal because of their good reflecting properties.

Good and bad emitters

Some surfaces also emit radiation better than others when they are hot. If you hold the backs of your hands on either side of a hot copper sheet which has one side polished and the other blackened (Figure 43.8), it will be found that the *dull black surface is a better emitter of radiation than the shiny one.*

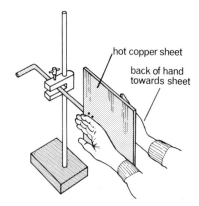

Figure 43.8

Teapots and kettles which are polished are poor emitters and keep their heat longer. In general, surfaces that are good absorbers of radiation are good emitters when hot.

Vacuum flask

A vacuum or Thermos flask keeps hot liquids hot or cold liquids cold. It is very difficult for heat to travel into or out of the flask.

Transfer by conduction and convection is minimized by making the flask a double-walled glass vessel with a vacuum between the walls (Figure 43.9). Radiation is reduced by silvering both

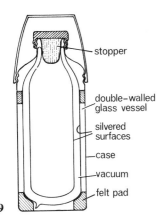

Figure 43.9

walls on the vacuum side. Then, if for example, a hot liquid is stored, the small amount of radiation from the hot inside wall is reflected back across the vacuum by the silvering on the outer wall. The slight heat loss which does occur is by conduction up the walls and through the stopper.

Questions

1. Explain why (*a*) newspaper wrapping keeps hot things hot and cold things cold, (*b*) fur coats would keep their owners warmer if they were worn inside out, (*c*) a string vest keeps a person warm even though it is a collection of holes bounded by string.

2. Why is a glass bottle likely to crack if boiling water is poured into it?

3. Convection takes place, A only in solids, B only in liquids, C only in gases, D in solids and liquids, E in liquids and gases?

4. We feel the heat from a fire by, A convection, B conduction, C regelation (refreezing), D diffusion, E radiation.

5. Three beakers are of the same size and shape: one beaker is painted dull black, one is dull white and one is gloss white. The beakers are filled with boiling water. In which beaker will the water cool most quickly? Give a reason.

6. The door canopy in Figure 43.10 shows in a striking way the difference between white and black surfaces when radiation falls on them. Explain why.

Figure 43.10

44 CHANGES OF STATE

When a solid is heated, it may melt and change its state from solid to liquid. If ice is heated it becomes water. The opposite process of freezing occurs when a liquid solidifies.

A pure substance melts at a definite temperature, called the *melting point*; it solidifies at the same temperature—the *freezing point*.

Experiment 1 Cooling curve of ethanamide

Half fill a test-tube with ethanamide (acetamide) and place it in a beaker of water (Figure 44.1a). Heat the water until all the ethanamide has melted.

Remove the test-tube and arrange it as in Figure 44.1b with a thermometer in the liquid ethanamide. Record the temperature every minute until it has fallen to 70 °C.

Plot a cooling curve of temperature against time.

● What is the freezing (melting) point of ethanamide?

Latent heat of fusion

The previous experiment shows that the temperature of liquid ethanamide falls until it starts to solidify (at 82 °C) and remains constant till it has all solidified. The cooling curve in Figure 44.2 is for a pure substance; the flat part AB occurs at the melting point when the substance is solidifying.

During solidification a substance loses heat to its surroundings but its temperature does not fall.

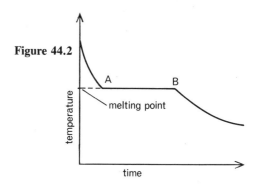

Figure 44.2

When a solid is melting, the heat supplied does not cause a temperature rise. For example, the temperature of a well-stirred ice–water mixture remains at 0 °C until all the ice is melted.

Heat which is *absorbed* by a solid during melting or *given out* by a liquid during solidification is called *latent heat of fusion* (see also page 189). *Latent* means hidden and *fusion* means melting. Latent heat does not cause a temperature change; it seems to disappear.

Latent heat of vaporization

Latent heat is also needed to change a liquid into a vapour. The reading of a thermometer in boiling water remains constant even though heat, called *latent heat of vaporization* (see also page 190), is still being absorbed by the water from whatever is heating it.

When steam condenses to form water, latent heat is given out. This is why a scald from steam may be more serious than one from boiling water.

Effect of pressure on melting point

Increasing the pressure on ice lowers its melting point. This may be shown by the demonstration in

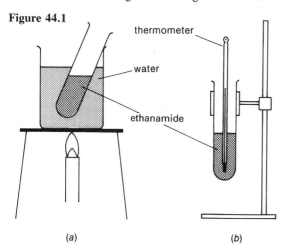

Figure 44.1

Figure 44.3 in which a weighted copper wire passes through a block of ice without cutting it in two.

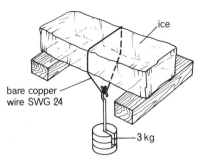

Figure 44.3

A large pressure is exerted on the ice below the wire. The *melting point* is lowered and the ice, being at 0 °C, melts since its temperature is now above its new melting point. The wire sinks through the water which is no longer under pressure and refreezes above the wire because the melting point returns to 0 °C. In refreezing the water gives out latent heat of fusion and this is *conducted* down through the wire to enable the ice below it to melt.

The effect is called *regelation* (refreezing). Regelation causes the snow to 'bind' together when a snowball is made. It may also account for the motion of glaciers.

Effect of impurities on melting point

The temperature of a well-stirred ice–water mixture is normally 0 °C but when an *impurity* such as salt is added, it may fall to −20 °C. The freezing mixture so formed can be used for cooling purposes.

The fall in temperature is due to the salt lowering the *melting point* of the ice. The ice, however, is still at 0 °C, i.e. above its new melting point. It therefore melts, absorbing latent heat from the mixture whose temperature falls till it freezes at the new melting point.

This effect explains the use of antifreeze in car radiators and also why brine (and sea sand) is spread on icy roads. The 'impurity' lowers the freezing point of the mixture and may prevent it freezing.

Effect of pressure on boiling point

The boiling point of water rises when the pressure above it is raised. Using the apparatus in Figure 44.4*a* the pressure can be increased by pinching the rubber tube for *just long enough* to see that the thermometer reading rises.

To show that decreasing the pressure lowers the boiling point, the water should be boiled for a few minutes so that the steam sweeps out most of the air. The heating is then stopped and the clip closed.

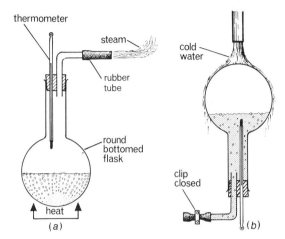

Figure 44.4

Cold water is run over the inverted flask (Figure 44.4*b*), so condensing the water vapour inside it and reducing the pressure above the water. The water starts to boil and if the cooling in this way is continued, it may go on boiling until about 40 °C.

Effect of impurities on boiling point

An 'impurity' such as salt when added to water raises the boiling point.

Evaporation

In evaporation a liquid changes to a vapour without ever reaching its boiling point. A pool of water in the road evaporates and does so most rapidly when there is (*i*) a wind, (*ii*) sun, and (*iii*) only a little water vapour present in the air.

Evaporation and boiling are compared below.

Evaporation	Boiling
1. Occurs at any temperature.	Occurs at a definite temperature—the boiling point.
2. Occurs at surface of liquid: no bubbles.	Occurs within liquid: bubbles appear.

Both processes need latent heat of vaporization. In evaporation this is obtained by the liquid from its surroundings, as may be shown by using the apparatus in Figure 44.5.

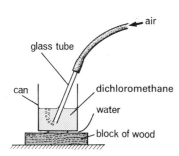

Figure 44.5

Dichloromethane is a *volatile liquid*, that is, it has a low boiling point and evaporates readily at room temperature, especially when air is blown through it. Latent heat is taken first from the dichloromethane itself and then from the water below the can. The water soon freezes causing the block and can to stick together.

Volatile liquids feel cold when spilt on the hand. They are used in perfumes.

Water evaporates from the skin when we sweat. This is the body's way of using unwanted heat and keeping a constant temperature. After vigorous exercise there is a risk of the body being over-cooled, especially in a draught; it is then less able to resist infection.

Cooling by evaporation

The molecules of a liquid have an average speed at a particular temperature but some are moving faster than others. Evaporation occurs when faster molecules escape from the surface of the liquid. The average speed, and therefore the average K.E. of the molecules left behind, decreases, so the temperature of the liquid falls.

A refrigerator causes cooling by evaporation.

Refrigerator

In a refrigerator, heat is taken in at one place and given out at another by the refrigerating substance as it is pumped round a circuit, Figure 44.6.

The coiled pipe round the *freezer* at the top of the refrigerator contains a volatile liquid. This

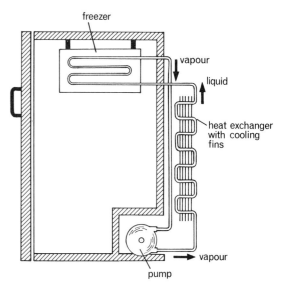

Figure 44.6

evaporates and takes latent heat from its surroundings, so causing cooling. The electrically-driven pump removes the vapour (so reducing the pressure, lowering the boiling point and encouraging evaporation or even boiling) and forces it into the *heat exchanger* (pipes with cooling fins outside the rear of the refrigerator). Here the vapour is compressed and liquefies, giving out latent heat of vaporization to the surrounding air. The liquid returns to the coils round the freezer and the cycle is repeated.

An adjustable thermostat switches the pump on and off, controlling the rate of evaporation and so the temperature in the refrigerator.

Moisture in the air

(*a*) *Dew point*. There is always water vapour in the air due to evaporation from the sea, rivers, lakes, etc. The amount which the air can hold depends on the temperature. If cooling occurs a temperature called the *dew point* is reached, at which *the water vapour present is sufficient to saturate the air*. Further cooling causes condensation of water vapour. If a cold clear night follows a warm day, the excess water vapour often forms a dew on the ground.

(*b*) *Relative humidity*. In a humid atmosphere we feel 'sticky' because moisture is evaporating slowly from our skin. The rate at which this occurs is decided not so much by the amount of water vapour present in the air but by how close it is to saturation, i.e. by its *relative humidity* (R.H.).

$$\text{R.H.} = \frac{\text{mass of water vapour in a certain volume of air}}{\text{mass of water vapour needed to saturate the same volume of air at the same temperature}}$$

This can be found from an instrument called a *hygrometer*.

Questions

1. Explain the following:
(a) On a very cold day good snowballs cannot be made.
(b) When walking on snow it 'cakes' and sticks to the sole of the shoe.
(c) A good cup of tea cannot be made on a high mountain.

2. Some water is stored in a bag made of a porous material, e.g. canvas, which is hung where it is exposed to a draught of air. Explain why the temperature of the water is lower than that of the air.

45 MEASURING HEAT

Specific heat capacity

If 1 kg of water and 1 kg of paraffin are heated in turn for the same time by the same heater, the temperature rise of the paraffin is about *twice* that of the water. Since the heater gives equal amounts of heat energy to each liquid, it seems that different substances require different amounts of heat to cause the same temperature rise in the same mass, say 1 °C in 1 kg.

The 'thirst' of a substance for heat is measured by its *specific heat capacity* (symbol c).

The specific heat capacity of a substance is the heat required to produce unit temperature rise in unit mass.

Heat, like other forms of energy, is measured in joules (J) and the unit of specific heat capacity is the joule per kilogram °C (J/kg °C).

In physics the word 'specific' means 'unit mass' is being considered.

The heat equation

If a substance has a specific heat capacity of 1000 J/kg °C then:

    1000 J   raise the temp. of 1 kg by 1 °C
∴   2 × 1000 J   ,,   ,,   ,,   ,,   2 kg by 1 °C
∴ 3 × 2 × 1000 J   ,,   ,,   ,,   ,,   2 kg by 3 °C

That is, 6000 J will raise the temperature of 2 kg of this substance by 3 °C. We have obtained this answer by multiplying together:

(*i*) the *mass* in kg
(*ii*) the *temperature rise* in °C
(*iii*) the *specific heat capacity* in J/kg °C.

If the temperature of the substance fell by 3 °C, the heat given out would also be 6000 J. In general, we can write the 'heat equation' as:

HEAT RECEIVED OR GIVEN OUT
   = MASS × TEMP. CHANGE × SP. HEAT CAPACITY

For example, if the temperature of a 5 kg mass of copper of specific heat capacity 400 J/kg °C rises from 15 °C to 25 °C, the heat received

= 5 × (25 − 15) × 400
= 5 × 10 × 400 = 20 000 J

Experiment 1 Specific heat capacities

As a 12 V (page 351) electric immersion heater is to be used for this experiment you need to know its power. A 40 W (page 359) heater converts 40 J of electrical energy into heat energy per second. If the power is not marked on the heater ask about it.*

a *Water.* Weigh out 1 kg of water into a container, e.g. an aluminium saucepan. Note the temperature of the water, insert the heater (Figure 45.1) and switch on the 12 V supply. Stir the water and after 5 minutes switch off, but continue stirring and note the *highest* temperature reached.

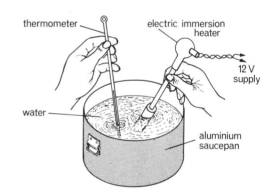

Figure 45.1

Assuming that the heat supplied by the heater equals the heat received by the water, work out the specific heat capacity of water in J/kg °C, as shown below.

Heat received by water (J)
   = power of heater (J/s) × time heater on (s)

---

* The power is found by immersing the heater in water, connecting it to a 12 V d.c. supply and measuring the current taken (usually 3–4 amperes). Then power in watts = volts × amperes.

Rearranging the 'heat equation' we get:

$$\text{sp. heat capacity of water} = \frac{\text{heat received by water (J)}}{\text{mass (kg)} \times \text{temp. rise (°C)}}$$

Suggest causes of error in this experiment.

b *Aluminium.* A cylinder weighing 1 kg and having two holes drilled in it is used. Place the immersion heater in the central hole and a thermometer in the other hole (Figure 45.2).

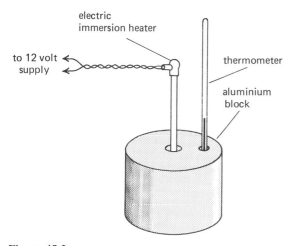

Figure 45.2

Note the temperature, connect the heater to a 12 V supply and switch it on for 5 minutes. When the temperature stops rising, record its highest value.

Calculate the specific heat capacity as before.

Worked examples.

1. *A tank holding 60 kg of water is heated by a 3 kW electric immersion heater. If the specific heat capacity of water is 4200 J/kg °C, estimate the time for the temperature to rise from 10 °C to 60 °C.*

A 3 kW (3000 W) heater supplies 3000 J of heat energy per second.

Let t = time taken in seconds to raise the temperature of the water by $(60 - 10) = 50$ °C,

∴ heat supplied to water in time $t = 3000 \times t$ J

From the 'heat equation', we can say:

heat received by water $= 60 \times 4200 \times 50$ J

Assuming heat supplied = heat received

$$3000 \times t = 60 \times 4200 \times 50$$
$$\therefore t = \frac{60 \times 4200 \times 50}{3000} = 4200 \text{ s } (70 \text{ min})$$

2. *A piece of aluminium of mass 0·5 kg is heated to 100 °C and then placed in 0·4 kg of water at 10 °C. If the resulting temperature of the mixture is 30 °C, what is the specific heat capacity of aluminium if that of water is 4200 J/kg °C?*

When two substances at different temperatures are mixed, heat flows from the one at the higher temperature to the one at the lower temperature until both are at the same temperature—the temperature of the mixture. If there is no loss of heat, then in this case:

heat given out by aluminium
= heat taken in by water

Using the 'heat equation' and letting c be the specific heat capacity of aluminium in J/kg °C, we have:

$$\text{heat given out} = 0.5 \times c \times (100 - 30) \text{ J}$$
$$\text{heat taken in} = 0.4 \times 4200 \times (30 - 10) \text{ J}$$
$$0.5 \times c \times 70 = 0.4 \times 4200 \times 20$$
$$\therefore c = \frac{4200 \times 8}{35} = 960 \text{ J/kg °C}$$

Specific latent heat of fusion

The specific latent heat of fusion of a substance is the quantity of heat needed to change unit mass from solid to liquid without temperature change.

It is measured in J/kg or J/g.

Experiment 2 Specific latent heat of fusion of ice

Place a 12 V electric immersion heater of known power in a filter funnel and pack small pieces of ice round it (Figure 45.3). Switch on the heater for 3

Figure 45.3

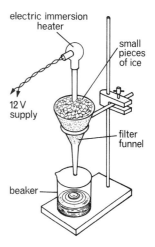

minutes and find the mass of water which collects in a beaker. Arrange the results as shown.

Power of immersion heater = W (J/s)
Time heat supplied = s
∴ Heat supplied to ice = J
Mass of beaker empty = g
Mass of beaker + melted ice = g
∴ Mass of melted ice = g

Calculate the heat needed to melt 1 g of ice.

- What are the causes of error in this experiment?

Specific latent heat of vaporization

The specific latent heat of vaporization of a substance is the quantity of heat needed to change unit mass from liquid to vapour without change of temperature.

An estimate of its value for water can be made using the apparatus of Figure 45.4. The mains-operated immersion heater is clamped so that it is

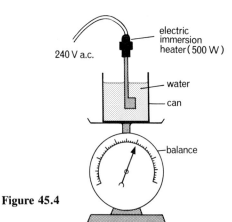

Figure 45.4

well covered by the water in the can. When the water is *boiling briskly* the reading on the balance is noted and a stop clock started. The time for 50 g of water to be boiled off is found.

Suppose a 500 W heater is used and that the time required is 4 minutes (240 s), then:

heat supplied in 240 s = 500 J/s × 240 s
= 120 000 J
∴ latent heat of vaporization = 120 000 J/50 g
= 2400 J/g

The accepted value is 2300 J/g = 2 300 000 J/kg = 2·3 × 10⁶ J/kg = 2·3 MJ/kg. Errors arise because the can and water lose heat to the surroundings.

Latent heat and the kinetic theory

(*a*) *Fusion.* The kinetic theory explains latent heat of fusion as being the energy which enables the molecules of a solid to change their vibratory motion about a fixed position to the greater range of movement they have as liquid molecules. Their P.E. increases but not their average K.E. as happens when heat causes a temperature rise.

(*b*) *Vaporization.* If liquid molecules are to overcome the forces holding them together and gain the freedom to move around independently as gas molecules, they need a large amount of energy. They receive this as latent heat of vaporization which, like latent heat of fusion, increases the P.E. of the molecules but not their K.E. It also gives the molecules the energy required to push back the surrounding atmosphere in the large expansion that occurs when a liquid vaporizes.

Worked examples.

The following values are required.

	Water	Ice	Aluminium
Sp. heat cap. (J/g °C)	4·2	2·0	0·90
Sp. lat. heat (J/g)	2300	340	

1. *How much heat is needed to change 20 g of ice at 0 °C to steam at 100 °C?*

There are three stages in the change.

Heat to change 20 g of *ice at 0 °C* to *water at 0 °C*

= mass of ice × sp. lat. heat of ice
= 20 × 340 = 6800 J

Heat to change 20 g *water at 0 °C* to *water at 100 °C*

= mass of water × sp. heat cap. of water × temp. rise
= 20 × 4·2 × 100 = 8400 J

Heat to change 20 g *water at 100 °C* to *steam at 100 °C*

= mass of water × sp. lat. heat of steam
= 20 × 2300 = 46 000 J

Total heat supplied therefore

= 6800 + 8400 + 46 000 = 61 200 J

2. *An aluminium can of mass 100 g contains 200 g of water. Both, initially at 15 °C, are placed in a refrigerator at −5·0 °C. Calculate the quantity of heat that has to be removed from the water and the can for their temperatures to fall to −5·0 °C.*

Heat lost by *can* in falling from 15 °C to −5·0 °C

= mass of can × sp. heat cap. aluminium × fall in temp.
= 100 × 0·90 × (15 − (−5)) = 100 × 0·90 × 20
= 1800 J

Heat lost by *water* in falling from 15 °C to 0 °C

= mass of water × sp. heat cap. water × temp. fall
= 200 × 4·2 × 15 = 12 600 J

Heat lost by *water* at 0 °C freezing to ice at 0 °C

= mass of water × sp. lat. heat
= 200 × 340 = 68 000 J

Heat lost by *ice* in falling from 0 °C to −5·0 °C

= mass of ice × sp. heat cap. of ice × temp. fall
= 200 × 2·0 × 5·0 = 2000 J

Total heat removed therefore

= 1800 + 12 600 + 68 000 + 2000 = 84 400 J

Questions

1. How much heat is needed to raise the temperature by 10 °C of 5 kg of a substance of specific heat capacity 300 J/kg °C?

2. 2000 J of energy is needed to heat 1 kg of paraffin through 1 °C. How much heat is needed to heat 3 kg of paraffin through 10 °C?

3. The same quantity of heat was given to different masses of three substances, A, B and C. The temperature rise in each case is shown in the table. Calculate the specific heat capacities of A, B and C.

Material	Mass (kg)	Heat given (J)	Temp. rise (°C)
A	1·0	2000	1·0
B	2·0	2000	5·0
C	0·5	2000	4·0

4. How much heat is given out when an iron ball of mass 2 kg and specific heat capacity 440 J/kg °C cools from 300 °C to 200 °C?

5. How many joules of heat energy are supplied by a 2 kW heater in (*a*) 10 s, (*b*) 1 minute?

6. An electric water-heater raises the temperature of 0·5 kg of water by 30 °C every minute. The specific heat capacity of water is 4200 J/kg °C. Assuming that no heat is lost, what is the power of the heater?

7. What mass of cold water at 10 °C must be added to 60 kg of hot water at 80 °C by someone who wants to have a bath at 50 °C? Neglect heat losses and take the specific heat capacity of water as 4200 J/kg °C.

For the following, use values given in *Worked examples* for latent heat

8. (*a*) How much heat will change 10 g of ice at 0 °C to water at 0 °C?
(*b*) What quantity of heat must be removed from 20 g of water at 0 °C to change it to ice at 0 °C?

9. (*a*) How much heat is needed to change 5 g of ice at 0 °C to water at 50 °C?
(*b*) If a refrigerator cools 200 g of water from 20 °C to its freezing point in 10 minutes, how much heat is removed per minute from the water?

10. How long will it take a 50 W heater to melt 100 g of ice at 0 °C?

11. Some small aluminium rivets of total mass 170 g and at 100 °C are emptied into a hole in a large block of ice at 0 °C.

(*a*) What will be the final temperature of the rivets?
(*b*) How much ice will melt?

12. (*a*) How much heat is needed to change 4 g of water at 100 °C to steam at 100 °C?
(*b*) Find the heat given out when 10 g of steam at 100 °C condenses and cools to water at 50 °C.

13. A 3 kW electric kettle is left on for 2 minutes after the water starts to boil. What mass of water is boiled off in this time?

191

46 LIGHT RAYS

You can see an object only if light from it enters your eyes. Some objects such as the sun, electric lamps and candles make their own light. We call these *luminous* sources.

Most things you see do not make their own light but reflect it from a luminous source. They are *non-luminous* objects. This page, you and the moon are examples. Figure 46.1 shows some others.

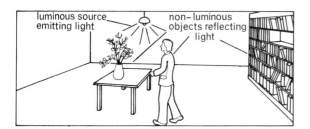

Figure 46.1

Luminous sources radiate light when their atoms become 'excited' as a result of receiving energy. In a light bulb, for example, the energy comes from electricity. The 'excited' atoms give off their light haphazardly in most luminous sources.

A light source that works differently is the *laser*, invented in 1960. In it the 'excited' atoms act together and emit a narrow, very bright beam of light which can cut a hole through a key 2 mm thick in one-thousandth of a second (Figure 46.2). Other uses are being found for the laser in industry and medicine.

PROPERTIES OF LIGHT RAYS

Rays and beams

Sunbeams streaming through trees (Figure 46.3) and light from a cinema projector on its way to the screen both suggest that *light travels in straight lines*. The beams are visible because dust particles in the air reflect light into our eyes.

Figure 46.2

Figure 46.3

The direction of the path in which light is travelling is called a *ray* and is represented in diagrams by a straight line with an arrow on it. A *beam* is a stream of light and is shown by a number of rays; it may be parallel, diverging (spreading out) or converging (getting narrower) (Figure 46.4).

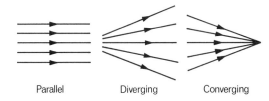

Parallel Diverging Converging

Figure 46.4

Experiment 1 The pinhole camera

One is shown in Figure 46.5a. Make a small pinhole in the centre of the black paper. Half-darken the room. Hold the box at arm's length so that the pinhole end is nearest to and about 1 m from a luminous object, e.g. a carbon filament

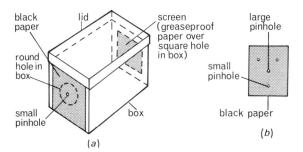

Figure 46.5

lamp or a candle. Look at the *image* on the screen (an image is a likeness of an object and need not be an exact copy).

- Can you see *three* ways in which the image differs from the object?
- What is the effect of moving the camera closer to the object?

Make the pinhole larger.

- What happens to the (*i*) brightness, (*ii*) sharpness, (*iii*) size of the image?

Make several small pinholes round the large hole (Figure 46.5b) and view the image again.

- What do you see?

The way a pinhole camera forms an image is shown in Figure 46.6.

Figure 46.6

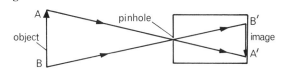

Shadows

Shadows are formed because light travels in straight lines. A very small source of light, called a *point* source, gives a sharp shadow which is equally dark all over. This may be shown as in Figure 46.7a where the small hole in the card acts as a point source.

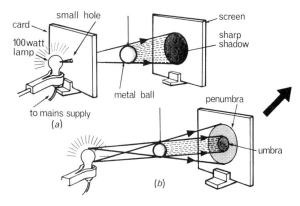

Figure 46.7

If the card is removed, the lamp acts as a large or *extended* source (Figure 46.7b). The shadow is then larger and has a central dark region, the *umbra*, surrounded by a ring of partial shadow, the *penumbra*. You can see by the rays that some light reaches the penumbra but none reaches the umbra.

Eclipses

There is an eclipse of the sun by the moon when the sun, moon and earth are in a straight line. The sun is an extended source (like the bulb in Figure 46.7b). People in the umbra of the moon's shadow,

Figure 46.8

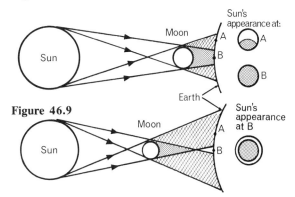

Figure 46.9

at B in Figure 46.8, see a *total* eclipse of the sun (that is, they can't see the sun at all). Those in the penumbra, at A, see a *partial* eclipse (part of the sun is still visible).

Sometimes the moon is farther away from the earth (it does not go round the earth in a perfect circle), and then the tip of the umbra does not reach the earth (Figure 46.9). When this happens people at A still see a partial eclipse, but those at B see an *annular* eclipse (the central region of the sun is hidden but not its outer parts).

A total eclipse seen from one place may last for up to 7 minutes. During this time, although it is day, the sky is dark, stars are visible, the temperature falls and birds stop singing.

Questions

1. How would the size and brightness of the image formed by a pinhole camera change if the camera was made longer?

2. What changes would occur in the image if the single pinhole in a camera was replaced by (*a*) four pinholes close together, (*b*) a hole 1 cm wide.

3. Draw a diagram to show a possible position of the moon for a lunar eclipse to be seen on earth.

47 REFLECTION OF LIGHT

Reflection

If we know how light behaves when it is reflected we can use a mirror to change the direction in which it is travelling. This happens in a periscope.

An ordinary mirror is made by depositing a thin layer of silver on one side of a piece of glass and protecting it with paint. The silver—at the *back* of the glass—acts as the reflecting surface.

Experiment 1 Reflection by a plane mirror

Draw a line AOB on a sheet of paper and using a protractor mark angles on it. Measure them from the perpendicular ON, which is at right angles to AOB. Set up a plane (flat) mirror with its reflecting surface on AOB.

a *Ray method.* Shine a narrow ray of light along say the 30° line, on to the mirror (Figure 47.1). Mark the position of the reflected ray, remove the mirror and measure the angle between the reflected ray and ON. Repeat for rays at other angles.

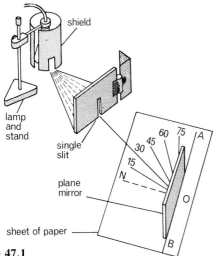

Figure 47.1

- What can you conclude?

b *Pin method.* Insert two pins P_1 and P_2 on the 30° line (Figure 47.2) to indicate a 'ray' of light falling

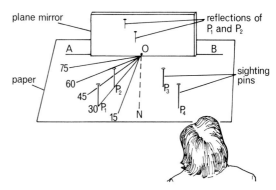

Figure 47.2

at this angle on the mirror. Look into the mirror and insert two sighting pins P_3 and P_4 so they are in line with the reflections (images) of P_1 and P_2. P_3P_4 gives the path of 'ray' P_1P_2 after it is reflected.

Remove P_3 and P_4 and mark their positions with crosses (lettered P_3 and P_4). Remove the mirror and draw a straight line through P_3 and P_4 to meet the mirror; this should be at O. Measure angle P_4ON. Repeat for other angles.

- What do you conclude?

Laws of reflection

Terms used in connection with reflection are shown in Figure 47.3. The perpendicular to the mirror at the point where the incident ray strikes it

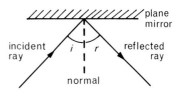

Figure 47.3

is called the *normal*. Note that the angle of incidence i is the angle between the incident ray and the *normal*; similarly for the angle of reflection r.

Figure 47.5

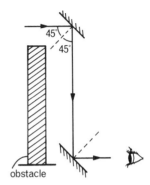

Figure 47.4

There are two laws of reflection.

The angle of incidence equals the angle of reflection.

The incident ray, the reflected ray and the normal all lie in the same plane. (This means that they can all be drawn on a flat sheet of paper.)

Periscope

A simple periscope consists of a tube containing two plane mirrors, fixed parallel to and facing one another. Each makes an angle of 45° with the line joining them (Figure 47.4). Light from the object is turned through 90° at each reflection and an observer is able to see over a crowd, for example, (Figure 47.5) or over the top of an obstacle.

In more elaborate periscopes like those used in submarines, prisms replace mirrors (see page 204).

Regular and diffuse reflection

If a parallel beam of light falls on a plane mirror it is reflected as a parallel beam (Figure 47.6a) and *regular* reflection is said to occur. Most surfaces, however, reflect light irregularly and the rays in an incident parallel beam are reflected in many directions (Figure 47.6b).

Irregular or *diffuse* reflection is due to the surface of the object not being perfectly smooth like a mirror. At each point on the surface the laws of reflection are obeyed, but the angle of incidence and so the angle of reflection varies from point to point. The reflected rays are scattered haphazardly. Most objects, being rough, are seen by diffuse reflection.

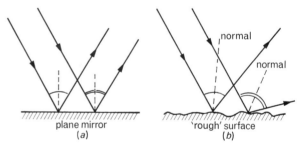

Figure 47.6

Plane mirrors

When you look into a plane mirror on the wall of a room you see an image of the room behind the mirror; it is as if there was another room. Restaurants sometimes have a large mirror on one wall just to make them look bigger. You may be able to say how much bigger after the next experiment.

The position of the image formed by a mirror depends on the position of the object.

Experiment 2 Position of image

Support a piece of thin glass on the bench, as in Figure 47.7. It must be *vertical* (at 90° to the bench). Place a small paper arrow O about 10 cm from the glass. The glass acts as a mirror and an image of O will be seen in it; the darker the bench top the brighter is the image.

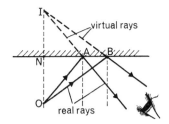

Figure 47.8

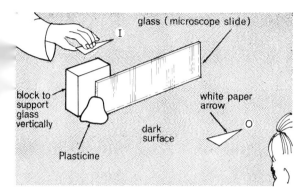

Figure 47.7

- How do the sizes of O and its image compare?

Lay another identical arrow I on the bench behind the glass; move it until it coincides with the image of O. Measure the distances of the points of O and I from the glass along the line joining them.

- How do they compare?

Try O at other distances.

Real and virtual images

A *real* image is one which can be produced on a screen and is formed by rays that actually pass through it.

A *virtual* image cannot be formed on a screen and is produced by rays which seem to come from it but do not pass through it. The image in a plane mirror is virtual. Rays from a point on an object are reflected at the mirror and appear to come from the point behind the mirror where the eye imagines the rays intersect when produced backwards (Figure 47.8). Rays IA and IB do not exist; they are virtual and are represented by dotted lines.

Lateral inversion

If you close your left eye your image in a plane mirror seems to close the right eye. In a mirror image, left and right are interchanged and the image is said to be *laterally inverted*. The effect occurs whenever an image is formed by one reflection and is very evident if print is viewed in a mirror (Figure 47.9).

- What happens if two reflections occur as in a periscope?

Figure 47.9

Properties of the image

The image in a plane mirror is:

(*i*) as far behind the mirror as the object is in front and the line joining the object and image is perpendicular to the mirror
(*ii*) the same size as the object
(*iii*) virtual
(*iv*) laterally inverted.

Use of plane mirrors

Apart from their everyday use, plane mirrors can improve the accuracy of scientific measurements.

A reading made on an instrument which has a pointer moving over a scale is correct only if your eye is directly over the pointer. In any other position there is an error, called the 'parallax' error. (There is parallax between two objects—here the pointer and the scale—if they appear to move in opposite directions when you move your head sideways. It arises when objects do not coincide; if they do coincide they move together.)

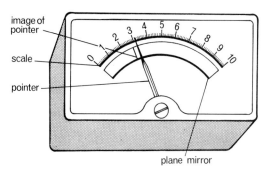

Figure 47.10

If a plane mirror is fitted in the scale, the correct position can then be found by moving your head until the image of the pointer in the mirror is hidden behind the pointer (Figure 47.10). (A similar error occurs when reading a ruler, due to its thickness, if you do not look at right angles to it.)

Two mirrors at right angles

Two mirrors at 90° to each other give *three* images of an object placed between them. In Figure 47.11, M_1 and M_2 are plane mirrors and O an object. Image I_1 is formed by one reflection at M_1; image I_2 by one reflection at M_2; and image I_3 by reflections at M_1 and M_2, as the diagram shows.

We can think of I_3 as being either the image of I_1 acting as an 'object' for mirror M_2 (extended to the left) or as the image of I_2 acting as an 'object' for mirror M_1 (extended downwards). In all cases an image is as far behind the mirror as the 'object' is in front.

Figure 47.11

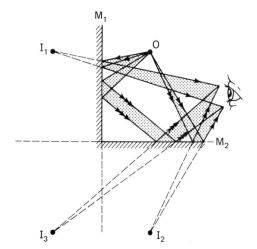

Curved mirrors

For some purposes curved mirrors are more useful than plane mirrors. You can see the images they give by looking into both sides of a polished spoon.

A *concave* mirror curves inwards like a cave (Figure 47.12a), a *convex* one curves outwards (Figure 47.12b). Many curved mirrors have spherical surfaces and the *principal axis* of such a mirror is the line joining the *pole* P or centre of the mirror to the *centre of curvature* C. C is the centre of the sphere of which the mirror is a part; it is in front of a concave mirror and behind a convex one.

The *radius of curvature r* is the distance CP.

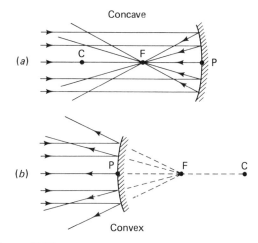

Figure 47.12

Principal focus

When a beam of light parallel to the principal axis is reflected from a concave mirror, the laws of reflection hold and it converges to a point on the axis, called the *principal focus* F. Since light does pass through it, it is a real focus and can be obtained on a screen. A convex mirror has a virtual principal focus behind the mirror, from which the reflected beam appears to diverge.

The *focal length f* is the distance FP and experiment and theory show that $FP = CP/2$ or $f = r/2$, i.e.

FOCAL LENGTH = HALF THE RADIUS OF CURVATURE

These statements apply only to small mirrors or to large mirrors if the beam is close to its axis. A large mirror does not form a point focus of a wide parallel beam but causes the reflected rays to form a curve called a *caustic*. One may be seen on the surface of tea in a cup (Figure 47.13). Why?

Figure 47.13

Ray diagrams

Facts about the images formed by spherical mirrors can be found by drawing *two* of the following rays.

a A ray parallel to the principal axis which is reflected through the principal focus F.
b A ray through the centre of curvature C which hits the mirror normally, is reflected back along its own path. (The radius of a sphere is perpendicular to the surface where it meets the surface.)
c A ray through the principal focus F which is reflected parallel to the principal axis.

In diagrams a curved mirror is represented by a *straight* line. A good-sized object can then be shown and rays from it regarded as forming a point focus, i.e. the mirror behaves as a small one. In numerical questions, *horizontal* distances have sometimes to be scaled down.

Images in a concave mirror

The ray diagrams in Figure 47.14 show the images for four object positions. In each case, two rays are drawn from the top A of an object OA and where they intersect after reflection gives the top B of the image IB. The foot I of each image is on the axis since ray OP hits the mirror normally and is reflected back along the axis. In 47.14 *d* the dotted rays and the image are virtual (not real).

Experiment 3 *f* of a concave mirror

We use the fact that when an object is placed at the centre of curvature C of a concave mirror, a real image is formed also at C (Figure 47.15*a*).

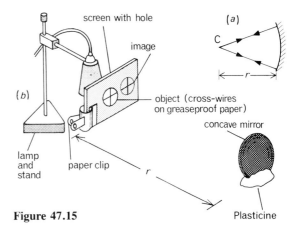

Figure 47.15

Move the mirror, arranged as in Figure 47.15*b*, until a *sharp* image of the cross-wire is obtained on the screen. Measure the distance *r* from the screen to the mirror, then $f = r/2$.

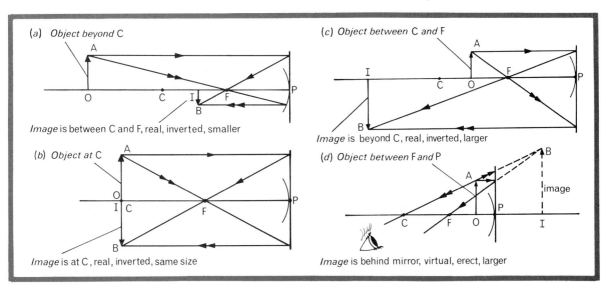

Figure 47.14

Uses of curved mirrors

(*a*) *Shaving mirror.* This is often concave. It forms a magnified, virtual, erect image of an object inside F (Figure 47.14*d*) but the view is limited.

(*b*) *Reflector.* When a point source of light is at the principal focus of a concave mirror, the reflected beam is parallel if the mirror is small. To obtain a wide parallel beam in, for example, a searchlight, the mirror must be parabolic (Figure 47.16).

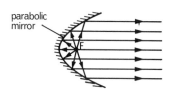

Figure 47.16

(*c*) *Driving mirror.* A convex mirror gives a wider field of view than a plane mirror of the same size (Figure 47.17*a* and *b*). For this reason, and because it always gives an erect (but smaller) image, it is used as a car driving mirror.

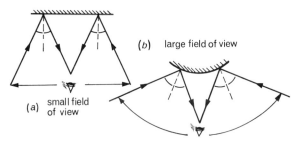

Figure 47.17

Questions

1. In Figure 47.18 a ray of light AO is shown falling on a mirror XY. ON is normal to the mirror.

Figure 47.18

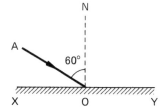

(*a*) What is the angle of incidence of the ray AO on XY?
(*b*) Copy the diagram and show the path of the ray AO after it is reflected from XY. Mark on your diagram the value of the angle of reflection.

2. In Figure 47.19 PQ and QR are two mirrors at right angles to each other. BO is a ray of light and NO a normal to PQ.

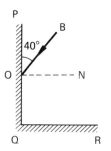

Figure 47.19

(*a*) What are the angles of reflection of the rays reflected from (*i*) PQ, (*ii*) QR?
(*b*) Copy the diagram and continue the ray BO to show the path it takes after reflection at both mirrors.

3. Figure 47.20 shows a trick in which a lighted candle can be made to appear to be burning in a jar of water.

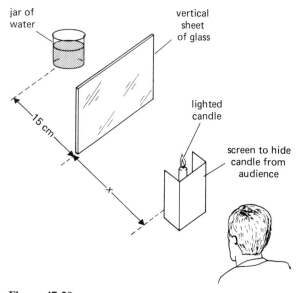

Figure 47.20

(*a*) What must be the value of the distance *x* for it to work?
(*b*) What does the glass do to the light to get this effect?

Figure 47.21

4. If a watch with dots instead of numbers is held up in front of a mirror at 11.15 (Figure 47.21) what will the time appear to be on the image of the watch?

5. Draw diagrams to represent (a) a convex mirror, (b) a concave mirror. Mark on each the principal axis, the principal focus F and the centre of curvature C if the focal length of (a) is 3 cm and of (b) 4 cm.

6. A concave mirror has a focal length of 3 cm and a real object 2 cm tall is placed 7 cm away from it. By means of an accurate full size diagram find where the image would be and measure its length.

48 REFRACTION OF LIGHT

Refraction

If you place a coin in an empty cup and move back until you *just* cannot see it, the result is surprising if someone *gently* pours in water. Try it.

Although light travels in straight lines in one transparent material, e.g. air, if it passes into a different material, e.g. water, it changes direction at the boundary between the two, i.e. it is bent. The *bending of light* when it passes from one material (called a medium) to another is called *refraction*. It causes effects like the coin trick.

Experiment 1 Refraction in glass

a *Ray method.* Shine a ray of light at an angle on to a glass block (which has its lower face painted white or frosted) (Figure 48.1). Draw the outline ABCD of the block on the sheet of paper under it. Mark the positions of the various rays in air and in glass.

Remove the block and draw the normals on the paper at the points where the ray enters AB (see Figure 48.1) and where it leaves CD.

- What *two* things happen to the light falling on AB?
- When the ray enters the glass at AB is it bent towards or away from the part of the normal in the block?
- How is it bent at CD?
- What can you say about the direction of the ray falling on AB and the direction of the ray leaving CD?
- What happens if the ray hits AB at right angles?

b *Pin method.* Draw round a glass block ABCD on a sheet of paper. Also mark a normal ON and a line making an angle of about 45° with ON (Figure 48.2). Stick two pins P_1 and P_2 on this line as far apart as possible. Look through the block and stick another two pins P_3 and P_4 in line with the images I_1 and I_2 of P_1 and P_2.

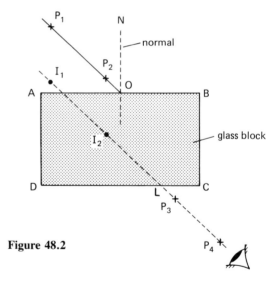

Figure 48.2

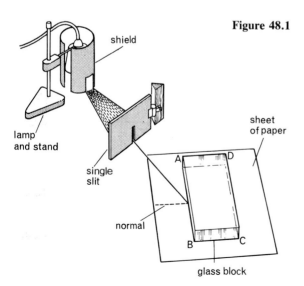

Figure 48.1

Remove all four pins and mark their positions with small crosses. Remove the block and draw a line through P_1 and P_2 to meet AB at O: P_1P_2 represents a ray *falling* on the block. Also draw a line through P_3 and P_4 to meet CD at L. P_3P_4 represents a ray *emerging* from the block. Join O to L: OL represents the refracted ray in the glass.

- When ray P_1P_2 enters the glass at AB is it bent towards or away from the part of the normal in the block?
- How is it bent at CD?
- What can you say about the direction of rays P_1P_2 and P_3P_4?

Laws of refraction

Terms used in refraction and the two laws are illustrated in Figure 48.3.

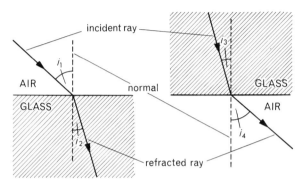

Figure 48.3

Light is bent towards the normal when it enters a denser medium at an angle and away from the normal when it enters a less dense medium at an angle.

The incident ray, the refracted ray and the normal all lie in the same plane.

Real and apparent depth

Rays of light from a point O on the bottom of a pool are refracted away from the normal at the water surface since they are passing into a less dense medium, i.e. air (Figure 48.4). On entering the eye they appear to come from a point I *above* O; I is the virtual image of O formed by refraction. The apparent depth of the pool is less than its real depth.

Figure 48.4

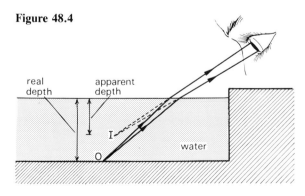

Refractive index (n)

If i_1 is the angle between the normal and a ray in *air* and i_2 is the angle between the normal and the same ray in a *material*, experiment shows that for all pairs of values of i_1 and i_2

$$\frac{\sin i_1}{\sin i_2} = \text{a constant}$$

This is *Snell's law* of refraction and is true whether the ray travels from air to the material or vice versa. The constant is called the *refractive index (n)* of the material. For glass n is about $1\cdot5$ (3/2) and for water $1\cdot33$ (4/3). The larger n is for a material the more it bends light.

- How will a ray be refracted when it passes from water to glass?

In Figure 48.3 $n = \sin i_1/\sin i_2 = \sin i_4/\sin i_3$.

Refraction by a prism

In a triangular glass prism (Figure 48.5a) the deviation of a ray due to refraction at the first surface is added to the deviation at the second surface (Figure 48.5b). The deviations do not cancel out as in a parallel-sided block where the emergent ray, although displaced, is parallel to the incident ray.

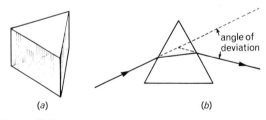

Figure 48.5

Total internal reflection

When light passes at small angles of incidence from a denser to a less dense medium, e.g. from glass to air, there is a strong refracted ray and a weak ray reflected back into the denser medium (Figure 48.6a). Increasing the angle of incidence increases the angle of refraction.

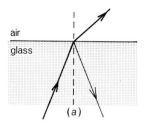 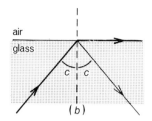 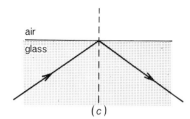

Figure 48.6

At a certain angle of incidence, called the *critical angle c*, the angle of refraction is 90° (Figure 48.6*b*). For angles of incidence greater than *c*, the refracted ray disappears and *all* the incident light is reflected inside the denser medium (Figure 48.6*c*). The light does not cross the boundary and is said to suffer *total internal reflection*.

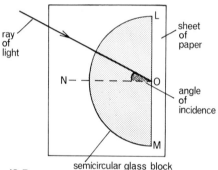

Figure 48.7

Experiment 2 Critical angle of glass

Place a semicircular glass block on a sheet of paper (Figure 48.7) and draw the outline LOMN where O is the centre and ON the normal at O to LOM. Direct a narrow ray (at an angle of about 30°) *along a radius towards* O.

- Why is the ray not refracted at the curved surface?

Note the refracted ray in the air beyond LOM and also the weak internally reflected ray in the glass.

Slowly rotate the paper so that the angle of incidence increases until total internal reflection *just* occurs. Mark the incident ray. Measure the angle of incidence; it equals the critical angle.

Multiple images in a mirror

An ordinary mirror silvered at the back forms several images of one object, due to multiple reflection inside the glass (Figure 48.8*a* and *b*).

These blur the main image I (which is formed by one reflection at the silvering), especially if the glass is thick. The trouble is absent in front-silvered mirrors but they are easily damaged.

Totally reflecting prisms

The defects of mirrors are overcome if 45° right-angled glass prisms are used. The critical angle of ordinary glass is about 42° and a ray falling normally on face PQ of such a prism (Figure 48.9*a*) hits face PR at 45°. Total internal reflection occurs and the ray is turned through 90°. Totally reflecting prisms replace mirrors in good periscopes.

Figure 48.8

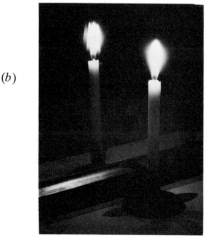

204

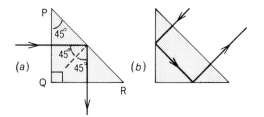

Figure 48.9

Light can also be reflected through 180° by a prism (Figure 48.9b); this happens in binoculars.

Mirages

They can often be seen on a hot day as a pool of water on the road some distance ahead. One explanation is that the light from the sky is gradually refracted away from the normal as it passes through layers of warm but less dense air near the hot road. Warm air has a slightly smaller refractive index than cool air and when the light meets a layer at the critical angle, it suffers total internal reflection (Figure 48.10). To an observer the reflection of the sky appears as a puddle in the road.

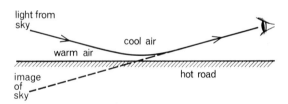

Figure 48.10

Light pipes

Light can be trapped by total internal reflection inside a bent glass rod and 'piped' along a curved path (Figure 48.11). A single, very thin glass fibre behaves in the same way. If several thousand are taped together a flexible light pipe is obtained that can be used to light up some awkward spot for inspection.

Figure 48.11

Refractive index and critical angle

From Figure 48.6b and the definition of refractive index

$$n = \frac{\text{sin of angle between ray in air and normal}}{\text{sin of angle between ray in glass and normal}}$$

$$\therefore n = \frac{\sin 90°}{\sin c} = \frac{1}{\sin c} \quad (\sin 90° = 1)$$

If $n = 3/2$, $\sin c = 2/3$ and $c = 42°$.

Questions

1. Figure 48.12 shows a ray of light entering a rectangular block of glass.

Figure 48.12

(a) Copy the diagram and draw the normal at the point of entry.
(b) Sketch the approximate path of the ray through the block and out of the other side.

2. Draw two rays from a point on a fish in a stream to show where someone on the bank must aim to spear the fish.

3. Figure 48.13 shows a ray of light XY passing through a glass prism. Which of the rays A to E is the correct emerging ray?

Figure 48.13

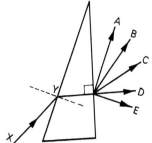

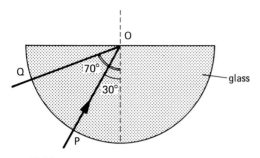

Figure 48.14

4. In Figure 48.14, two rays of light are shown entering a semicircular glass block.

(a) Why are they not bent at P and Q?
(b) Copy the diagram and draw approximate paths to show what happens to each ray after it reaches O.

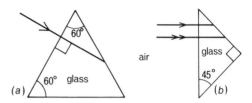

Figure 48.15

5. Copy Figure 48.15a and b and complete the paths of the rays.

49 LENSES

Lenses are used in many optical instruments (page 210); they often have spherical surfaces and there are two types. A *convex* lens is thickest in the centre and is also called a *converging* lens because it bends light inwards (Figure 49.1a). You may have used one as a magnifying glass (Figure 49.2a)

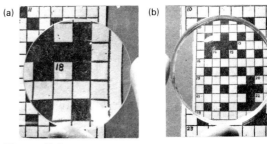

Figure 49.2

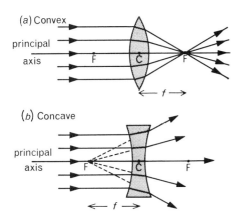

Figure 49.1

or as a burning glass. A *concave* or *diverging* lens is thinnest in the centre and spreads light out (Figure 49.1b); it always gives a diminished image (Figure 49.2b).

The centre of a lens is its *optical centre* C; the line through C at right angles to the lens is the *principal axis*.

Principal focus

When a beam of light parallel to the principal axis passes through a convex lens it is refracted so as to converge to a point on the axis called the *principal focus* F. It is a real focus. A concave lens has a virtual principal focus behind the lens, from which the refracted beam seems to diverge.

Since light can fall on both faces of a lens it has two principal foci, one on each side, equidistant from C. The distance CF is the *focal* length of the lens; it is an important property of the lens.

Ray diagrams

Information about the images formed by a lens can be obtained by drawing *two* of the following rays.

A ray parallel to the principal axis which is refracted through the principal focus F.

A ray through the optical centre C which is undeviated. (The central part of a lens acts as a small parallel-sided block which slightly displaces but does not deviate a ray passing through it; for a thin lens the displacement can be ignored.)

A ray through the principal focus F which is refracted parallel to the principal axis.

In diagrams a thin lens is represented by a *straight* line at which all the refraction is considered to occur. In numerical questions *horizontal* distances have sometimes to be scaled down (see Question 2, page 209).

Images formed by a convex lens

In the formation of images by lenses two important points on the principal axis are F and 2F; 2F is at a distance of twice the focal length from C.

In each ray diagram in Figure 49.3 two rays are drawn from the top A of an object OA and where they intersect after refraction gives the top B of the image IB. The foot I of each image is on the axis since ray OC passes through the lens undeviated. In *d* the dotted rays, and the image, are virtual.

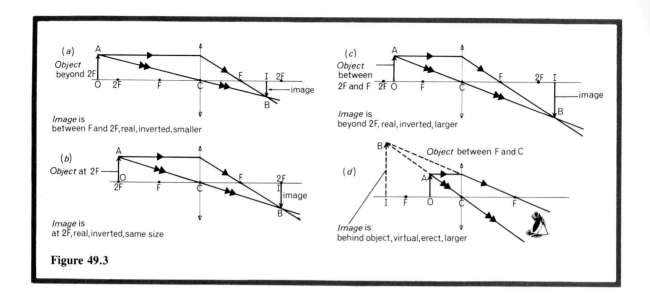

Figure 49.3

Experiment 1 f of a convex lens

a *Distant object method.* We use the fact that rays from a *point* on a very distant object, i.e. at infinity, are nearly parallel (Figure 49.4a).

Move the lens, arranged as in Figure 49.4b, until a *sharp* image of a window at the other side of the room is obtained on the screen.

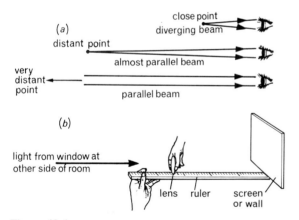

Figure 49.4

- Why is the distance between the lens and the screen roughly f?

b *Plane mirror method.* Using the apparatus in Figure 49.5a move the lens until a *sharp* image of the object, i.e. the illuminated cross-wire, is formed on the screen beside the object. When this happens, light from the object must travel back along nearly the same path and hit the mirror normally (Figure 49.5b).

- Why is the object then at the lens' principal focus F?

Magnification

The linear magnification m is given by:

$$m = \frac{\text{height of image}}{\text{height of object}}$$

From Figure 49.3d it can be shown that in all cases triangles OAC and BIC are similar and so

$$m = \frac{IB}{OA} = \frac{IC}{OC} = \frac{\text{distance of image from lens}}{\text{distance of object from lens}}$$

Figure 49.5

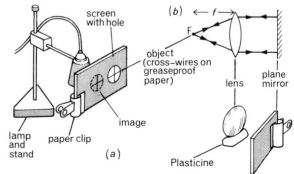

Questions

1. Copy the diagrams in Figure 49.6 and complete the path of each light ray after it passes through the lens.

2. An object AO 1 cm high is placed 25 cm in front of a converging lens of focal length 10 cm. Find the position, size and nature of the image formed by making a scale drawing showing the paths of rays from A. (HINT: have O on the principal axis and use a scale of 1 cm to represent 5 cm for horizontal distances.)

3. What is the magnification produced by a lens if it forms a real image on a screen 20 cm behind the lens of an object 5 cm in front of the lens?

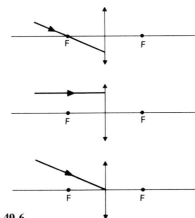

Figure 49.6

50 OPTICAL INSTRUMENTS

Camera

A camera is a light-tight box in which a convex lens forms a real image on a film (Figure 50.1). The film contains chemicals that change on exposure to light; it is 'developed' to give a negative. From the negative a photograph is made by 'printing'.

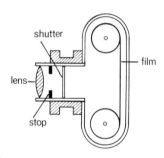

Figure 50.1

(a) *Focusing.* In simple cameras the lens is fixed and all distant objects, i.e. beyond about 2 m, are in reasonable focus.

- Roughly how far from the film will the lens be if its focal length is 5 cm?

In other cameras exact focusing of an object at a certain distance is done by altering the lens position. For near objects it is moved away from the film, the correct setting being shown by a scale on the focusing ring.

(b) *Shutter.* When a photograph is taken, the shutter is opened for a certain time and exposes the film to light entering the camera. Sometimes exposure times can be varied and are given in fractions of a second, e.g. 1/1000, 1/60, etc. Fast-moving objects require short exposures.

(c) *Stop.* The brightness of the image on the film depends on the amount of light passing through the lens when the shutter is opened and is controlled by the size of the hole (aperture) in the stop. In some cameras this is fixed but in others (Figure 50.2) it can be made larger for a dull scene and smaller for a bright one.

Figure 50.2

The aperture may be marked in *f-numbers*. The diameter of an aperture with f-number 8 is one eighth of the focal length of the lens and so the *larger* the f-number the *smaller* the aperture. The numbers are chosen so that on passing from one to the next higher, e.g. from 8 to 11, the area of the aperture is halved.

The human eye

This works in a similar way to the camera. For a full description, see page 327.

Projector

A projector forms a real image on a screen of a slide in a slide projector, and on a film in a cine projector. The image is usually so highly magnified that very strong but even illumination of the slide or film is needed if the image is also to be bright. This is achieved by directing light from a small but powerful lamp on to the 'object' by means of a concave mirror and a condenser lens system arranged as in Figure 50.3. The image is produced by the projection lens which can be moved in and out of its mounting to focus the picture.

- How must the slide be placed in a projector like that in Figure 50.4 to give an erect image the right way round?

Figure 50.3

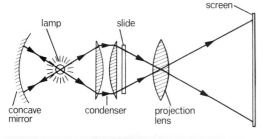

Figure 50.4

Magnifying glass

A watchmaker's magnifying glass is shown in use in Figure 50.5. The sleepers on a railway track are all the same length but those nearby seem longer. This is because they enclose a larger angle at your eye than more distant ones: as a result their image on the retina is larger so making them appear bigger.

Figure 50.5

A convex lens gives an enlarged, upright virtual image of an object placed inside its principal focus F (Figure 50.6a). It acts as a magnifying glass since the angle β made at the eye by the image, formed at the near point, is greater than the angle α made by the object when it is viewed directly at the near point without the magnifying glass (Figure 50.6b).

The fatter (more curved) a convex lens is, the shorter is its focal length and the more does it magnify. Too much curvature however distorts the image.

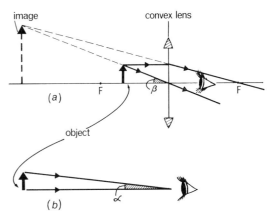

Figure 50.6

Questions

1. The camera shown in Figure 50.7a is focused on an object close to it. Is the camera in Figure 50.7b focused on an object which is closer still or farther away?

Figure 50.7

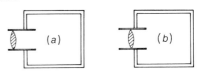

2. If a projector is moved farther away from the screen on which it was giving a sharp image of a slide, state (a) *three* changes that occur in the image, (b) how the projection lens must be adjusted to refocus the image.

3. (a) Three converging lenses are available having focal lengths of 4 cm, 40 cm and 4 m, respectively. Which one would you choose as a magnifying glass?
(b) An object 2 cm high is viewed through a converging lens of focal length 8 cm. The object is 4 cm from the lens. By means of a ray diagram find the position, nature and magnification of the image.

51 DISPERSION AND COLOUR

Spectrum of white light

Colourful clothes, colour TV and the flashing coloured lights in a discotheque all help to make life brighter. It was Newton who, in 1666, set us on the road to understanding how colours may arise. He produced them by allowing sunlight (which is white) to fall on a triangular glass prism (Figure 51.1). The band of colours he obtained is called a

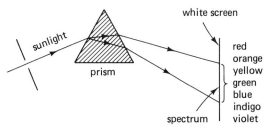

Figure 51.1

spectrum. He concluded that (*i*) white light is a mixture of many colours which the prism separates out because (*ii*) different colours are refracted by different amounts. The splitting up of white light into a spectrum is called *dispersion*.

Experiment 1 A pure spectrum

A pure spectrum is one in which the colours do not overlap, as they do when a prism alone is used. It needs a lens to focus each colour as in Figure 51.2a.

Arrange a lens L (Figure 51.2b) so that it forms an image of the vertical filament of a lamp on a screen at S_1, 1 m away. The filament acts as a narrow source of white light. Insert a 60° prism P and move the screen to S_2, keeping it at the same distance from L, to receive the spectrum; rotate P until the spectrum is pure.

Place different colour filters between P and S_2 and note the results.

Recombining the spectrum

The colours of the spectrum can be recombined to form white light by:

(*i*) arranging a second prism so that the light is deviated in the opposite direction (Figure 51.3a), or

(*ii*) using an electric motor to rotate at high speed a disc with the spectral colours painted on its sectors (Figure 51.3b). (The whiteness obtained is slightly grey because paints are not pure colours.)

Colour of an object

The colour of an object depends on (*i*) the colour of the light falling on it, and (*ii*) the colour(s) it transmits or reflects.

(*a*) *Filters*. A filter lets through light of certain colours only and is made of glass or celluloid. For example, a red filter transmits mostly red light and absorbs other colours; it therefore produces red light when white light shines through it.

Figure 51.2

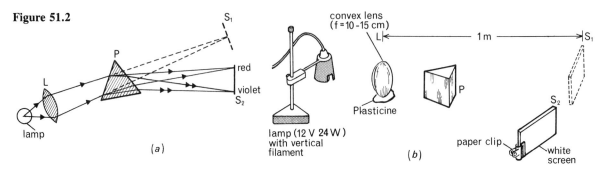

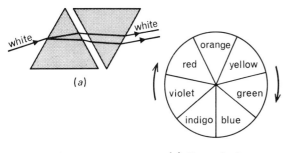

Figure 51.3 (b) Newton's disc

(b) *Opaque objects*. These do not allow light to pass through them but are seen by reflected light. A white object reflects all colours and appears white in white light, red in red light, blue in blue light, etc. A blue object appears blue in white light because the red, orange, yellow, green and violet colours in white light are absorbed and only blue reflected. It also looks blue in blue light but in red light it appears black since no light is reflected and blackness indicates the absence of colour.

Mixing coloured lights

In science red, green and blue are *primary* colours (they are not the artist's primary colours) because none of them can be produced from other colours of light. However, they give other colours when suitably mixed.

The primary colours can be mixed by shining beams of red, green and blue light on to a white screen so that they partially overlap (Figure 51.4a). The results are summarized in the 'colour triangle' of Figure 51.4b.

The colours formed by adding two primaries are called *secondary* colours; they are yellow, cyan (turquoise or peacock blue) and magenta. The three primary colours give white light, as do the three secondaries.

Figure 51.4

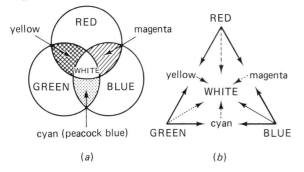

- Why would you also expect a primary colour and the secondary opposite it in the colour triangle to give white?

Any *two* colours producing white light are *complementary* colours, e.g. blue and yellow.

Mixing coloured pigments

Pigments are materials which give colour to paints and dyes by reflecting light of certain colours only and absorbing all other colours. Most pigments are impure, i.e. they reflect more than one colour. When they are mixed the colour reflected is the one common to all. For example, blue and yellow paints give *green* because blue reflects indigo and green (its neighbours in the spectrum) as well as blue, whilst yellow reflects green, yellow and orange (Figure 51.5). Only *green* is reflected by both.

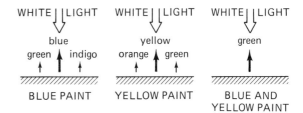

Figure 51.5

Mixing coloured pigments is a process of *subtraction*; coloured lights are mixed by *addition*.

Questions

1. In Figure 51.6a ray AB is of red light and in Figure 51.6b ray PQ is a mixture of red and blue light. Copy each diagram, mark and label (a) the path taken by the light through the prism, (b) what is seen on the screen.

Figure 51.6

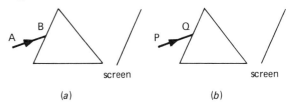

2. A white handkerchief viewed through a piece of red glass looks red. Explain why.

3. A book which looks blue in white light is viewed in (a) red light, (b) magenta light. What colour does it appear in each case?

52 WAVES

There are many kinds of waves. A wave can be produced by fixing one end of a rope (or spring) and jerking the other end (Figure 52.1). The

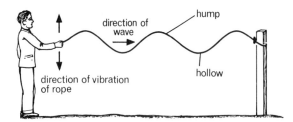

Figure 52.1

humps and hollows forming the wave, travel along the rope but each part of the rope moves to and fro *at right angles* to the direction of the wave. Such a wave is called a *transverse wave*.

Describing waves

Figure 52.2 helps to explain the terms used.

(a) *Wavelength*, represented by the Greek letter λ (lambda), is the distance between successive crests.

(b) *Frequency* f is the number of complete waves generated per second. If the end of a rope is jerked up and down twice in a second, two waves are produced in this time. The frequency of the wave is 2 vibrations per second or 2 *hertz* (2 Hz; the hertz being the unit of frequency) as is the frequency of jerking of the end of the rope. That is, the frequencies of the wave and its source are equal.

The frequency of a wave is also the number of crests passing a chosen point per second.

(c) *Speed* v of the wave is the distance moved by a crest or any point on the wave in 1 s.

(d) *Amplitude* a is the height of a crest or the depth of a trough measured from the undisturbed position of the object carrying the wave, e.g. a rope.

The wave equation

There is a useful connection between f, λ and v which is true for all types of wave.

Suppose waves of wavelength $\lambda = 20$ cm travel on a long rope and three crests pass a certain point every second. The frequency $f = 3$ Hz. If Figure 52.3 represents this wave motion then if crest A is

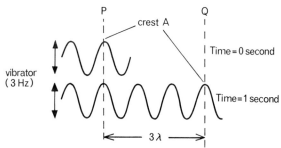

Figure 52.3

at P at a particular time, 1 s later it will be at Q, a distance from P of three wavelengths, i.e. $3 \times 20 = 60$ cm. The speed of the wave $v = 60$ cm per second (60 cm/s). Therefore:

$$\text{speed of wave} = \text{frequency} \times \text{wavelength}$$

or,

$$v = f\lambda$$

Experiment 1 The ripple tank

The behaviour of water waves can be studied in a ripple tank. This consists of a transparent tray con-

Figure 52.2

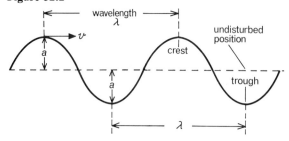

214

taining water, having a lamp above and a white screen below (Figure 52.4).

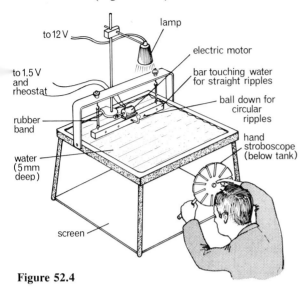

Figure 52.4

Pulses (i.e. short bursts) of ripples are obtained by dipping a finger in the water to obtain circular ones and a ruler for straight ones. *Continuous* ripples are generated by an electric motor on a bar which gives straight ripples if it just touches the water and circular ripples if the bar is raised and a small ball fitted to it so as to be in the water.

Continuous ripples are studied more easily if they are *apparently* stopped ('frozen') by viewing the screen through a disc, with equally spaced slits, that can be spun by hand, i.e. a stroboscope.

a *Reflection*. In Figure 52.5 *straight* water waves are represented falling on a metal strip placed in the tank at an angle of 60°, i.e. the angle i between

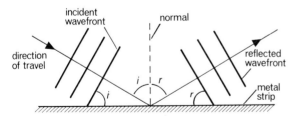

Figure 52.5

the direction of travel of the waves and the normal to the strip is 60°, as is the angle between the wavefront and the strip. The angle of reflection r is 60°. Incidence at other angles shows that the angles of reflection and incidence are always equal.

b *Refraction*. If a glass plate is placed in the tank so that the water is about 1 mm deep over it but 5 mm elsewhere, continuous straight waves in the shallow region are found to have a shorter wavelength than those in the deeper parts (Figure 52.6a). Both sets of waves have the frequency of

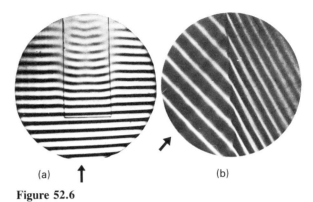

Figure 52.6

the vibrating bar and since $v = f\lambda$, then if λ has decreased so has v, since f is fixed. Hence *waves travel more slowly in shallow water*.

When the plate is at an angle to the waves (Figure 52.6b) their direction of travel in the shallow region is bent towards the normal (Figure 52.7), i.e. refraction occurs.

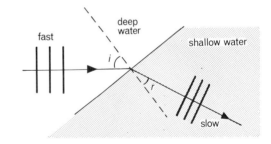

Figure 52.7

Questions

1. The lines in Figure 52.8 are crests of straight ripples.
   (a) What is their wavelength?
   (b) If ripple A occupied 5 s ago the position now occupied by ripple F, what is the frequency of the ripples?
   (c) What is the speed of the ripples?

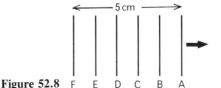

Figure 52.8 F E D C B A

2. A straight ripple ABC is shown in Figure 52.9 moving towards a wall XY. Draw one diagram to show the

Figure 52.9

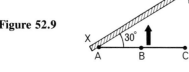

position of the ripple when B reaches the wall and another when C reaches it. On each diagram mark the angles which are 30°.

3. One side of a ripple tank ABCD is raised slightly (Figure 52.10) and a ripple started at P by a finger. After a second the shape of the ripple is as shown.

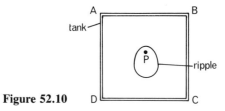

Figure 52.10

(*a*) Why is it not circular?
(*b*) Which side of the tank has been raised?

53 THE ELECTROMAGNETIC SPECTRUM

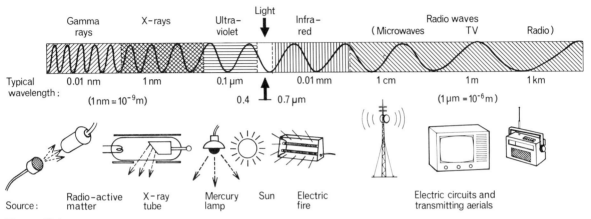

Figure 53.1

The members of the electromagnetic spectrum (Figure 53.1) are wave-like radiations which carry energy. They all travel at 300 million metres per second (3×10^8 m/s) in air. This is the speed of light. Their different wavelengths λ and frequencies f are related to their speed v by $v = f\lambda$.

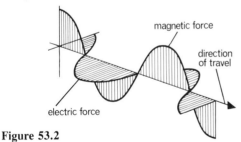

Figure 53.2

Whilst each is produced differently, all have an electrical origin. Because of this and their ability to travel in a vacuum (e.g. from the sun to the earth), they are regarded as a combination of travelling electric and magnetic forces which vary in value and are directed at right angles to each other and to the direction of travel, i.e. they are transverse-type waves (Figure 53.2).

Infrared radiation

The presence of infrared (i.r.) in the radiation from the sun or from the filament of an electric lamp can be detected by placing a phototransistor just beyond the red end of the spectrum formed by a prism (Figure 53.3). Our bodies detect i.r. (also called 'radiant heat' or 'heat radiation') by its heating effect.

Anything which is hot but not glowing, i.e. below 500 °C, emits i.r. alone. At about 500 °C a body becomes red-hot and emits red light as well as i.r.; the heating element of an electric fire is an

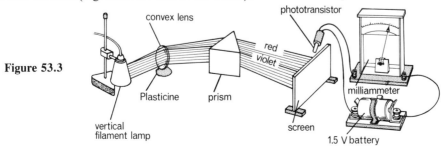

Figure 53.3

217

example. At about 1000 °C things such as lamp filaments are white-hot and radiate i.r. and white light, i.e. all the colours of the spectrum.

Infrared is also detected by special photographic films, and pictures can be taken in the dark like that in Figure 53.4 of a car; the white parts are hottest. Infrared lamps are used to dry the paint on cars during manufacture and in the treatment of muscular complaints.

Ultraviolet radiation

Ultraviolet (u.v.) rays have shorter wavelengths than light and can be detected just beyond the violet end of the spectrum (due to the sun or a filament lamp) by fluorescent paper. They cause sun-tan and produce vitamins in the skin but an overdose can be harmful, especially to the eyes.

Ultraviolet causes teeth, finger nails, fluorescent paints and clothes washed in detergents to fluoresce, i.e. they glow by reradiating as light the energy they absorb as u.v. The shells of fresh eggs fluoresce with a reddish colour, those of 'bad' eggs appear violet.

Figure 53.4

A u.v. lamp used for scientific or medical purposes contains mercury vapour and this emits u.v. when an electric current passes through it. Fluorescent tubes also contain mercury vapour and their inner surfaces are coated with powders which radiate light when struck by u.v.

Radio waves

Radio waves vary in wavelength from a few millimetres to several kilometres. The shortest are called microwaves and are radiated by 'saucer-shaped' aerials like those in Figure 53.5. They are

Figure 53.5

used in radar and in a new method of cooking which cooks food right through quickly. VHF and TV signals have wavelengths of a few metres and travel in straight lines from sender to receiver. Long distance radio waves rely on 'reflection' from the *ionosphere*; this consists of layers of electrically charged gases between 80 and 400 km above the earth. For satellite communication, signals with wavelengths less than 10 m or so must be used since only they can penetrate the ionosphere.

X-rays

X-rays have smaller wavelengths than u.v. They can penetrate solid objects and affect a film; Figure 53.6 is an X-ray photograph of someone shaving with an electric razor. Very penetrating X-rays are used in hospitals to kill cancer cells. They also damage healthy cells, so careful shielding of the X-ray tube with lead is needed. Less penetrating X-rays have longer wavelengths and penetrate flesh but not bone.

In industry X-rays are used to inspect welded joints for faults.

Gamma rays come from radioactive substances and are more penetrating and dangerous than X-rays.

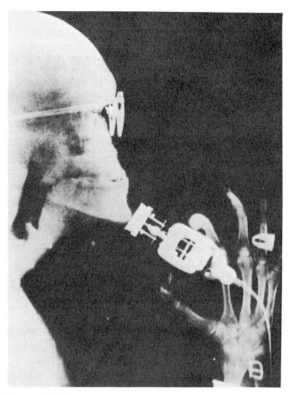

Figure 53.6

Questions

1. Name *three* properties common to all electromagnetic waves.

2. Name one type of electromagnetic wave which (*a*) causes sun-tan, (*b*) bounces off the earth's upper atmosphere, (*c*) passes through a thin sheet of lead, (*d*) is used to take photographs in haze.

54 SOUND WAVES

Sources of sound such as a drum, a guitar and the human voice have some part which *vibrates*. The sound travels through the air to our ears and we hear it. That the air is necessary may be shown by pumping out a glass jar containing a ringing electric bell (Figure 54.1); the sound disappears though the striker can still be seen hitting the gong.

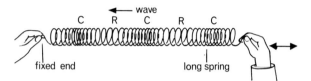

Figure 54.2

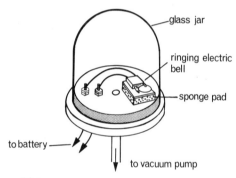

Figure 54.1

Evidently sound cannot travel in a vacuum like light.

Other materials, including solids and liquids, transmit sound.

SOUND WAVES AND THEIR PROPERTIES

Longitudinal waves

In a longitudinal wave the particles of the transmitting medium vibrate in the *same direction* as that in which the wave is travelling and not at right angles to it as in a transverse wave. A longitudinal wave can be sent along a spring, stretched out on the bench and fixed at one end if the free end is repeatedly pushed and pulled sharply. Compressions C (where the coils are closer together) and rarefactions R (where the coils are farther apart) (Figure 54.2), travel along the spring.

A sound wave produced for example by a loudspeaker consists of a train of compressions and rarefactions in the air (Figure 54.3). The speaker has a cone which is made to vibrate in and out by an electric current. When the cone moves out the air in front is compressed; when it moves in the air is rarefied (goes 'thinner'). The wave progresses through the air but the air as a whole does not move. The air particles vibrate backwards and forwards a little as the wave passes, and this causes the sound you hear when the wave reaches your ear.

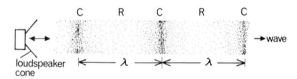

Figure 54.3

The number of compressions produced per second is the frequency f of the sound wave (and equals the frequency of the vibrating cone); the distance between successive compressions is the wavelength λ. As for transverse waves the speed $v = f\lambda$.

Human beings hear only sounds with frequencies from about 20 Hz to 20 000 Hz (20 kHz). These are the *limits of audibility*; they decrease with age.

Reflections and echoes

Sound waves are reflected well from hard, flat surfaces such as walls or cliffs and obey the same laws of reflection as light. The reflected sound forms an *echo*.

If the reflecting surface is nearer than 15 m, the echo joins up with the original sound which then seems to be prolonged. This is called *reverberation*. Some reverberation is desirable in a concert hall to stop it sounding 'dead', but too much causes 'confusion'.

Speed of sound

If you stand about 100 m from a high wall and clap your hands, echoes are obtained. When the clapping rate is such that each clap coincides with the echo of the previous one, the sound has travelled to the wall and back in the time between two claps, i.e. an interval. By timing thirty intervals with a stop-watch and knowing the distance to the wall, the speed of sound can be found roughly.

It is about 330 m/s in air at 0 °C and increases at higher temperatures. In steel it is 6000 m/s.

Musical notes

Irregular vibrations cause *noise*, regular vibrations such as occur in the instruments of an orchestra, produce *musical notes* which have three properties—pitch, loudness and quality.

(*a*) *Pitch.* The pitch of a note depends on the frequency of the sound wave reaching the ear, i.e. on the frequency of the source of sound. A high-pitched note has a high frequency and a short wavelength. The frequency of middle C is 256 vibrations per second or 256 Hz and that of upper C is 512 Hz. Notes are an *octave* apart if the frequency of one is twice that of the other.

Notes of known frequency can be produced in the laboratory by a signal generator supplying alternating electric current (a.c.) to a loudspeaker. The cone of the speaker vibrates at the frequency of the a.c. which can be varied and read off a scale on the generator. A set of tuning forks with frequencies marked on them can also be used. A tuning fork (Figure 54.4) has two steel prongs which vibrate when struck; the prongs move in and out *together*, generating compressions and rarefactions.

(*b*) *Loudness.* A note is louder when more sound energy enters our ears per second than before and is caused by the source vibrating with a larger amplitude. If a violin string is bowed more strongly, its amplitude of vibration increases as does that of the resulting sound wave and the note heard is louder because more energy has been used to produce it.

(*c*) *Quality.* The same note on different instruments sounds different; we say the notes differ in *quality* or *timbre*. The difference arises because no instruments (except a tuning fork and a signal generator) emit a 'pure' note, i.e. of one frequency. Notes consist of a main or *fundamental* frequency mixed with others, called *overtones*, which are usually weaker and have frequencies that are exact multiples of the fundamental. The number and strength of the overtones decides the quality of a note. A violin has more and stronger higher overtones than a piano. Overtones of 256 Hz (middle C) are 512 Hz, 768 Hz, and so on.

String and wind instruments

(*a*) *String.* In a string instrument such as a guitar, the 'string' is a tightly-stretched wire or length of gut. When it is plucked it vibrates and the frequency (pitch) of the note produced depends on its:

(*i*) length—short wires emit high notes
(*ii*) mass per unit length—thin wires give high notes
(*iii*) tension—tight wires produce high notes.

(*b*) *Wind.* In wind instruments, musical notes are produced by vibrating columns of air. In an organ, for example, the vibration is started by a jet of air hitting a sharp edge. The shorter the air column the higher the note.

Questions

1. If 5 s elapse between a lightning flash and the clap of thunder how far away is the storm? (Speed of sound = 330 m/s.)

2. How could you raise the pitch of the note from a certain guitar string without changing its length?

Figure 54.4

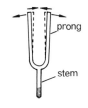

Formulas, Equations and Chemical Compounds

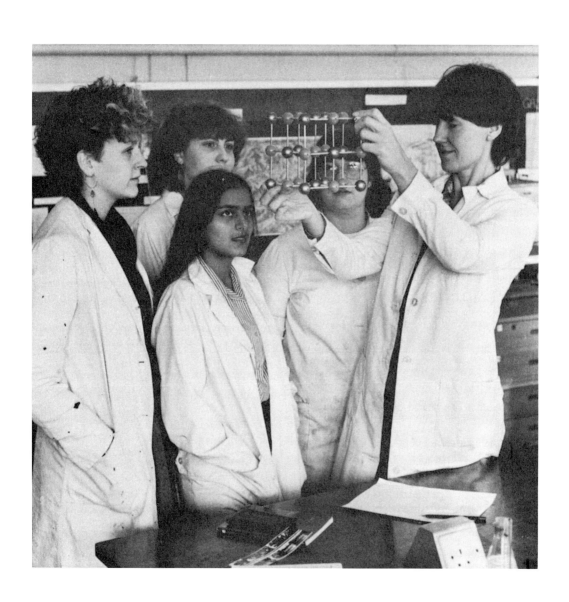

55 MORE ABOUT ATOMS

The particles in elements

A lump of iron is made of very many particles. Because iron is an element, these particles are *atoms*. For the present, you can imagine that all these atoms are the same. Scientists have found that iron atoms have a diameter of about 0·000 000 1 mm (1×10^{-7} mm) and a mass of about 0·000 000 000 000 000 000 000 09 g (9×10^{-23} g). They are very small indeed!

The atoms of one element are different from the atoms of all other elements. This is why the properties of different elements are different. Table 55.1 shows the masses and the diameters of the atoms of some elements.

Table 55.1

Elements	Mass in g	Diameter in mm
Aluminium	$4\cdot5 \times 10^{-23}$	$2\cdot86 \times 10^{-7}$
Calcium	$6\cdot7 \times 10^{-23}$	$3\cdot94 \times 10^{-7}$
Carbon	$2\cdot0 \times 10^{-23}$	$1\cdot54 \times 10^{-7}$
Copper	11×10^{-23}	$2\cdot56 \times 10^{-7}$
Gold	33×10^{-23}	$2\cdot88 \times 10^{-7}$
Lead	34×10^{-23}	$3\cdot50 \times 10^{-7}$
Sulphur	$5\cdot3 \times 10^{-23}$	$2\cdot08 \times 10^{-7}$

Note: 10^{-23} is $1/10^{23}$ or 1/100000 million million million; 10^{-7} is $1/10^{7}$ or 1/10 million.

A few elements are made of *molecules* rather than separate atoms.

A molecule of an element is a separate particle containing a small number of the same atoms strongly bound together.

Elements made of separate atoms (like iron) are said to be *monatomic*, and those made of molecules with two of the same atoms joined together are said to be *diatomic* (Figure 55.1). There are also some elements which have more than two of the same atoms joined together in each of their molecules: for example, sulphur molecules are made of eight sulphur atoms.

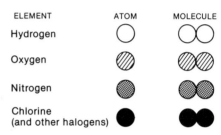

Figure 55.1

The molecules of an element can sometimes be split into the separate atoms, but this is usually difficult because the bonds (page 246) between the atoms are strong.

The particles in compounds

Atoms of different elements can often join together to form molecules. Compounds that are liquids or gases at room temperature are made of molecules and some examples are shown in Figure 55.2.

Carbon dioxide is made of molecules each having one carbon atom and two oxygen atoms joined together. Water is a liquid, and its

Figure 55.2

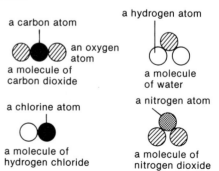

molecules are made of two hydrogen atoms and one oxygen atom joined together. But notice from Figure 55.2 that, unlike a molecule of carbon dioxide, a water molecule does not have its atoms in a straight line.

Not all compounds are made of molecules. One example is iron sulphide. This has two kinds of atom, iron atoms and sulphur atoms, but it does *not* contain molecules of iron sulphide. Instead the solid is made of iron atoms and sulphur atoms packed closely together. Figure 55.3 shows that each iron atom or each sulphur atom 'touches' more than one of the other kind of atom.

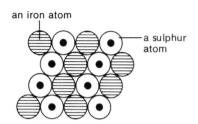

Figure 55.3

In fact, many of the compounds you meet in your science course are not made of molecules. Chapter 56 deals with the special kind of particles that make up all electrolytes.

Inside the atom

At one time, scientists thought of atoms as very small, hard spheres which could not be broken down into smaller parts. They now think of atoms as being made from a collection of much smaller particles, the *sub-atomic particles*.

One kind of sub-atomic particle is called the *electron*. The first evidence for the electron was obtained by the British physicist J. J. Thomson in 1897. A diagram of the apparatus he used is given in Figure 55.4.

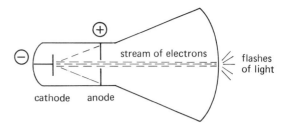

Figure 55.4

The tube contained a gas at a very low pressure and the cathode was heated by passing electricity through it. When a high voltage was applied between the cathode and the anode, an electric current flowed across the tube and a faint green glow appeared on a screen beyond the anode. On a closer look, this glow was found to be made of thousands of tiny flashes of light.

Thomson decided that the flashes were due to particles with a *negative charge* given off from the hot cathode and attracted towards the positive anode. Some of them passed through the hole in the anode and travelled on to hit the screen.

The same kind of particles are produced no matter which metal is used for the cathode and which gas is in the tube. The particles are electrons and Table 55.2 shows their mass and charge. Obviously their mass is much lower than the mass of any atom.

Atoms are electrically neutral, that is, they have no net negative or positive charge. If atoms contain electrons, then they must also contain enough particles with a *positive* charge to balance the negative charge of the electrons. These positively charged particles are called *protons*. Table 55.2 shows that the mass of the proton is the same as the mass of a hydrogen atom, and its charge is the same in size as the charge on the electron.

Because the negative charge carried by the electron is equal to the positive charge carried by the proton, all neutral atoms must be made of equal numbers of electrons and protons. The number of

Table 55.2

Name of sub-atomic particle	Mass of particle compared with mass of hydrogen atom (H=1)	Mass in g	Relative charge	Charge in C
Electron	1/1840	9.11×10^{-28}	-1	-1.60×10^{-19}
Proton	1	1.67×10^{-24}	$+1$	$+1.60 \times 10^{-19}$
Neutron	1	1.67×10^{-24}	0	0

protons in an atom of an element is called the *atomic number* of that element. You can easily find the atomic number of any element because it is the same as the numbered position of the element in the periodic table (inside back cover).

Further experiments have shown that there is one more type of sub-atomic particle. This is the *neutron*. It has the same mass as a proton, but carries no charge. All atoms except hydrogen contain neutrons.

At first, it was assumed that the electrons, protons and neutrons were packed together to make a solid sphere. This picture had to be changed when the results of the following experiment were examined.

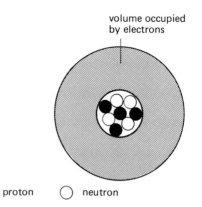

● proton ○ neutron

Figure 55.5 Note that the nucleus is really a very much smaller part of the atom than this diagram suggests.

Figure 55.6

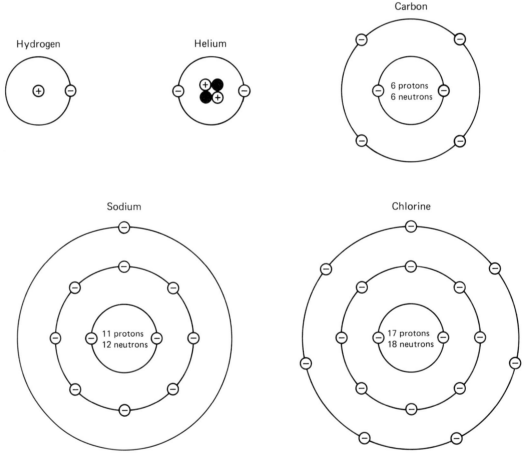

⊕ proton

⊖ electron

● neutron

The Geiger–Marsden experiment

Fast moving positively charged alpha particles (page 380) were 'fired' at some very thin gold foil. It was found that:

a Most of the alpha particles passed through with little or no change in direction.
b A very few alpha particles suffered very large changes in direction. Some even bounced straight back.

Working with these results, Rutherford reasoned as follows:

a Even the very thin gold foil used is several hundred atoms thick. It follows that *most of each atom is empty space*.
b A positively charged particle can only 'bounce' straight back if a large force repels it. This repelling force can only happen if atoms contain a *separate* positive charge, i.e. the *protons and electrons in an atom are widely separated*.

From this, he deduced a model of the atom which looks like that shown in Figure 55.5. Almost all the mass is concentrated in the *nucleus*, which contains protons and neutrons.

The electrons orbit around the nucleus, rather like planets moving round the sun. Like our planetary system, an atom is mostly empty space. To get an idea of the scale, imagine that the nucleus is the size of a tennis ball. The nearest electron will be about 4 km away!

Figure 55.6 shows simplified pictures of some atoms we have met previously. Notice that in each, the number of electrons orbiting the nucleus is the same as the number of protons in the nucleus. It has also been shown that the electron orbits as a series of increasing levels. These levels are a simple way of indicating that electrons have different *energies*. The more energy an electron has, the higher the level, i.e. the bigger the orbit. If more energy is supplied to an atom (e.g. by heating or passing an electrical discharge) electrons can move to higher energy levels. When enough energy is supplied, some electrons can actually escape from their atoms. In Chapter 91 we will study one application of this ability of electrons to escape from their atoms.

Questions

1. 'Argon is a monatomic gas and nitrogen is a diatomic gas.'
Explain what this statement means using diagrams to represent their molecules.

2. Carbon monoxide and nitrogen dioxide are two gases made of molecules. Use their names to decide how many atoms of each kind there are in each of the molecules.

3. Complete the following passage about atoms by filling in the gaps with the words listed below. Each term may be used more than once.

charge, electron, energy level, neutron, nucleus, particle, proton

Atoms consist of a central .... which is made of two kinds of sub-atomic .... One kind is called the .... and carries a positive .... while the other kind is called the .... and carries no ....
The third kind of sub-atomic .... is the .... Each of these is found in an .... surrounding the centre of the atom, and carries a negative .... of the same size as the one on the ....
In Rutherford's famous experiments, some positively charged particles 'fired' at gold foil rebounded if they came close to the .... of any of the atoms.

56 CHARGED PARTICLES

More electrolysis experiments

A green solution of copper chromate is being electrolysed in the apparatus shown in Figure 56.1. The gelatin makes the solution set like jelly and stops the solutions from mixing by diffusion. After a few minutes of this electrolysis, the space above the green jelly on the cathode side goes blue and the space on the anode side goes yellow. After 1–2 hours, definite bands can be seen, blue near the cathode and yellow near the anode.

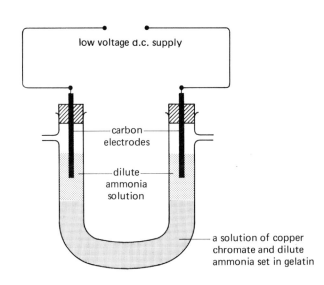

Figure 56.1

The green solution of copper chromate has separated into a blue part which moves towards the cathode and a yellow part which moves towards the anode. If small equal amounts of the blue and yellow solutions are mixed, green copper chromate is produced again.

These results can be explained by using the idea of charged particles. The particles that cause the blue colour must have a positive charge because they are attracted to the cathode which is negative. The particles that cause the yellow colour must have a negative charge because they are attracted to the anode which is positive.

Since a green solution of copper chromate is electrically neutral, the charges on the blue and yellow parts must exactly balance each other when the two parts are mixed in certain proportions.

So copper chromate is made of two kinds of charged particles, a positive kind and a negative kind. These charged particles are called *ions*: *cations* have a positive charge and move to the cathode, and *anions* have a negative charge and move to the anode (Figure 56.2).

Solutions of copper sulphate and copper nitrate are blue but solutions of sodium chromate and potassium chromate are yellow. Thus it seems likely that the blue colour is due to copper cations and the yellow colour to chromate anions.

Another way of investigating the movement of coloured ions during electrolysis is shown in Figure 56.3. The strip of filter paper is dampened with water and a crystal of potassium permanganate is

228

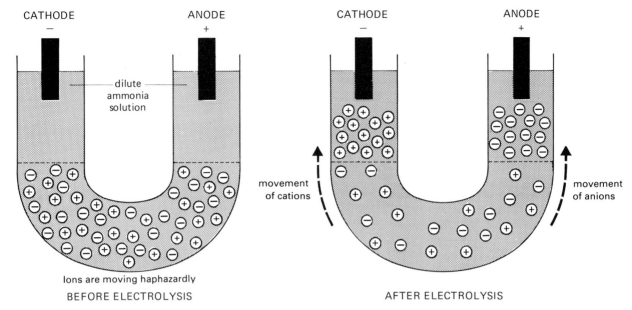

Figure 56.2

placed in the middle of it. A patch of purple slowly moves towards the anode when the electricity is passed between the crocodile clips acting as electrodes.

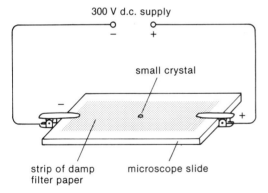

Figure 56.3

Potassium permanganate is purple. At first sight, it looks as if the whole substance moves towards the anode. But sodium permanganate is also purple and does the same thing. It is more likely that only the permanganate part of the substance moves towards the anode. So potassium permanganate is made of a permanganate anion and a potassium cation. The potassium cation must be moving towards the cathode but it cannot be seen, because it is colourless.

Explaining electrolysis

Look at Figure 56.4 which shows a way of describing what happens in the various parts of the circuit during an electrolysis experiment.

In the wires, there is a flow of *electrons* (page 225). The flow is from the battery to the cathode and from the anode back to the battery.

In the electrolyte, there is a flow of ions. Cations move towards the cathode and anions move towards the anode.

So positively charged ions arrive at the cathode from the electrolyte and negatively charged electrons move to the cathode from the battery. The cations combine with the electrons to form neutral particles:

At the cathode:

cations + electrons ⟶ atoms
(positively (taken from (formed at
charged ions) the cathode) the cathode)

If the electrolyte is molten lead bromide, lead cations combine with electrons to form lead atoms and you see a deposit of lead on the cathode. If the electrolyte is dilute sulphuric acid or sodium chloride solution, hydrogen cations combine with electrons to form hydrogen atoms. These atoms combine to form hydrogen molecules and you see hydrogen gas given off.

At the anode, negatively charged ions arrive from the electrolyte and electrons flow from the

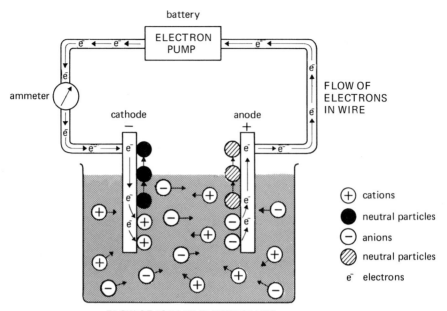

Figure 56.4   FLOW OF IONS IN ELECTROLYTE

anode to the battery. When anions reach the anode, they lose electrons to the anode and the ions become neutral atoms:

At the anode:

anions ⟶ atoms + electrons
(negatively (formed at (given to
charged ions) the anode) the anode)

If the electrolyte is molten lead bromide, bromide anions arrive at the anode, lose electrons to the anode and form bromine atoms. The atoms then form molecules and you see bromine gas given off.

Charges on ions

A neutral atom has an equal number of protons and electrons (page 225). Many ions are just charged atoms. The charge comes about because there are either more protons than electrons in the

Figure 56.5

CATIONS

(extra protons in atom)

| 4 protons in nucleus | + | 3 electrons around nucleus | ⟶ | X^+, cation with one extra proton |

| 4 protons in nucleus | + | 2 electrons around nucleus | ⟶ | X^{2+}, cation with two extra protons |

ANIONS

(extra electrons in atom)

| 8 protons in nucleus | + | 9 electrons around nucleus | ⟶ | Y^-, anion with one extra electron |

| 8 protons in nucleus | + | 10 electrons around nucleus | ⟶ | Y^{2-}, anion with one extra electron |

atom or more electrons than protons.

Some examples of charged atoms are given in Figure 56.5. One extra proton gives a charge on the cation of $+1$, two extra protons give a charge of $+2$, and so on. One extra electron gives a charge on the anion of -1, two extra electrons give a charge of -2, and so on.

There is a list of ions together with their charges in Table 57.6 (page 234).

Electrolytes and ions

All acids, alkalis and salts are electrolytes when they are dissolved in water or in the molten state (page 58). But they cannot be electrolysed when they are in the solid state. The reason for this is that electrolysis can only happen *if the ions are free to move*.

A solid salt such as sodium chloride is made of ions but, like the particles in any solid, they are held close together. The only movement possible is vibration about these fixed positions (page 16). When the solid melts, the ions are still close to one another but they can now move in the spaces between each other (Figure 56.6). So molten salts can be electrolysed.

During the dissolving of a salt in water, the water molecules are attracted towards the ions on the edges of the crystals. This attraction helps to dislodge these ions which then move off into the water to form a solution. Again, this produces ions that are free to move through the electrolyte and so salts in solution can be electrolysed.

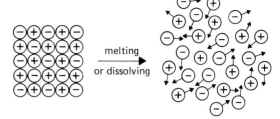

a regular arrangement of anions and cations in part of a crystal of a solid salt (only shown in two-dimensions)

mobile ions in the molten salt or in a solution of the salt in water

$\oplus$ cation

$\ominus$ anion

Figure 56.6

Questions

1. Use the idea of protons and electrons to explain the difference between a sodium atom and a sodium cation with a $+1$ charge.

2. Solid salt and solid sugar do not conduct electricity. A solution of sugar in water does not conduct electricity either, but a salt solution can be electrolysed. Explain these facts.

3. Why do changes in an electrolyte (such as the evolution of a gas) caused by the passage of an electric current take place only at the electrodes and not in the body of the electrolyte?

4. When a deep purple solution of copper permanganate is electrolysed in a U-tube, a blue colour appears around the cathode and a lighter purple colour around the anode.

(*a*) What kind of charge must the particles that cause the blue colour have?
(*b*) What kind of charge must the particles that cause the light purple colour have?
(*c*) Name the ions that cause the blue colour, and the ions that cause the light purple colour.
(*d*) Given that the formula of copper permanganate is $Cu(MnO_4)_2$, what can you say about the *relative* sizes of the charges on the two kinds of ion?

57 CHEMICAL FORMULAS

What are formulas?

The *atomic symbols* for some of the more common elements are given on the inside back cover. These are useful as a kind of 'chemical shorthand'. Some symbols are just the first letter, the first two letters or the first and third letters of the element's name.

Others do not look at all like the element's name. This is because they are based on their name in another language such as Latin: for example, the Latin name for sodium is *Na*trium and for potassium it is *K*alium.

Scientists often use atomic symbols to write down the *formulas* of compounds. Here are the formulas of two copper compounds:

$$CuO \qquad CuCl_2$$
black copper oxide   green copper chloride

Each formula is made of the atomic symbol for the metal put in front of the atomic symbol for the non-metal. But in the formula for green copper chloride there is something else: a number 2 comes after the Cl symbol.

A formula can tell you two things:

— which elements are joined together in the compound
— how much of each element there is in the compound compared with the other elements

Black copper oxide is made from copper and oxygen. The formula CuO also tells you that there is 1 'lot' of O atoms for every 1 'lot' of Cu atoms. But the formula $CuCl_2$ tells you that there are 2 'lots' of Cl atoms for every 1 'lot' of Cu atoms.

For most elements, the symbol is the same as the formula: the formula for iron is Fe and for carbon is C. A few non-metals are different because they are made of molecules with two or more of the same atoms (page 224). For example, the formulas of hydrogen, oxygen and sulphur molecules are H_2, O_2 and S_8, respectively.

Patterns in the formulas of compounds

There are three kinds of formulas for metal chlorides in Table 57.1, depending on how many

Table 57.1 Metal chlorides

LiCl	$MgCl_2$	$AlCl_3$
NaCl	$CaCl_2$	$FeCl_3$
KCl	$CuCl_2$	
AgCl	$FeCl_2$	
	$PbCl_2$	
	$ZnCl_2$	

'lots' of Cl atoms are joined with 1 'lot' of metal atoms. For some metals, such as iron, there are two chlorides with different 'lots' of Cl atoms joined with 1 'lot' of Fe.

The formulas for metal bromides and metal iodides look very like metal chlorides (Table 57.2).

Table 57.2 Metal bromides and iodides

BROMIDES	KBr	$FeBr_2$	$FeBr_3$
	AgBr	$PbBr_2$	
IODIDES	KI	FeI_2	FeI_3
	AgI	PbI_2	

There is also a pattern in the formulas of metal oxides (Table 57.3), but it is not the same as the pattern for chlorides. You could try to sort out for yourself these formulas into three classes. The formulas for sulphides are like those for oxides except that S is used in place of O.

Table 57.3 Metal oxides

Al_2O_3	CaO	Cu_2O	CuO	FeO	Fe_1O_3
Li_2O	Na_2				

Remembering all these formulas is a problem, but the periodic table (inside back cover) can help you. Look at Table 57.4, which gives the formulas of chlorides and oxides of some elements on the left-hand side of the table. For this part of the table at least, the number of 'lots' of Cl atoms is the same as the group number.

The pattern for the formulas of the oxides is a little more complicated. It can be understood if you

Table 57.4

Group (column) number	I	II	III	IV
Element (3rd row)	Na	Mg	Al	Si
Formula of chloride	NaCl	$MgCl_2$	$AlCl_3$	$SiCl_4$
Formula of oxide	Na_2O	*	Al_2O_3	SiO_2
Element (4th row)	K	Ca	Ga (gallium)	Ge (germanium)
Formula of chloride	KCl	$CaCl_2$	$GaCl_3$	$GeCl_4$
Formula of oxide	K_2O	CaO	Ga_2O_3	GeO_2

* You will find this formula if you carry out the experiment on page 235.

think of each 'lot' of O atoms as being worth 2 'lots' of Cl atoms. So, for example, 4 'lots' of Cl atoms but only 2 'lots' of O atoms are joined with 1 'lot' of C atoms to make CCl_4 and CO_2, respectively.

The formulas in Table 57.5 are for compounds with three or more kinds of atoms. The 'non-metal' parts of these are all groups of atoms such as hydrogencarbonate (HCO_3), hydroxide (OH), nitrate (NO_3), carbonate (CO_3) and sulphate (SO_4). The HCO_3 group contains 1 'lot' of H atoms, 1 'lot' of C atoms and 3 'lots' of O atoms, and so on.

Table 57.5

Group (column) number	I	II	III
Element (3rd row)	Na	Mg	Al
Formula of:			
hydrogencarbonate	$NaHCO_3$	$Mg(HCO_3)_2$	–
hydroxide	NaOH	$Mg(OH)_2$	$Al(OH)_3$
nitrate	$NaNO_3$	$Mg(NO_3)_2$	$Al(NO_3)_3$
Formula of:			
carbonate	Na_2CO_3	$MgCO_3$	–
sulphate	Na_2SO_4	$MgSO_4$	$Al_2(SO_4)_3$

The hydrogencarbonate, hydroxide and nitrate groups give similar formulas to chloride (Cl in Table 57.1). The only difference is that you have to put brackets around these groups of atoms when they are followed by a number 2 or 3. The formula $Mg(NO_3)_2$ means that 1 'lot' of Mg atoms is joined with 2 'lots' of NO_3 groups.

The carbonate and sulphate groups give similar formulas to oxides (O in Table 57.3). Again, you have to use brackets in the cases where the group of atoms is followed by a number.

Ions and the formulas of electrolytes

The formulas of electrolytes (acids, alkalis and salts) can always be worked out if the charges on their cations and anions are known. Table 57.6 gives these charges for some common ions. Some ions are 'compound ions', that is, they contain more than one atom: these are the 'groups of atoms' mentioned above.

An electrolyte has no *overall* charge. The total positive charge on the cations must balance the total negative charge on the anions. Figure 57.1 shows how this balance can be reached for three different electrolytes.

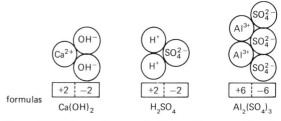

Figure 57.1 These diagrams just show the proportions in which the cations and anions are joined together: they do not represent particles of $Ca(OH)_2$, H_2SO_4 or $Al_2(SO_4)_3$.

Table 57.6

CATIONS					
Charge +1		Charge +2		Charge +3	
Ammonium	NH_4^+	Barium	Ba^{2+}	Aluminium	Al^{3+}
Copper(I)	Cu^+	Calcium	Ca^{2+}	Iron(III)	Fe^{3+}
Hydrogen	H^+	Copper(II)	Cu^{2+}		
Potassium	K^+	Iron(II)	Fe^{2+}		
Silver	Ag^+	Lead	Pb^{2+}		
Sodium	Na^+	Magnesium	Mg^{2+}		
		Zinc	Zn^{2+}		

ANIONS			
Charge −1		Charge −2	
Bromide	Br^-	Carbonate	CO_3^{2-}
Chloride	Cl^-	Oxide	O^{2-}
Hydrogen carbonate	HCO_3^-	Sulphate	SO_4^{2-}
		Sulphide	S^{2-}
Hydrogen sulphate	HSO_4^-	Sulphite	SO_3^{2-}
Hydroxide	OH^-		
Iodide	I^-		
Nitrate	NO_3^-		

If the formula of aluminium sulphate were $AlSO_4$, there would be an overall charge of $+3$, -2 or $+1$. So this cannot be the correct formula. The formula $Al_2(SO_4)_3$ shows that two charges of size $+3$ (total $+6$) and three charges of size -2 (total -6) are needed.

Some elements appear under two separate columns showing that they can form two kinds of ions with different charges. In these cases, the ions are written or spoken about by putting a number after the name of the element. For example, Cu^+ is called a copper (I) ion and Fe^{3+} an iron (III) ion.

In naming some compounds, it is usual to include this number so that there is no confusion over which cation is present in the compound. So red copper oxide formula (Cu_2O) is called copper(I) oxide and black copper oxide formula (CuO) is called copper(II) oxide.

There is no need to use a number where there is only one kind of ion. So Ca^{2+} is not called a calcium(II) ion because there is no Ca^+ or Ca^{3+} to confuse it with.

Moles of atoms

From the next experiment, you can work out the formula of magnesium oxide for yourself. To do this, you first need to find how many 'lots' of Mg atoms and O atoms join to make magnesium oxide.

Atoms are very small indeed (see inside back cover), and the mass of a single atom of an element cannot be found directly using a balance. Instead, scientists compare the masses of different atoms using an instrument called the *mass spectrometer*.

Imagine that the apparatus in Figure 57.2 is used to separate some ball-bearings into three kinds according to their masses. As they roll past the powerful horseshoe magnet at the same constant speed, they are bent into differently curved paths depending on their masses. The ball-bearings with

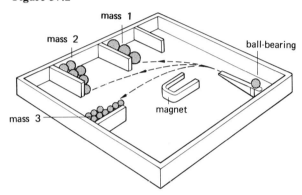

Figure 57.2

the largest mass are only bent slightly while the ones with the smallest mass are bent the most.

The mass spectrometer works rather like this. The atoms of the elements are first ionized (made into ions, in this case cations). They are then passed through a magnetic field (Figure 57.3). If the charge on all the ions is the same and if they are all moving at about the same speed, the amount they are bent depends on their masses.

The atom of hydrogen has the lowest mass of all atoms because it is made of only one proton (and one electron). So the masses of other atoms used to be given as so many times greater than this lowest mass. Table 57.7 shows that a magnesium atom has twenty-four times the mass of a hydrogen atom, and an oxygen atom has sixteen times the mass.

Table 57.7 also shows that the mass of a magnesium atom is one and a half times the mass of an oxygen atom, and that the mass of a copper atom is four times that of an oxygen atom.

The masses of atoms can be compared with the mass of any particular atom you choose. But nowadays, all atomic masses are compared with one-twelfth of the mass of a *carbon* atom. This makes no difference to the numbers given in Table 57.7.

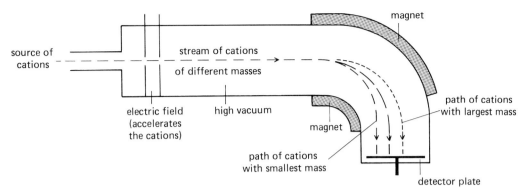

Figure 57.3

There is a list of the *relative atomic masses* of elements in the periodic table on the inside back cover.

Scientists talk about 1 *mole* of Mg atoms instead of 1 'lot' of Mg atoms. The mass of 1 mole of Mg atoms is 24 g and the mass of 1 mole of O atoms is 16 g. The list of relative atomic masses can be used to find the mass of 1 mole of atoms of other elements.

Table 57.7

Atomic symbol of element	Relative atomic mass
H	1
C	12
O	16
Mg	24
S	32
Cl	35
Cu	64

All these relative atomic masses are given to the nearest whole number.

Finding a formula by experiment

Magnesium oxide can be made by burning magnesium ribbon in air. The word equation for this chemical reaction is:

magnesium + oxygen ⟶ magnesium oxide

Before you do the experiment, think of some of the results you might get. Suppose you find that 48 g of magnesium joins with 16 g of oxygen. This would mean that 2 moles of Mg atoms are combined with 1 mole of O atoms in magnesium oxide. The formula of magnesium oxide would then be Mg_2O. If you find that 24 g of magnesium joins with 16 g of oxygen, then you should be able to work out that the formula would be MgO.

Experiment 1 Finding the combining masses of magnesium and oxygen

Put exactly 0.24 g (about 24 cm length) of clean magnesium ribbon into a crucible of known mass. Heat the crucible strongly on a pipeclay triangle until the magnesium starts to burn (Figure 57.4).

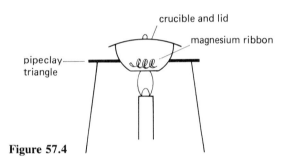

Figure 57.4

Use tongs to lift the crucible lid every so often to let air get in, but do not let any of the white ash (magnesium oxide) get out. Make sure that all the magnesium has reacted by gently blowing on the hot ribbon every so often, each time replacing the lid very quickly afterwards.

Put the crucible back on the balance after it has cooled down and find the mass of the crucible together with the white ash.

- What is the mass of the magnesium oxide that is formed?
- What must be the mass of the oxygen that has joined with the 0.24 g of magnesium?
- Now work through these steps:

1 Work out, by proportion, the mass of oxygen that would join with 24 g of magnesium.
2 About how many moles of O atoms are there in this mass of oxygen?
3 How many moles of Mg atoms are there in 24 g of magnesium?
4 Compare the number of moles of O atoms with

the number of moles of Mg atoms and then write down the formula of magnesium oxide.

Moles of compounds

The formulas of copper(I) oxide and copper(II) oxide are Cu_2O and CuO, respectively. Figure 57.5 shows how the masses of 1 mole of each of these compounds can be worked out.

If these answers are written without the unit (grams), then the numbers are called the *relative molecular mass* of the compound. The relative molecular mass of copper(I) oxide is 144, and for copper(II) oxide it is 80.

Some formulas have brackets, and working out the relative molecular mass for one of these compounds is a little more difficult (Figure 57.6).

Some more calculations

To find the formula of a compound, we first need to know the proportions by mass of all its elements. Experiments like the one above show that these proportions are the same for any sample of a particular compound: this is the Law of Constant Composition (page 13).

Here is an example:

An 80 g sample of copper(II) oxide contains 64 g of copper and 16 g of oxygen. A different 40 g sample of copper(II) oxide contains 32 g of copper and 8 g of oxygen, and so on.

The *percentage composition by mass* of copper(II) oxide is $64/80 \times 100\%$ Cu (80% Cu) and $16/80 \times 100\%$ O (20% O).

Figure 57.6

	CALCIUM HYDROXIDE	
Formula	$Ca(OH)_2$	
Number of moles of each atom or group of atoms	1 mole of Ca atoms	2 moles of OH 'groups'
Masses of these moles	40 g	2×17 or 34 g
Total mass	74 g	

So the mass of 1 mole of $Ca(OH)_2$ is 74 g and its relative molecular mass is 74.

Below you can see how to work out the formula of a compound from its percentage composition by mass.

Worked example.

Copper(II) oxide contains 80 per cent copper by mass. What is its formula? (Relative atomic masses: Cu = 64, O = 16).

100 g of copper(II) oxide contains 80 g of copper and 20 g of oxygen.

STEP 1 Change the *grams* of the elements to *moles*.

a 64 g of copper is 1 mole of Cu atoms
1 g of copper is 1/64 mole of Cu atoms
So 80 g of copper is 80/64 or 1·25 moles of Cu atoms.

Figure 57.5

	COPPER(I) OXIDE		COPPER(II) OXIDE	
Formula	Cu_2O		CuO	
Number of moles of each kind of atom	2 moles of Cu atoms	1 mole of O atoms	1 mole of Cu atoms	1 mole of O atoms
Masses of these moles	2×64 or 128 g	16 g	64 g	16 g
Total mass	144 g		80 g	

So the mass of 1 mole of Cu_2O is 144 g.

So the mass of 1 mole of CuO is 80 g.

b 16 g of oxygen is 1 mole of O atoms
1 g of oxygen is 1/16 mole of O atoms
So 20 g of oxygen is 20/16 or 1·25 moles of O atoms.

STEP 2 Work out the simplest whole number ratio between the moles of the elements.
In copper(II) oxide, 1·25 moles of Cu atoms combine with 1·25 moles of O atoms.
So the simplest whole number ratio of moles Cu: moles O is 1:1.
The formula of copper(II) oxide is *CuO*.

The formula obtained by this kind of calculation is sometimes called an *empirical formula*. Many substances only have one formula and this is always the empirical formula obtained from the percentage composition by mass. But some have another kind of formula as well.

The formula of sulphur can be written as S or S_8. The S formula just tells you that sulphur is an element of atoms of sulphur only. The S_8 formula tells you that sulphur is made of molecules, each containing eight sulphur atoms. The second formula is a *molecular formula*.

Some compounds also have both an empirical formula and a molecular formula.

Worked example.

A gas called ethane contains 80 per cent carbon and 20 per cent hydrogen. What is its formula? (Relative atomic masses: C = 12, H = 1.)

STEP 1
a 80 g of carbon is 80/12 or 6·7 moles of C atoms
b 20 g of hydrogen is 20/1 or 20 moles of H atoms

STEP 2
The simplest whole number ratio of moles C: moles H is 1:3.
The empirical formula of ethane is CH_3.

Other experiments have shown that the relative molecular mass of ethane is 30. So 1 mole of ethane has a mass of 30 g.
But the mass of 1 mole of CH_3 is (12 + 3) or 15 g: the empirical formula cannot be the same as the molecular formula. Because 30 is 15 times 2, the molecular formula must be the empirical formula 'times 2'. In other words, the molecular formula is C_2H_6, and ethane is made of C_2H_6 molecules (Figure 57.7), not molecules of CH_3.

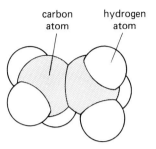

Figure 57.7 A model of an ethane molecule, C_2H_6.

Questions

1. Using Table 57.6 write down the formula of each of these electrolytes: (*a*) sodium iodide, (*b*) calcium hydroxide, (*c*) copper(I) chloride, (*d*) iron(II) sulphate, (*e*) lithium hydroxide, (*f*) lead iodide, (*g*) sodium hydrogencarbonate, (*h*) potassium carbonate, (*i*) ammonium sulphate, (*j*) iron(III) nitrate.

2. 7·4 g of zinc powder is added to 25·4 g of iodine, and drops of ethanol are added to the mixture until no more chemical reaction takes place. The product, zinc iodide, dissolves in ethanol, and the excess of zinc powder is filtered off and dried. It is found that 0·9 g of zinc is left over after the chemical reaction is finished. (Relative atomic masses: Zn = 65, I = 127.)

   (*a*) Work out the mass of zinc that has reacted with 25·4 g of iodine.
   (*b*) What fraction of a mole of Zn atoms is this mass?
   (*c*) What fraction of a mole of I atoms is 25·4 g?
   (*d*) Work out the formula of zinc iodide.
   (*e*) Predict the likely formulas of zinc chloride and zinc bromide from this result.

3. Calcium bromide contains 20 per cent calcium by mass. What is its formula? (Relative atomic masses: Ca = 40, Br = 80.)

4. Using a gas called butane, C_4H_{10}, explain the difference between a molecular formula and an empirical formula.

58 EQUATIONS

WHAT DO EQUATIONS MEAN?

In this book, chemical equations are written in two ways—as word equations and as symbol equations. Word equations just give the names of the substances that are reacting (the *reactants*) and the substances that are produced (the *products*). Symbol equations also show the reactants and products, but they contain some important extra information.

Consider the symbol equation for the chemical reaction between magnesium and steam (page 35):

$$Mg(s) + H_2O(g) \longrightarrow MgO(s) + H_2(g)$$

where 's' stands for 'solid' and 'g' for 'gas'. Each of the four formulas in this equation are more than just a shorthand way of writing the name of the substance. Each stands for a mole of the substance. So the equation shows that 1 mole of Mg reacts with 1 mole of H_2O to give 1 mole of MgO and 1 mole of H_2.

Not all equations are as simple as this one. Look at the equation for the chemical reaction between magnesium and dilute hydrochloric acid:

$$Mg(s) + 2HCl(aq) \longrightarrow MgCl_2(aq) + H_2(g)$$

where 'aq' stands for 'aqueous' and shows that HCl and $MgCl_2$ are both in solution with water. An important difference between this equation and the one above is that there is a number 2 before one of the formulas. This means that 2 moles of HCl are reacting with 1 mole of Mg to form 1 mole of $MgCl_2$ and 1 mole of H_2.

The numbers placed before formulas in equations show the numbers of moles of the substances that are involved in the chemical reaction.

Balancing symbol equations

The Law of Conservation of Mass (page 8) says that, in chemical reactions, the total mass of the reactants must equal the total mass of the products: no new matter is produced and no old matter is destroyed. Since matter is made of atoms, this is like saying that atoms cannot be created or destroyed during a reaction.

To *balance* a symbol equation means making sure that the quantities of each kind of atom are the same on either side of the equation. Consider the burning of magnesium in oxygen to form magnesium oxide. Before being balanced, the symbol equation would be:

UNBALANCED $Mg(s) + O_2(g) \longrightarrow MgO(s)$

After balancing, it becomes:

BALANCED $2Mg(s) + O_2(g) \longrightarrow 2MgO(s)$

Figure 58.1 shows that the balanced equation is the correct one because the numbers of moles of Mg atoms are the same on either side and so are the numbers of moles of O atoms.

Another example is the burning of natural gas (mostly methane, CH_4) to form carbon dioxide and water. The unbalanced and balanced equations are:

UNBALANCED $CH_4(g) + O_2(g) \longrightarrow CO_2(g) + H_2O(l)$

BALANCED $CH_4(g) + 2O_2(g) \longrightarrow CO_2(g) + 2H_2O(l)$

where 'l' stands for 'liquid'.

Figure 58.2 shows that the second equation is fully balanced for moles of C atoms, H atoms and O atoms.

FROM EXPERIMENT TO EQUATION

In the first place, equations are always found through experiments. Suppose that you wanted to know something about the equation for the chemical reaction between copper metal and silver nitrate solution. Silver metal is displaced by the copper, which itself goes into solution. An experiment shows that 0·64 g of copper gives 2·16 g of silver.

64 g of copper is 1 mole of Cu atoms
So 0·64 g of copper is 1/100 mole of Cu atoms.

Figure 58.1

	LEFT-HAND SIDE			RIGHT-HAND SIDE		
	Reactants	Moles of Mg atoms	Moles of O atoms	Products	Moles of Mg atoms	Moles of O atoms
UNBALANCED	1 mole of Mg atoms	1		1 mole of MgO	1	①
	1 mole of O_2 molecules		②			
BALANCED	2 moles of Mg atoms	2		2 moles of MgO	2	2
	1 mole of O_2 molecules		2			

◯ shows where there is an imbalance in moles of atoms.

Figure 58.2

	LEFT-HAND SIDE				RIGHT-HAND SIDE			
	Reactants	Moles of C atoms	Moles of H atoms	Moles of O atoms	Products	Moles of C atoms	Moles of H atoms	Moles of O atoms
UNBALANCED	1 mole of CH_4	1	④		1 mole of CO_2	1		2
	1 mole of O_2 molecules			②	1 mole of H_2O		②	1 Total=③
BALANCED	1 mole of CH_4	1	4		1 mole of CO_2	1		2
	2 moles of O_2 molecules			4	2 moles of H_2O		4	2 Total=4

◯ shows where there is an imbalance in moles of atoms.

108 g of silver is 1 mole of Ag atoms.
So 2·16 g of silver is the mass of 2/100 (1/50) mole of Ag atoms.

These results show that 1/100 mole of Cu atoms displaces 2/100 mole of Ag atoms, and 1 mole of Cu atoms displaces 2 moles of Ag atoms.
Part of the equation for this chemical reaction can now be written:

Cu(s) + ... ⟶ 2Ag(s) + ...

Silver nitrate solution is the other reactant. The formula of silver nitrate is $AgNO_3$ and 2 moles of $AgNO_3$ are needed to produce 2 moles of Ag atoms. So the equation now becomes:

Cu(s) + 2AgNO₃(aq) ⟶ 2Ag(s) + ...

Chemical analysis shows that the only other product is copper(II) nitrate, $Cu(NO_3)_2$. This is also the product you would expect by looking at the symbols on the left-hand side of the equation that are missing on the right-hand side: these symbols are Cu and $2NO_3$. So the fully balanced equation is:

Cu(s) + 2AgNO₃(aq) ⟶ 2Ag(s) + Cu(NO₃)₂(aq)

Figure 58.3

	LEFT-HAND SIDE				RIGHT-HAND SIDE		
Reactants	Moles of Cu atoms	Moles of Ag atoms	Moles of NO_3 group	Products	Moles of Cu atoms	Moles of Ag atoms	Moles of NO_3 group
1 mole of Cu atoms	1			2 moles of Ag atoms		2	
2 moles of $AgNO_3$		2	2	1 mole of $Cu(NO_3)_2$	1		2

Figure 58.3 shows why this is balanced. Notice that you only need to check that the numbers of moles of NO_3 groups (actually NO_3^- anions) are the same on either side. It is not necessary to check on the individual elements, nitrogen and oxygen.

Reactions involving gases

For the chemical reaction between copper and silver nitrate solution, it is easy to find the numbers of moles of Cu atoms and Ag atoms by measuring the masses of these two elements.

For gases, this would be more difficult because their masses for the volumes collected in school laboratory experiments are small. Fortunately, there is another way of finding the number of moles of gas taking part in a reaction.

1 mole of any gas at a given temperature and pressure has about the same volume.

At sea-level, this volume is around 24 litres (24 000 cm³), but Table 58.1 shows that it is higher in places well above sea-level where the air pressure is much lower than 760 mmHg.

Suppose you carry out the experiment shown in Figure 58.4 in which some powdered calcium carbonate is being reacted with dilute hydrochloric acid to give carbon dioxide. The gas collects in the glass syringe and its volume is measured at the end of the reaction. It is found that 0·10 g of calcium carbonate gives 24 cm³ of carbon dioxide.

100 g of $CaCO_3$ is 1 mole of $CaCO_3$ (Figure 58.5)
So 0·10 g is 1/1000 mole of $CaCO_3$.
24 000 cm³ is the volume of 1 mole of CO_2 at the same temperature and pressure
So 24 cm³ is the volume of 1/1000 mole of CO_2.

Table 58.1

Air pressure in mmHg	550	600	650	700	725	750	775
Volume of 1 mole of gas in litres	33	30	28	26	25	24	23

The volumes occupied by 1 mole of any gas at 'room' temperature for different air pressures. Add 1 litre to each of these volumes if the temperature of your laboratory is over 25 °C. The normal air pressure at sea-level is 760 mmHg.

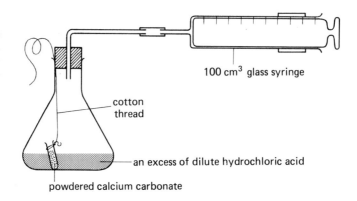

Figure 58.4 To start the reaction, the flask is tilted so that the acid mixes with the calcium carbonate.

Figure 58.5

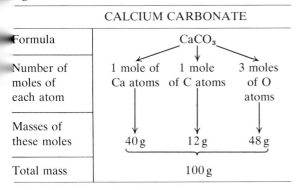

	CALCIUM CARBONATE		
Formula	$CaCO_3$		
Number of moles of each atom	1 mole of Ca atoms	1 mole of C atoms	3 moles of O atoms
Masses of these moles	40 g	12 g	48 g
Total mass	100 g		

These results show that 1/1000 mole of $CaCO_3$ gives 1/1000 mole of CO_2, and 1 mole of $CaCO_3$ gives 1 mole of CO_2.

Part of the equation for this chemical reaction can now be written:

$CaCO_3(s) + \ldots \longrightarrow \ldots + CO_2(g) + \ldots$

The other reactant is hydrochloric acid (HCl) and experiment shows that the solution left over at the end of the chemical reaction contains only calcium chloride ($CaCl_2$). So the equation now becomes:

$CaCO_3(s) + HCl(aq) \longrightarrow CaCl_2(aq) + CO_2(g) + \ldots$

There is obviously something wrong with this equation as it stands. The 2 moles of Cl atoms on the right-hand side need to be balanced by 2 moles on the left-hand side. This can be done by putting a 2 in front of HCl(aq). But even more of a problem is that there are no H atoms on the right-hand side to match those on the left. And the moles of O atoms on either side do not balance either.

The fully balanced equation (Figure 58.6) has water (H_2O) as the third product:

$CaCO_3(s) + 2HCl(aq) \longrightarrow CaCl_2(aq) + CO_2(g) + H_2O(l)$

IONS IN EQUATIONS

Acid–alkali reactions

The symbol equation for the neutralization of dilute hydrochloric acid by sodium hydroxide solution can be written as:

$NaOH(aq) + HCl(aq) \longrightarrow NaCl(aq) + H_2O(l)$

But acids, alkalis and salts are electrolytes and are made of ions. There is another way of writing the equation which uses ions rather than whole formulas.

Figure 58.7 represents the neutralization in the form of a diagram. The H^+ and OH^- ions join to form H_2O molecules while the Na^+ and Cl^- ions just stay in the solution without being changed in any way.

Figure 58.6

LEFT-HAND SIDE					
Reactants	Moles of Ca atoms	Moles of C atoms	Moles of O atoms	Moles of H atoms	Moles of Cl atoms
1 mole of $CaCO_3$	1	1	3		
2 moles of HCl				2	2

RIGHT-HAND SIDE					
Products	Moles of Ca atoms	Moles of C atoms	Moles of O atoms	Moles of H atoms	Moles of Cl atoms
1 mole of $CaCl_2$	1				2
1 mole of CO_2		1	2		
1 mole of H_2O			1	2	

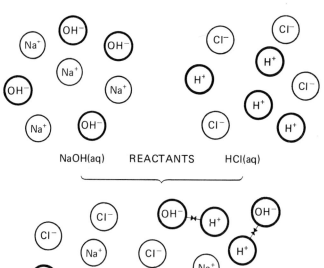

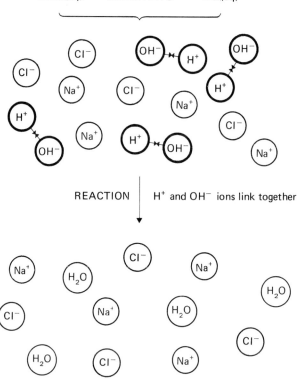

Figure 58.7 Note that only those molecules of water produced in the reaction are shown.

So the *ionic equation* for this chemical reaction is:

$$H^+(aq) + OH^-(aq) \longrightarrow H_2O(l)$$

There is no need to include the Na^+ and Cl^- ions because nothing happens to them.

This equation is the same for *any* acid–alkali reaction because all acids contain H^+ ions and all alkalis contain OH^- ions.

Precipitation reactions

Ionic equations can also be written for chemical reactions in which an insoluble salt is precipitated when two soluble salts are mixed (page 69).

Figure 58.8 represents the reaction between calcium chloride solution and sodium sulphate solution in the form of a diagram. Again, the diagram can be used to write the ionic equation. The Ca^{2+} and SO_4^{2-} ions join to form $CaSO_4$ (which is precipitated) while the Na^+ and Cl^- ions just stay in the solution without being changed in any way.

So the ionic equation for this chemical reaction is:

$$Ca^{2+}(aq) + SO_4^{2-}(aq) \longrightarrow CaSO_4(s)$$

There is no need to include the Na^+ and Cl^- ions because nothing happens to them.

Balancing ionic equations

All symbol equations for the chemical reactions of electrolytes can be written using ions. Balancing these ionic equations is a little more complicated

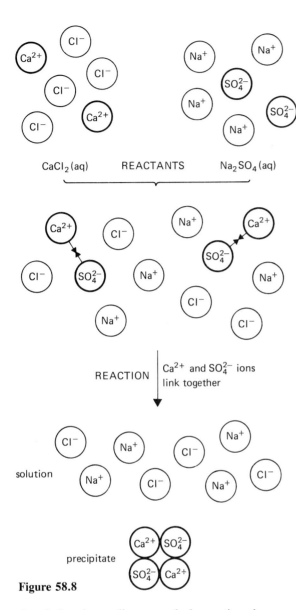

Figure 58.8

than balancing ordinary symbol equations because there are two things to check, not just one. Ordinary symbol equations need only to be balanced for moles of atoms. Ionic equations need to be balanced for moles of atoms and for electrical charges.

Consider the precipitation of lead iodide (PbI_2) by mixing a solution containing lead ions (Pb^{2+}) and a solution containing iodide (I^-) ions. The unbalanced ionic equation is:

$$UNBALANCED \quad Pb^{2+}(aq) + I^-(aq) \longrightarrow PbI_2(s)$$

This equation is not balanced for moles of I. It is also not balanced for electrical charges: on the right-hand side there is no charge while on the left-hand side the +2 and −1 charges give a total charge of +1.

The fully balanced equation is:

$$BALANCED \quad Pb^{2+}(aq) + 2I^-(aq) \longrightarrow PbI_2(s)$$

The moles of atoms are now balanced, and so are the electrical charges. The +2 and −2 charges on the left-hand side give a total charge of zero, which is the total charge on the right-hand side.

Questions

1. Balance these equations for moles of atoms of each kind. (NOTE: there are no other products apart from those given.)

(a) $Zn(s) + O_2(g) \longrightarrow ZnO(s)$
(b) $ZnO(s) + HCl(aq) \longrightarrow ZnCl_2(aq) + H_2O(l)$
(c) $ZnCl_2(aq) + AgNO_3(aq) \longrightarrow Zn(NO_3)_2(aq) + AgCl(s)$
(d) $Zn(NO_3)_2(s) \longrightarrow ZnO(s) + NO_2(g) + O_2(g)$
(e) $ZnO(s) + Al(s) \longrightarrow Zn(s) + Al_2O_3(s)$

2. Balance these ionic equations for moles of atoms of each kind and for charges. (NOTE: there are no other products apart from those given.) One of the equations is already fully balanced: make sure you identify this and leave it as it is.

(a) $Mg(s) + H^+(aq) \longrightarrow Mg^{2+}(aq) + H_2(g)$
(b) $Ba^{2+}(aq) + SO_4^{2-}(aq) \longrightarrow BaSO_4(s)$
(c) $Pb^{2+}(aq) + Br^-(aq) \longrightarrow PbBr_2(s)$
(d) $Mg(s) + Ag^+(aq) \longrightarrow Mg^{2+}(aq) + Ag(s)$
(e) $Cl_2(aq) + I^-(aq) \longrightarrow I_2(aq) + Cl^-(aq)$

3. Explain why both the neutralization of nitric acid by sodium hydroxide solution and the neutralization of hydrochloric acid by potassium hydroxide solution can be represented by the single ionic equation:

$$H^+(aq) + OH^-(aq) \longrightarrow H_2O(l)$$

59 CALCULATIONS USING EQUATIONS

Mass–mass calculations

A symbol equation shows how many moles of each substance react or are formed. From the number of moles of each substance, each of their masses can be worked out. This information about a chemical reaction allows a chemist to work out the proportions in which the reactants should be mixed and also the amount of product (the *yield*) expected.

Suppose that you want to know how much chromium metal you could expect to get from 100 kg of chromium(III) oxide. The chemical reaction would be carried out using aluminium as the reducing agent (page 35). The symbol equation for this chemical reaction is:

$$2Al(s) + Cr_2O_3(s) \longrightarrow 2Cr(s) + Al_2O_3(s)$$
2 moles   1 mole   2 moles   1 mole

The masses (in grams) of each of the reactants and products are:

   2×27 g   152 g 2×52 g   102 g

So 152 g of chromium(III) oxide gives a yield of 104 g of chromium, or 152 kg of chromium(III) oxide gives 104 kg of chromium.

The next step is to find how many kilograms of chromium are produced by 100 kg of chromium(III) oxide.

152 kg of chromium(III) oxide gives 104 kg of chromium
So 1 kg of chromium(III) oxide gives 104/152 kg of chromium and 100 kg of chromium(III) oxide gives $100 \times 104/152$ or 68·4 kg of chromium.

The steps in this calculation are:

1 Use the balanced equation to write down the numbers of moles of reactants and products.
2 Change these moles to grams using the masses of 1 mole of the reactants and products (these are found by working out their relative molecular masses, page 236).
3 Use the information in the question to change the masses in **2** to the required masses.

Mass–volume calculations

The calculation is a little different if you need to find the yield of a gas produced in a chemical reaction. Here it is a volume that is important rather than a mass.

Suppose that you want to know how much carbon dioxide you could expect to get by heating 10 tonnes of limestone (calcium carbonate). The symbol equation for this chemical reaction is:

$$CaCO_3(s) \longrightarrow CaO(s) + CO_2(g)$$
1 mole   1 mole   1 mole

The masses (in grams) of the solids are:

   100 g   56 g

Also, 1 mole of any gas has a volume of 24 litres at room temperature and pressure (page 240).

So 100 g of limestone gives a yield of 24 litres of carbon dioxide at 'ordinary' temperature and pressure, and 100 tonnes (100 000 kg or 100 000 000 g) give 24 000 000 litres of carbon dioxide.

Lastly, you have to work out how many litres of carbon dioxide are produced by 10 tonnes of limestone.

100 tonnes of limestone give 24 000 000 litres of carbon dioxide.
So 1 tonne of limestone gives 24 000 000/100 litres of carbon dioxide and 10 tonnes of limestone give $10 \times 24\,000\,000/100$ or *2 400 000 litres of carbon dioxide*.

The steps in this calculation are:

1 Use the balanced equation to write down the numbers of moles of reactants and products.
2 Change the moles to grams for solids and liquids (using the masses of 1 mole of each), and to litres for gas (using the volume of 1 mole of any gas).
3 Use the information in the question to change the masses and volumes in **2** to the required masses and volumes.

Questions

1. The equation for the reaction between zinc and dilute hydrochloric acid is:

$$Zn(s) + 2HCl(aq) \longrightarrow ZnCl_2(aq) + H_2(g)$$

What volume of hydrogen would be given off when 13 g of zinc is added to an excess of the acid? (Relative atomic mass: $Zn = 65$; assume that 1 mole of any gas has a volume of 24 litres under the conditions of the experiment.)

2. The equation for the electrolysis of molten sodium chloride is:

$$2NaCl(l) \xrightarrow{electricity} 2Na(l) + Cl_2(g)$$

Work out:

(a) the mass of sodium metal, and
(b) the volume of chlorine at room temperature and pressure that would be produced by the breakdown of 234 kg of sodium chloride.

(Relative atomic masses: $Na = 23$, $Cl = 35.5$; assume that 1 mole of any gas has a volume of 24 litres at room temperature and pressure.)

3. A gas called ethene (ethylene) has a molecular formula of C_2H_4. It burns in oxygen to form carbon dioxide and water.

(a) What is the relative molecular mass of ethene?
(b) What is the empirical formula of ethene?
(c) What kind of bonds, ionic or covalent, must hold the atoms together in a molecule of ethane? (See Chapter 60.)
(d) Write (i) the word equation and (ii) the fully balanced symbol equation for the burning of ethene in oxygen.
(e) Use the balanced equation to find how many litres of carbon dioxide would be produced by 14 g of C_2H_4 at room temperature and pressure.

(Relative atomic masses: $C = 12$, $H = 1$; assume that 1 mole of any gas has a volume of 24 litres at room temperature and pressure.)

60 CHEMICAL BONDING

What are bonds?

To stretch a solid, energy must be supplied to move particles apart against strong attractive forces (see page 16). This energy may be in the form of mechanical energy (page 159) or heat energy (page 159).

The atoms which make up a molecule are also difficult to separate from each other. Again, energy is needed to break down molecules into separate atoms. So there must be attractive forces between atoms in a molecule.

The forces that hold particles together—ions or atoms in solids, or atoms in molecules—are called *bonds*. To understand why bonds exist, we look again at the structure of an atom.

When atoms are close to each other, it is their outermost parts that interact. What happens to these parts explains how bonds are formed.

Electrons in atoms

We assume that an atom (page 226) consists of a small dense positively charged nucleus surrounded by a space containing negatively charged electrons. There is strong experimental evidence that the electrons are arranged in definite patterns—'shells' or 'layers' called *energy levels* arranged at different distances from the nucleus. For very simple atoms like hydrogen, there is only one energy level occupied by electrons whereas for more complicated atoms like sodium and chlorine, there are several (Figure 60.1).

Electrons in different energy levels have different energies while those in any one energy level have similar (though not always exactly the same) energies. There are forces of attraction between the electrons and the nucleus.

For each kind of atom, the total number of electrons and also the number of electrons in the outermost energy level (the *outer electrons*) are known. These numbers are given in Table 60.1 for the first twenty elements in the periodic table.

There is a pattern in the numbers of outer electrons when the elements are arranged in the order of their atomic numbers (Figure 60.2).

Table 60.1

Element	Atomic number (total number of electrons)	Number of outer electrons	Outer energy level	Number of inner electrons
Hydrogen	1	1	1st	0
Helium	2	2	1st	0
Lithium	3	1		2
Beryllium	4	2		2
Boron	5	3		2
Carbon	6	4		2
Nitrogen	7	5	2nd	2
Oxygen	8	6		2
Fluorine	9	7		2
Neon	10	8		2
Sodium	11	1		10
Magnesium	12	2		10
Aluminium	13	3		10
Silicon	14	4		10
Phosphorus	15	5	3rd	10
Sulphur	16	6		10
Chlorine	17	7		10
Argon	18	8		10
Potassium	19	1	4th	18
Calcium	20	2		18

The properties of an element are controlled by the number of outer electrons in its atom because these are the electrons that are used to make bonds between elements (see below). So lithium, sodium and potassium show similar properties because

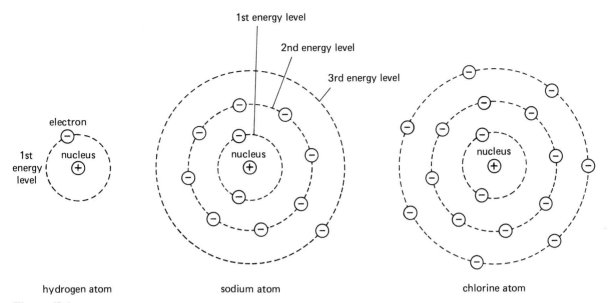

Figure 60.1

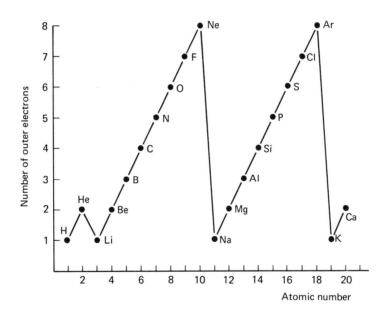

Figure 60.2

their atoms all have one outer electron. Fluorine and chlorine show very different properties from the alkali metals because they have a different number of outer electrons. But fluorine and chlorine are very alike because their atoms all have the same number (seven) of outer electrons.

So the way that elements in the periodic table are arranged in families (page 47) is the result of the regular way in which the number of outer electrons varies with atomic number.

For many elements the number of outer electrons in their atoms is equal to the number of the group in the periodic table in which the element is found.

247

Bonds between ions

When sodium and chlorine react to form sodium chloride (Na⁺Cl⁻), sodium atoms must each lose one electron to form sodium ions and chlorine atoms must each gain one electron to form chloride ions. The numbers of electrons in these atoms and ions are given in Table 60.2.

Table 60.3 gives the total numbers of electrons and the numbers of outer electrons in the atoms of the noble gases. It is clear that the sodium ion has the same number of electrons as the neon atom and the chloride ion has the same number of electrons as the argon atom.

It is very difficult to decompose sodium chloride into its elements. In other words, it is very difficult to change sodium ions to sodium atoms and chloride ions to chlorine atoms. Neon and argon must also be very *stable* because they belong to the very unreactive family of elements called the noble gases (page 20). So sodium and chloride ions are stable because they have the same (stable) number of electrons as neon and argon atoms.

Figure 60.3 shows a way of describing how sodium and chloride ions are formed from their atoms. Each sodium atom loses one electron to a chlorine atom.

In solid sodium chloride, each sodium ion is surrounded by chloride ions and each chloride ion by sodium ions (Figure 60.4). So adjacent ions attract each other because of their opposite charges. The strong electrostatic forces of attraction between oppositely charged ions are called *ionic bonds*. In a crystal of sodium chloride, these bonds extend all through the *structure*.

The particles (ions) in all solid salts are held together by ionic bonds.

Table 60.2

Atom or ion	Atomic number	Total number of electrons in one atom or ion
Na	11	11
Na⁺	11	10
Cl	17	17
Cl⁻	17	18

Table 60.3

Element	Atomic number	Total number of electrons in one atom	Number of outer electrons
He	2	2	2
Ne	10	10	8
Ar	18	18	8
Kr	36	36	8
Xe	54	54	8
Rn	86	86	8

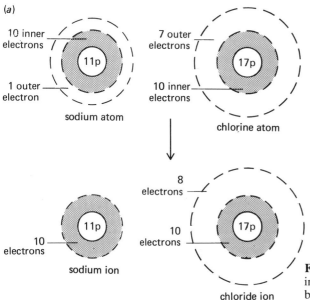

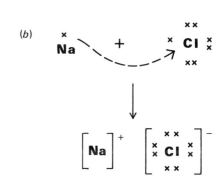

Figure 60.3 Note that the inner electrons are not included in diagram (*b*). They are not important here because they are not involved in forming the bond.

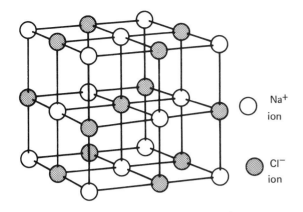

Figure 60.4 Ball-and-spoke model of the structure of sodium chloride.

Bonds between atoms

Hydrogen molecules (H₂) are the simplest of all molecules. A single hydrogen atom has one proton and one electron. Figure 60.5 shows a way of describing what happens to the two electrons when the atoms join to form a hydrogen molecule.

When the two atoms are a long way from one another, each electron is attracted only by the proton in its own atom. As the two atoms get nearer, each electron is now attracted by both protons while the two electrons repel one another and the two protons repel one another.

When the atoms are close enough, a strong bond is formed. The region in which the electrons are most likely to be found is where they are most strongly attracted by both protons at the same time. This region is shaded in the diagram.

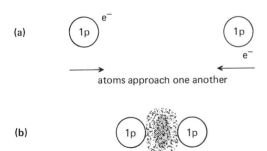

Figure 60.5

There is another way of thinking about the bond between two hydrogen atoms. Both atoms are trying to get the same number of electrons as a helium atom. They can do this by each gaining one extra electron. But it is impossible for both atoms to gain complete control of both electrons. Instead, they have to share the two electrons between them (Figure 60.6).

Figure 60.6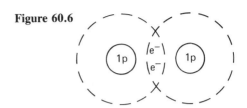

So, in effect, both atoms now have two outer electrons even though this pair of electrons is being shared.

The atoms are held together as a molecule because both nuclei are attracting and holding on to the same thing—the shared pair of electrons. The two approaches (Figures 60.5 and 60.6) give a similar picture of the hydrogen molecule.

It is very difficult to decompose hydrogen molecules to atoms and so the bond between the atoms must be strong. This is so because the atoms in a molecule have the same number of electrons as a helium atom, and the atoms of noble gases have a very stable arrangement of electrons.

This kind of bond between atoms is called a *covalent bond*. The difference between this bond and an ionic bond is that it is formed by the *sharing* of a pair of electrons between atoms rather than the *transfer* of one or more electrons from one atom to another.

This difference is like the difference between the merger of two industrial companies (the sharing) and the take-over of one company by another (the transfer) (Figure 60.7). In a merger each company may be thought of as owning the other, just as in a covalent bond each atom may be thought of as having the pair of electrons between them.

Figure 60.7

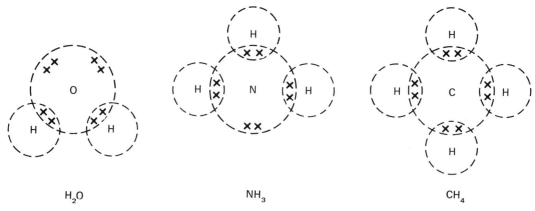

Figure 60.8 Note that all inner electrons have been missed out of these diagrams because they are not involved in forming the bonds.

The atoms in the molecules of compounds are also held together by covalent bonds. Figure 60.8 shows the arrangement of outer electrons in molecules of water, ammonia and methane. In each case, the other (inner) electrons are missed out because they do not take part in forming covalent bonds.

Questions

1. Explain the difference between an ionic bond and a covalent bond using two simple examples.

2. Why is it wrong to speak of *molecules* of sodium chloride? What should we say instead?

3. Figure 60.9 shows how the bonds are formed in (a) magnesium chloride, and (b) chlorine molecules. In each case, say which kind of bond is found in the substance and explain how it is formed.

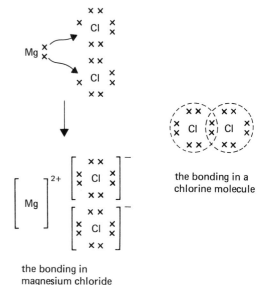

Figure 60.9 Note that only the outer electrons in each atom are shown.

Chemicals from Nature

61 METALS FROM ROCKS

METALS AND ORES

Ores (page 25) are usually compounds of metals, though a few are elements. Gold and silver are often found in rocks as the elements. But metals higher than gold and silver in the reactivity series are almost always found in compounds.

Whenever a metal is *extracted* from one of its compounds, the chemical reaction is called *reduction*, even if the compound is not an oxide. This is a new use of the word *reduction* compared with the way it is used on page 35.

Figure 61.1 shows the most common metals that are found, either alone or joined with other elements in the earth's rocks. Most of the well-known

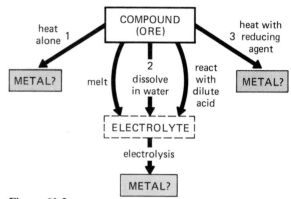

Figure 61.2

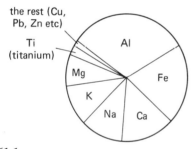

Figure 61.1

metals, such as copper, lead and zinc, are really quite rare. At the rate we are using these rarer metals, some scientists think that we shall have used up all the known deposits of their ores in another few decades.

GETTING THE COPPER OUT OF A COPPER ORE

Two ores of copper are malachite and copper pyrites. Malachite contains only one metal whereas copper pyrites contains some iron as well as copper. Malachite is a carbonate and copper pyrites is a sulphide.

Compounds can be broken down in three main ways (Figure 61.2). When malachite is heated, it breaks down into carbon dioxide, water and copper(II) oxide. So heating alone does not form copper metal.

Malachite itself is not an electrolyte: it does not dissolve in water and it does not melt when heated. But malachite can be reacted with a dilute acid to make a solution. This can then be electrolysed.

The third way of breaking down compounds is also shown in Figure 61.2. Hydrogen can reduce copper(II) oxide to copper (page 36): it can also reduce heated malachite to copper. But a more convenient and safer reducing agent for you to use is charcoal (carbon).

Experiment 1 Trying to reduce some metal oxides with charcoal

You are given some calcium oxide, copper(II) oxide, lead(II) oxide and iron(III) oxide. Use the apparatus shown in Figure 61.3 to find out which of these oxides can be reduced by heating them with charcoal.

If reduction takes place, there may be a change in the colour of the oxide as the metal forms.

- When carbon removes the oxygen from the oxide, which gas is likely to be formed?
- What effect does this gas have on lime water?

Charcoal easily reduces the oxides of metals low in the reactivity series because these metals are not strongly bound to oxygen. But the oxides of metals

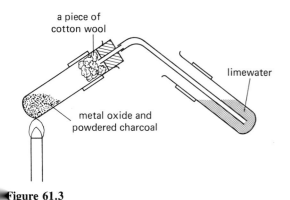

Figure 61.3

higher in the reactivity series have to be reduced in other ways.

Experiment 2 Getting copper out of malachite

You are given some powdered malachite. From what you have read in this section, try to find *two* ways of getting the copper from it. Ask your teacher for any other chemicals and also any apparatus you need.

METAL EXTRACTIONS AND THE REACTIVITY SERIES

Table 61.1 describes how four well-known metals are extracted from their common ores. The order of these metals is their order in the reactivity series.

Metals low in the reactivity series can sometimes be extracted by heating them alone (method 1 in Figure 61.2). Carbon, in the form of coke, is a common reducing agent used in the extraction of metals which are found in the middle or the lower half of the reactivity series. And for the most reactive metals, the main extraction process is the electrolysis of their molten ores.

Extracting iron

In the laboratory it is more difficult to extract iron from an iron ore than to extract copper from malachite. This is because iron is higher than copper in the reactivity series and so must be more strongly bound to other elements in its ores.

Even so, iron can be extracted in industry using coke if the temperature of the mixture is kept around 1200 °C. Air is pumped into the furnace (Figure 61.5) and this reacts with the hot coke to form carbon monoxide. This gas is a very good reducing agent and it is this, not the carbon, that reduces the iron ore to iron.

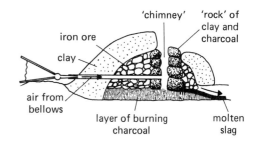

Figure 61.4

If the iron ore is haematite (iron(III) oxide), the word and symbol equations for this extraction are:

iron(III) oxide + carbon monoxide ⟶ iron + carbon dioxide

$Fe_2O_3(s) + 3CO(g) \longrightarrow 2\,Fe(s) + 3CO_2(g)$

Table 61.1

Metal	Ore	Method of extraction
Sodium	Rock salt (sodium chloride)	Electrolysis of molten ore
Aluminium	Bauxite (aluminium oxide)	Electrolysis of molten ore
Iron	Haematite (iron(III) oxide)	Reduction of ore using coke (carbon)
Copper	Copper pyrites (copper iron sulphide)	Heating ore alone in air

Coke is made by strongly heating coal in the absence of air.

Figure 61.5

Iron has been used by man for a very long time. The first furnaces were much smaller than those shown in Figure 61.5 and they were packed with charcoal instead of coke. Older village people in some parts of Africa still know how to extract iron. Figure 61.4 shows what the inside of an old furnace looks like.

The charcoal is the fuel that heats the furnace. But it also reacts to form carbon monoxide (the reducing agent) with the air pumped in by the bellows. Molten slag is formed during the chemical reaction and this runs through the bed of burning charcoal and out of the furnace. At the end of this chemical reaction, the furnace can be opened to get out the hot solid iron. When this has been cooled under water, it is ready to be beaten into the shape of a tool. Each furnace can make a few kilograms at a time.

Extracting reactive metals

Most reducing agents are not 'powerful' enough to reduce ores of reactive metals because the metals are so strongly bound to the other elements in the ores. Instead, the molten ores can be broken down by electrolysis.

Sodium is extracted by the electrolysis of molten sodium chloride, and aluminium by the electrolysis of molten aluminium oxide. Large currents are needed for these electrolyses because the electrical energy has not only to break down the compounds but also to keep the electrolytes in the molten state. A cheap supply of electricity, usually from water-power, is a big advantage.

It is usual to add a *flux*, a substance that lowers the melting point of the ore without interfering with the electrolysis. Calcium chloride is used as a flux in the extraction of sodium and this lowers the melting point of sodium chloride from 801 °C to about 600 °C. Cryolite (another aluminium mineral) is used as a flux in the extraction of aluminium and this lowers the melting point of aluminium oxide from around 2000 °C to 950 °C.

A problem with extractions like these is that, at the working temperatures, the metals formed at the cathode could easily react with the products at the anode (chlorine or oxygen) to reform the original compounds. So it is important to keep the molten metal separate from the anode gas. In the extraction of sodium, this is done by placing a wire gauze between the cathode and anode compartments.

Extractions by electrolysis always produce very pure metals. For example, the aluminium made in this way is 99·9 per cent pure.

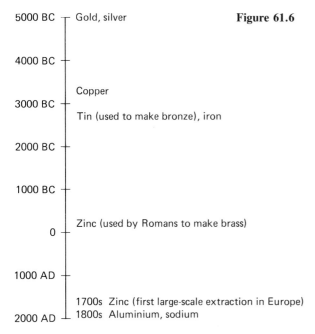

Figure 61.6

Questions

1. (a) Name *two* gases and *one* solid that can be used as reducing agents in the extraction of metals from their ores.
   (b) Which one of these three would you use to reduce malachite to copper in the laboratory? Give reasons for your answer.
   (c) What are the hazards (if any) associated with the use of each of the reducing agents you named in (a)?

2. Figure 61.6 shows some dates in the history of the extraction of metals. What is the pattern in this that links with the reactivity series for metals?

3. Here is some information about the extraction of four metals.

Metal	Method of extraction
Chromium	Reduction of oxide by aluminium
Lead	Reduction of oxide by coke
Magnesium	Electrolysis of molten chloride
Titanium	Reduction of chloride by magnesium

   (a) Which two of these metals are extracted using a more reactive metal as a reducing agent?
   (b) What other extraction method could possibly have been used for these two metals?
   (c) What is the third extraction method given in the table?
   (d) Using the information in the table and your knowledge of the reactivity series, write down an order of reactivity for the four metals, putting the most reactive one first.

62 CHEMICALS FROM LIMESTONE

The flow chart (Figure 62.1) summarizes the chemical reactions that are mentioned in this chapter. Study it carefully as you work through the chapter.

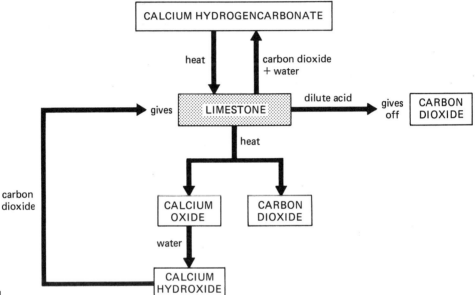

Figure 62.1

What is limestone?

Limestone is a sedimentary rock (page 27). Some limestones were formed from layers of shells on the sea floor and some others from the coral reefs along ancient tropical coastlines. In either case, the bits of material were pressed down by huge thicknesses of newer materials and, usually, a 'cement' was formed between the grains and shells.

Table 62.1 shows a chemical analysis of a limestone. Dilute hydrochloric acid 'dissolves' the calcium carbonate, releasing carbon dioxide.

calcium + hydrochloric → calcium
carbonate   acid   chloride
$CaCO_3(s) + 2HCl(aq) \longrightarrow CaCl_2(aq)$
                     + carbon + water
                       dioxide
                     $+ CO_2 + H_2O(l)$

This result can be used to find out if a sample of limestone contains anything else besides the carbonate. The extra material could be something that reacts with the acid without giving off carbon dioxide, or it could be something that is left over as a solid at the end of the reaction.

Experiment 1 Finding the purity of a limestone

a Use the apparatus shown in Figure 58.4 (page 240). Add exactly 0·30 g of pure calcium carbonate from a bottle to the small test-tube in the flask. When the reaction stops and no more gas collects in the syringe, read its volume on the scale.

b Do the experiment again, this time using 0·30 g of finely crushed limestone.

- What volume of carbon dioxide do you now collect?

Filter the mixture after all the gas has been given off.

- Is there any solid residue on the filter paper?

Table 62.1

Test	Result	Conclusion
Dilute hydrochloric acid is added to limestone	The limestone 'fizzes': a gas is given off. This gas (carbon dioxide) makes lime water go cloudy	There is a CARBONATE in limestone
Some crushed limestone is held on a nichrome wire in a Bunsen flame	The flame goes brick red for a time	There is a CALCIUM compound in limestone

- From the results in (*a*) and (*b*), do you think your limestone seems to be pure calcium carbonate?

More about carbon dioxide

Experiment 2 Making and testing carbon dioxide

The apparatus for this experiment is the same as for making hydrogen (Figure 9.5, page 35), but the chemicals used are calcium carbonate (as limestone (marble) chips) and dilute hydrochloric acid. Because carbon dioxide is only slightly soluble in water, the gas can be collected in the same way as hydrogen.

Make at least five test-tubes full of the gas. Close each tube with a bung. Test the gas with:

(*i*) damp universal indicator paper or damp blue litmus paper
(*ii*) a few cm³ of lime water
(*iii*) a burning splint plunged into the tube.

Try also to find a way of showing that carbon dioxide has a higher density than air.

Carbon dioxide is an acidic gas. When it dissolves in water, it forms carbonic acid. Figure 62.2 shows that the gas dissolves easily in sodium hydroxide solution: acids always react with alkalis (page 53).

Carbon dioxide forms a white solid when it is passed into lime water. If this solid is filtered off and tested with some dilute acid, carbon dioxide is reformed. This shows that the white solid is a carbonate. Because it is formed from calcium hydroxide (lime water), it must be calcium carbonate.

$$Ca(OH)_2(aq) + CO_2(g) \longrightarrow CaCO_3(s) + H_2O(l)$$

Using limestone

Quarries (Figure 62.3) can often be found where the rock just below the surface is limestone. These are man-made holes from which the rock has been blasted out and removed.

Limestone has many uses. The rock may be cut into blocks and used as a building stone or it may be broken up into small chippings and used in a

Figure 62.2

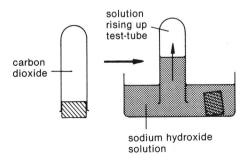

Figure 62.3

tarmac road surface. Also, the purer limestones are used in industry to make other chemicals.

Experiment 3 Making quicklime and slaked lime from limestone

Use the apparatus shown in Figure 62.4 to heat a small lump of limestone (a marble chip) very strongly.

- What can you see during heating?

After about 15 minutes, let the solid cool. This is quicklime. Put the quicklime on a watch-glass and add one drop of water to it. (CARE: do not let the quicklime come into contact with your skin or your eyes because it can cause burns.)

- What happens?

Add more water to the solid, drop by drop, until the chemical reaction stops. The dry solid is now slaked lime.

Put a *little* of the slaked lime in a test-tube and shake it with water. Let the mixture settle and then find the pH value of the solution. Also pass carbon dioxide through some of the *clear* solution.

- What happens?
- What is the chemical name of slaked lime?

Figure 62.5 shows what can be made starting from limestone. Match this diagram with the summary diagram given on page 256.

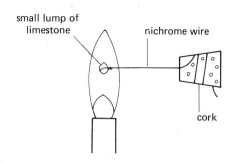

Figure 62.4

The word and symbol equations for the action of heat on limestone and for the *slaking* of quicklime with water are:

calcium $\xrightarrow{\text{heat}}$ calcium + carbon
carbonate   oxide   dioxide
$CaCO_3(s) \longrightarrow CaO(s) + CO_2(g)$

calcium oxide + water $\longrightarrow$ calcium hydroxide
(quicklime)   (slaked lime)
$CaO(s) \quad + H_2O(l) \longrightarrow Ca(OH)_2(s)$

Recycling limestone

Some of the limestone in Figure 62.6 has dissolved away. You can tell this because the cracks between the 'blocks' are much wider than they were when the rock was first formed. But in Figure 62.7 there

Figure 62.5

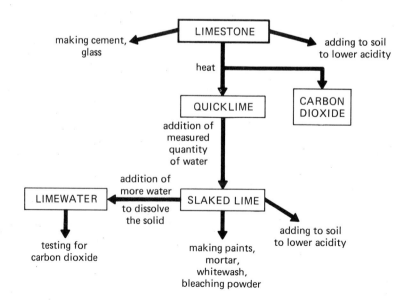

Figure 62.6

Figure 62.7

is an example of a 'new limestone' being created: the stalactites and stalagmites that are growing from the roof and the floor of this limestone cave are made of calcium carbonate.

Experiment 4 Dissolving and precipitating limestone

a Find the mass of a small lump of limestone (a marble chip) and then add it to a bottle of soda water (a solution of carbon dioxide in water under pressure). Screw the top of the bottle back on quickly so that the gas cannot escape. Keep the bottle in a cool place for at least a week.

Now remove the lump of limestone, dry it and find its new mass.

- Has any of the limestone dissolved?

b Add a few cm³ of the soda water left at the end of the experiment in part **a** to a boiling tube and heat it so that it boils gently.

- What happens?

Do the same thing to a *fresh* sample of soda water to show that the effect you observe must be linked to the dissolving of the limestone in part **a**.
c Use the first sample of boiled soda water from part **b**, that is, the one that was used in part **a** as well. Bubble carbon dioxide from a generator into this cooled mixture until a change takes place.

- What happens to the white cloudiness?

You are now back to the same clear solution that you obtained at the end of part **a**.

Limestone dissolves in water containing carbon dioxide (that is, carbonic acid) because it reacts to form calcium hydrogencarbonate.

$$\text{calcium carbonate (limestone)} + \text{carbon dioxide (carbonic acid)} + \text{water} \longrightarrow \text{calcium hydrogencarbonate}$$

$$CaCO_3(s) + CO_2(g) + H_2O(l) \longrightarrow Ca(HCO_3)_2(aq)$$

In the case of the limestone in Figure 62.6, the carbon dioxide is found in both the rain-water and the soil water that seeps through its cracks.

The chemical reaction shown by the above equations can be made to go the other way simply by boiling the solution of calcium hydrogencarbonate. The white cloudiness that forms is a precipitate of calcium carbonate.

$$\text{calcium hydrogencarbonate} \xrightarrow{heat} \text{calcium carbonate} + \text{carbon dioxide} + \text{water}$$

$$Ca(HCO_3)_2(aq) \longrightarrow CaCO_3(s) + CO_2(g) + H_2O(l)$$

In nature, solutions of calcium hydrogencarbonate are not boiled to form calcium carbonate. Instead, the stalactites and stalagmites in Figure 62.7 are formed when drops of solution trickling through the cave begin to evaporate or lose part of their carbon dioxide. Also, many animals living in the sea are able to extract calcium carbonate from water containing calcium hydrogencarbonate to build up their shells.

Questions

1. Limestone country often looks very dry: there are few surface streams and rivers and a lot of the water flows through underground caves. Explain how this comes about.

2. The rock called *chalk* is a very pure limestone. Describe an experiment which could be used to prove this.

3. (a) Give *two* pieces of evidence which show that limestone is a sedimentary rock.
   (b) Name *two* uses of limestone that do not involve changing it to another chemical.
   (c) Write the symbol equation for the chemical reaction of limestone with dilute nitric acid.
   (d) Suppose that 100 g of limestone gives off 20 litres of carbon dioxide at a certain temperature and pressure when it is added to an excess of dilute nitric acid. Is this limestone pure or impure calcium carbonate? Explain your answer. (Relative atomic masses: Ca = 40, C = 12, H = 1; the volume occupied by 1 mole of any gas at the given temperature and pressure is 24 litres.)

4. Slaked lime is used to make lime mortar, a material that is sometimes used in brick and stone buildings. When lime mortar is fresh, it gives no chemical reaction with dilute hydrochloric acid, but older lime mortar gives off carbon dioxide.

   (a) What compound is there likely to be in older lime mortar?
   (b) Explain, using a symbol equation, how the slaked lime in fresh lime mortar could be changed to this compound by one of the gases in the air.

63 CHEMICALS FROM SALT

The flow chart (Figure 63.1) summarizes the chemical reactions that are mentioned in this chapter. Study it carefully as you work through the chapter.

Salt from seas and lakes

When sea water is evaporated, solid salts (page 65) are always left over. A cubic metre of sea water contains about 27 kg of sodium chloride, about 5 kg of magnesium salts, and about 1 kg of calcium sulphate. Since over two-thirds of the earth's surface is covered by seas, this adds up to a huge amount of chemicals.

In many tropical countries, salt for cooking comes from the sea. A large flat area of land close to the sea is always needed. It is here that the sea water evaporates in the heat of the sun. The area is divided into shallow 'ponds' (Figure 63.2) and the sea water moves slowly from one 'pond' to another, getting more and more concentrated. In the final 'pond', the salt crystals are scraped together and piled into heaps to dry.

The salt made in this way is usually quite pure

Figure 63.1

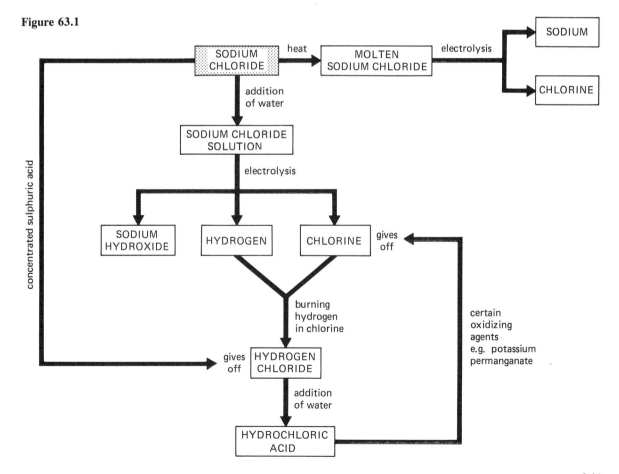

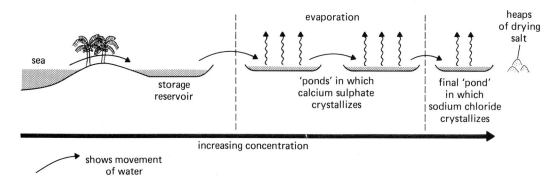

Figure 63.2

because the less soluble salts (like calcium sulphate) are crystallized out before the sea water gets to the final 'pond', and the more soluble salts (like the magnesium compounds) mostly stay in the water left over in the 'pond'.

Salt can also be obtained from the waters of some lakes. Most lakes are made of 'fresh' water and only have low concentrations of dissolved salts. But some of the lakes in East Africa, like Lake Katwe in Uganda and Lake Magadi in Kenya, are very salty. This is because the streams running into the lakes contain salts, and the concentration of the salts gets much higher as the lake water evaporates in the heat of the sun. Both these lakes are used as sources of sodium chloride.

About 250 million years ago the climate in Britain was very hot and dry as it is today around Lake Magadi. There was a very salty lake over one part and an arm of the sea over another. Sometimes, but not very often, enough rain fell to give the lake a new supply of water. Also, the arm of the sea occasionally received an inflow of less salty water from the nearby ocean.

Over many years, large thicknesses of salt crystals formed on the floors of the lake and the sea. Since then, these crystals have been covered by sand and clay and pressed down and hardened to form underground deposits of rock salt (see Figure 1.9).

Electrolysing sodium chloride solution

If the electrolysis of sodium chloride solution is carried out in a U-tube, the pH value of the solution in the cathode arm rises from its original value of 7 (Figure 63.3). This is because there is an increase in the concentration of hydroxide ions (the ions present in all solutions of alkalis in water) around the cathode as the hydrogen ions from the water combine with electrons to form hydrogen atoms (and then molecules, page 224). Chloride ions move to the anode where they lose electrons to form chlorine atoms (and then molecules).

In industry, this electrolysis is carried out using a very concentrated solution of sodium chloride called brine. The anodes are made of carbon but the cathode is a stream of mercury flowing across the bottom of the electrolysis cell. Although this electrolysis cell does not work in quite the same way as the one in Figure 63.3, the three products are still the same.

All three products have important uses—hydrogen in the manufacture of ammonia (page 273); chlorine to make insecticides, plastics and solvents; and sodium hydroxide (caustic soda) to make paper and soap.

Hydrogen chloride

Experiment 1 Making and testing hydrogen chloride

Figure 63.4 shows how hydrogen chloride can be made from sodium chloride and moderately concentrated sulphuric acid. (CARE: concentrated sulphuric acid is very corrosive.)

Collect four test-tubes of gas. Assume that a steady flow of gas for about 10 seconds will make sure that the tube is full. Close each tube with a bung.

Test the gas with:

(i) damp universal indicator
(ii) a burning splint plunged into the tube
(iii) an open tube containing a few cm³ of ammonia solution.

Figure 63.3

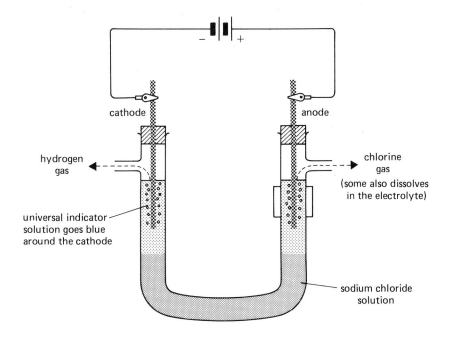

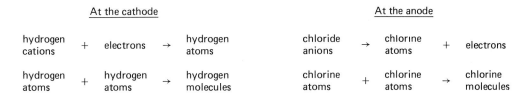

Use the method shown in Figure 63.5 to prove that hydrogen chloride is very soluble in water.

Hydrogen chloride is an acidic gas: it dissolves in water to form hydrochloric acid. This is why it reacts with ammonia, an alkaline gas (page 52), to form a white 'smoke' of ammonium chloride (a salt).

ammonia + hydrogen ⟶ ammonium
              chloride   chloride
$NH_3(g)$ + $HCl(g)$ ⟶ $NH_4Cl(s)$

Figure 63.4

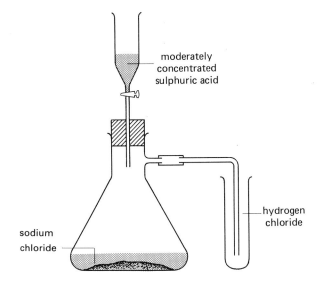

Figure 63.5

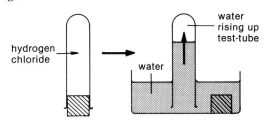

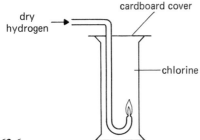

Figure 63.6

Hydrogen chloride can be synthesized from its elements (Figure 63.6). The reaction can be written in the form of a word or symbol equation, or it can be represented by diagrams of molecules (Figure 63.7).

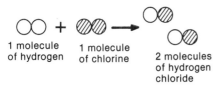

Figure 63.7

hydrogen + chlorine ⟶ hydrogen chloride
$H_2(g) + Cl_2(g) \longrightarrow 2HCl(g)$

Oxidizing hydrochloric acid

Concentrated hydrochloric acid can react with some oxidizing agents (substances that can release some or all of their oxygen to another substance). So the acid can act as a reducing agent by taking up this oxygen.

Figure 63.8 shows a way of describing what happens when concentrated hydrochloric acid is oxidized. The hydrogen part joins with the oxygen part to form water, so leaving chlorine behind. You can tell whether concentrated hydrochloric acid has been oxidized by finding out if any chlorine gas is given off the mixture.

Potassium permanganate ($KMnO_4$) is a suitable oxidizing agent for concentrated hydrochloric acid.

Chlorine

Experiment 2 Making and testing chlorine

Use the apparatus shown in Figure 63.4, but replace the sodium chloride by potassium permanganate and the moderately concentrated sulphuric acid by concentrated hydrochloric acid. (CARE: never add concentrated sulphuric acid to potassium permanganate; also note that concentrated hydrochloric acid is corrosive.)

Collect five boiling tubes of gas. You should be able to tell when the tubes are full by the colour of their contents (chlorine is greenish-yellow). Close each tube with a bung.

Test the gas with:

(i) damp universal indicator paper or damp blue litmus paper
(ii) a burning splint plunged into the tube
(iii) a few cm³ of sodium hydroxide solution
(iv) a lighted wax taper plunged into the tube
(v) a hot tuft of iron wool, held by some tongs, pushed into the tube.

(Some of these tests are shown in Figure 13.3, page 49).

Chlorine reacts with water to form two acids—hydrochloric acid and hypochlorous acid (HOCl). The hydrochloric acid turns universal indicator paper or blue litmus paper red while the hypochlorous acid is a bleaching agent. So if the indicator paper is left in the gas for long enough, all the colour is removed by the bleaching action of hypochlorous acid.

The chemical reactions of chlorine with elements can be compared with the burning of these elements in oxygen. For example, hydrogen burns in chlorine to form hydrogen chloride (see above) just as hydrogen burns in oxygen to form water (page 34).

Fuels can also burn in chlorine, as they do in oxygen. The paraffin wax in a candle or a wax taper is a compound of carbon and hydrogen (page 281). When this burns in chlorine, carbon (soot) and hydrogen chloride are formed.

Figure 63.8

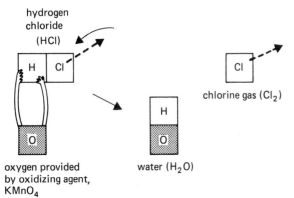

Figure 63.9

compound of C and H + chlorine ⟶ carbon + hydrogen chloride

A fuel burns in oxygen to form carbon (soot) when the supply of air is limited. (In a plentiful supply of air, the two products are carbon dioxide and water, page 281.)

compound of C and H + oxygen (limited supply) ⟶ carbon + hydrogen oxide (water)

In both these chemical reactions, the fuel is broken down and the hydrogen is reacted with the chlorine or oxygen. The burning in oxygen is called oxidation because oxygen is added to the fuel (page 35). But another way of looking at the same chemical reaction is to say that hydrogen is removed from the fuel by the oxygen.

New definitions of oxidation and reduction can now be given.

Oxidation in a chemical reaction takes place when hydrogen is removed from a substance.

Reduction in a chemical reaction takes place when hydrogen is added to a substance.

So the burning of fuels in chlorine can also be called oxidation. And, like oxygen, chlorine is an oxidizing agent.

You should now be able to explain the oxidation of concentrated hydrochloric acid by potassium permanganate (see above) using the idea of the transfer of hydrogen from one substance to another, rather than the transfer of oxygen.

Questions

1. (*a*) Why does the sea contain much higher concentrations of dissolved salts than the rivers that flow into them?
   (*b*) Why do a few lakes such as the 'Dead Sea' in the Middle East (Figure 63.9) contain higher concentrations of dissolved salts than the sea itself?

2. Hydrogen chloride reacts with zinc to form a white solid and hydrogen.

(*a*) Write a symbol equation for this chemical reaction.
(*b*) Draw a diagram of the apparatus you would use in order to carry out the chemical reaction and to collect the hydrogen free from any excess of hydrogen chloride.

3. When a mixture of hydrogen and chlorine is left for a few days, the green colour gradually fades and an acidic gas is produced. But if a similar mixture is passed over a hot platinum wire, the green colour fades very quickly indeed. (WARNING: this experiment could produce a dangerous explosion.)

(*a*) What is happening to the mixture as the green colour fades?
(*b*) What effect does the platinum wire have on the mixture?
(*c*) What word can be used to describe the effect of the wire?
(*d*) For which gas is the chemical reaction a synthesis?
(*e*) Describe another way in which this gas could be synthesized.

4. Starting with solid sodium chloride as your only compound containing chlorine, describe *two* ways in which you could make a few test-tubes of chlorine gas.

64 CHEMICALS FROM SULPHUR

Sulphur is one of the few elements that can be found in rocks by itself. It can also be found in minerals where it is joined with one or more metals, for example, galena (lead sulphide, PbS) and copper pyrites (copper iron sulphide, $CuFeS_2$). There are photographs of these two minerals on page 23.

The flow chart (Figure 64.1) summarizes the chemical reactions mentioned in this chapter starting from either rock sulphur or metal sulphides. Study it carefully as you work through the chapter.

The usual test for sulphur dioxide is to hold a piece of filter paper soaked in acidified potassium dichromate solution in a stream of the gas. The colour of the solution changes from orange to very pale green. Other gases with choking smells, for example, hydrogen chloride (page 262), have no effect on acidified potassium dichromate solution.

Another way of preparing sulphur dioxide is by heating concentrated sulphuric acid with copper turnings and collecting the gas by passing it down a gas jar or boiling tube. In this chemical reaction,

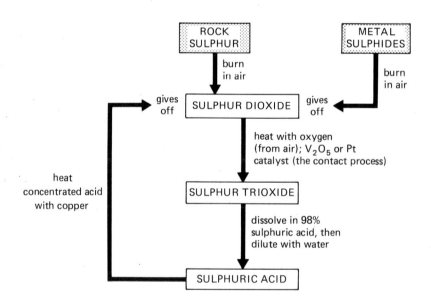

Figure 64.1

Sulphur dioxide

When sulphur or metal sulphides burn in pure oxygen (or air), an acidic gas with a choking smell is produced. This is called sulphur dioxide.

sulphur + oxygen ⟶ sulphur dioxide
$S(s) + O_2(g) \longrightarrow SO_2(g)$

concentrated sulphuric acid is acting as an oxidizing agent (page 35). It changes the copper to copper(II) sulphate, so releasing sulphur dioxide.

copper + sulphuric acid $\xrightarrow{\text{heat}}$ copper(II) + sulphur + water
 (concentrated)   sulphate   dioxide

$Cu(s) + 2H_2SO_4(aq) \longrightarrow CuSO_4(aq) + SO_2(g) + 2H_2O(l)$

Sulphuric acid

Sulphuric acid is a very important chemical because it can be used as the starting point for a whole range of useful substances—fertilizers (page 99), paints, man-made fibres, soaps and detergents, dyes, explosives and many other things.

The main stage in its manufacture, the so-called *contact process*, is the reaction between sulphur dioxide and oxygen in the presence of a catalyst to produce sulphur trioxide, SO_3. A good yield of sulphur trioxide at a fast enough rate can be obtained using a fairly high temperature, atmospheric pressure and a catalyst of vanadium(V) oxide, V_2O_5.

$$\text{sulphur dioxide} + \text{oxygen} \xrightarrow[\substack{V_2O_5 \\ \text{catalyst}}]{\text{heat}} \text{sulphur trioxide}$$

$$2SO_2(g) + O_2(g) \longrightarrow 2SO_3(g)$$

Figure 64.2 shows how sulphur trioxide can be made from sulphur dioxide in the laboratory. It is essential that both the sulphur dioxide and the oxygen are completely dry because the sulphur trioxide produced reacts very vigorously with water.

Sulphur dioxide and oxygen (in the approximate ratio of 2:1 by volume) are passed over hot platinized mineral wool, which acts as the catalyst. The gases are then passed through a cooled tube at 0 °C to form silky white crystals of sulphur trioxide.

A platinum catalyst is not used in the industrial process because it too easily adsorbs impurities from the gases. This *'poisoning'* of the catalyst makes it far less effective in speeding up the reaction.

The crystals of sulphur trioxide fume in damp air as they react with the water vapour. It is very dangerous to add them directly to water because the chemical reaction gives out a lot of heat. So in the industrial process, the gas is first dissolved in very concentrated sulphuric acid before being diluted with water to the required concentration. The overall chemical reaction can be represented by these equations:

$$\text{sulphur trioxide} + \text{water} \longrightarrow \text{sulphuric acid}$$
$$SO_3(g) + H_2O(l) \longrightarrow H_2SO_4(aq)$$

The equations show that the dense white fumes that form when sulphur trioxide meets damp air are small droplets of sulphuric acid. This is why you must never try to smell sulphur trioxide: the formation of sulphuric acid in your lungs would be very harmful.

Sulphur compounds in the air

The air always contains nitrogen, oxygen, water vapour, carbon dioxide and the noble gases (page 20). In addition, the air over populated areas of the world also contains *pollutants*. These are gases that are added to the air by the activities of man.

Sulphur compounds, especially sulphur dioxide, are among the chief air pollutants. Certain fuels

Figure 64.2

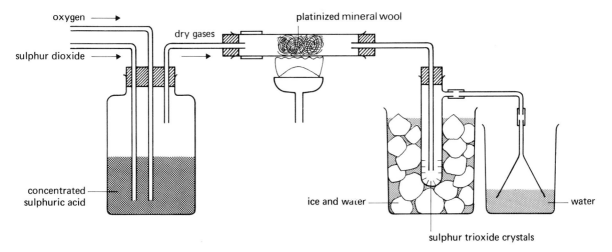

contain sulphur and when they burn, sulphur dioxide is given off: coal and fuel oil contain up to 3 per cent of sulphur and petrol about 0·05 per cent. Sulphur dioxide is also produced in the extraction of copper when the sulphide ore is heated in air (page 253).

The amounts of these pollutants in the air are not very high. On average, there is a little less than one-millionth of a gram (10^{-6} g) of sulphur dioxide in a cubic metre of air. Over cities there is usually much more and the concentration may reach one-thousandth of a gram (10^{-3} g) per cubic metre close to large sources of pollution.

These concentrations may not seem high, but it takes very little sulphur dioxide to damage the leaves of plants and stunt their growth, or to make people's respiratory illnesses much worse. Also, some sulphur dioxide in the air is oxidized to sulphuric acid. As well as being harmful to plants and animals, this acid attacks the stonework of buildings, especially limestone (Figure 64.3), and causes it to crumble.

It has also been found that sulphur dioxide in the air slowly affects the paper and binding of old books stored in the libraries of industrial cities. The paper goes yellow and loses some of its strength so that it starts to break up.

Experiment 1 The effect of sulphur dioxide on metals

Set up the two pieces of apparatus shown in Figure 64.4. The only difference between them is that one has ordinary air inside the box, whereas the other has air containing a lot of sulphur dioxide. Leave the metal foils in the boxes for at least a week.

- What happens to the foils in the box containing sulphur dioxide? Compare the appearance of these foils with those in the other box.

Figure 64.3

The corrosion (page 45) of metals in air containing sulphur dioxide is thought to be caused by sulphuric acid. This is formed on the surface of the metal and reacts with its protective layer. Once the fresh metal is exposed to the air, corrosion can take place much more quickly.

Figure 64.4

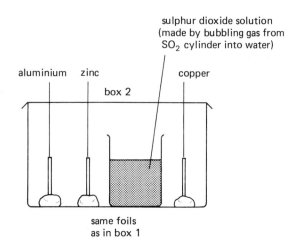

The protective layer on most metals is an oxide or a carbonate of the metal. Both these classes of compounds react with acids to form salts, and so the substances formed by this kind of corrosion are sulphates.

Questions

1. Both carbon dioxide and sulphur dioxide turn lime water cloudy. Describe one way (apart from smell) that you could use to distinguish between these two gases.

2. A country has reserves of crude oil and copper pyrites. Draw a flow diagram to show how the fertilizer 'sulphate of ammonia' (ammonium sulphate) could be made on a large scale starting from these two materials, and air and water.

3. Describe how you would make crystals of copper(II) sulphate starting from copper metal and concentrated sulphuric acid. Write a symbol equation for the chemical reaction.

4. List some of the ways in which man has tried to reduce pollution of the air.

65 CHEMICALS AND FOODS

The flow chart (Figure 65.1) summarizes the chemical reactions that are mentioned in this chapter. Study it carefully as you work through the chapter.

other elements (page 10). When they are broken down by strong heating, all the other elements are given off in liquids and gases, leaving only some of the carbon behind.

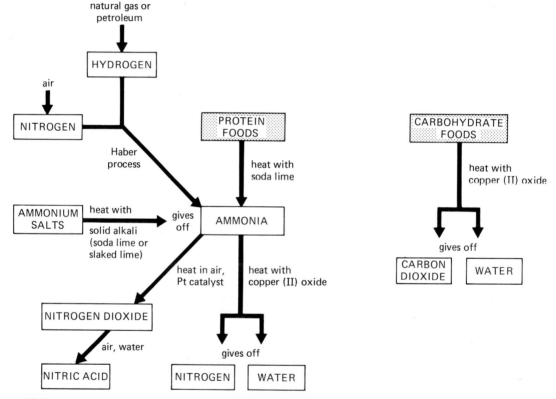

Figure 65.1

Breaking down foods

Food is for eating. But to find out more about the elements in food, you must break it down.

All foods go black if they are heated strongly. This also happens to both wood and paper: wood forms charcoal (page 7). Substances that come from living material, and this includes foods, wood and paper, are compounds of carbon and a few

Experiment 1 Heating some foods with copper(II) oxide

Use the apparatus shown in Figure 65.2. The copper(II) oxide must be dry.
Use a food chosen from this list: beans, breadcrumbs, wheat or maize flour, and sugar.

- What happens to the lime water when the mixture is heated?

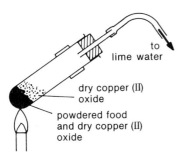

Figure 65.2

- Which element does this show the food *must* contain?
- Which other element *might* come from the food as well as from the copper(II) oxide?

Test the drops of the colourless liquid that form in the test-tube above the mixture with a piece of cotton wool sprinkled with white copper sulphate.

- Which element in the liquid *must* come from the food?
- Which one *might* come from the food as well as from the copper(II) oxide?

In this experiment, the foods are broken down by oxidation (page 35). The oxidizing agent is copper(II) oxide. It releases the oxygen that makes the compounds in the foods break down to form other compounds with much smaller molecules. In doing this, the copper(II) oxide itself changes to copper metal.

Some foods can be broken down in another way.

Experiment 2 Heating some foods with soda lime

Use one of these foods: cheese, chicken or meat, gelatin, and milk powder.

Mix two spatula measures of the food in a test-tube with the same volume of soda lime. Add some more soda lime on top of the mixture. Heat the test-tube strongly near to an open window. If a gas with a nasty smell is given off, test it with:
(i) damp universal indicator paper
(ii) a burning splint held near to the mouth of the tube.

When some foods are heated with soda lime, an alkaline gas called ammonia is given off. Foods that have this property always contain *proteins* (page 292). Many foods contain some proteins, but sometimes amounts are small.

Many foods also contain large amounts of *carbohydrates* and these substances do not give off ammonia when they are heated with soda lime. They are made of carbon, hydrogen and oxygen (as their name suggests) and give off carbon dioxide and water vapour when they are heated with copper(II) oxide.

Ammonia

The gas that is given off when proteins are heated with soda lime is called ammonia. This is a compound of nitrogen and hydrogen with the formula NH_3. Because soda lime itself does not contain any nitrogen, this element must come from the proteins. So proteins are made of nitrogen joined with other elements, usually just carbon, hydrogen and oxygen.

Experiment 3 Making and testing ammonia

Heat the mixture of solid ammonium sulphate and calcium hydroxide, to which a little water has been added (Figure 65.3). As the gas passes through the calcium oxide it is dried. Make about four test-tubes full of ammonia and close each one with a bung.

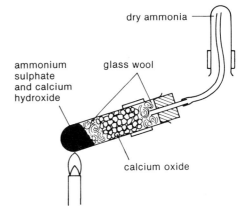

Figure 65.3

Test the gas as in Experiment 2, but also find a way of showing that ammonia is very soluble in water.

- What does the way in which the gas is collected tell you about the density of ammonia compared to that of air?

Also, hold an open test-tube of ammonia close to a test-tube containing a few cm³ of concentrated hydrochloric acid.

- What happens?

The word and symbol equations for this preparation of ammonia are:

ammonium + calcium $\xrightarrow{\text{heat}}$ calcium + ammonia + water
sulphate   hydroxide   sulphate

$(NH_4)_2SO_4(s) + Ca(OH)_2(s) \rightarrow CaSO_4(s) + 2NH_3(g) + 2H_2O(g)$

In fact, any ammonium salt reacts with any alkali to give off ammonia, and the chemical reaction can be compared with that between a protein (which is a nitrogen compound) and soda lime (which is an alkali).

Ammonia is the only common alkaline gas. It dissolves readily in water to form a solution with a pH value of 10 or 11. This property explains why ammonia reacts with hydrogen chloride (page 263).

Breaking down ammonia by oxidation

Figure 65.4 shows an experiment in which copper(II) oxide is used to break down ammonia by oxidation. Even though the ammonia and the copper(II) oxide are dry, some drops of water can be seen in the combustion tube. These must be formed by the reaction.

The oxygen in the water comes from the copper(II) oxide while the hydrogen comes from the ammonia. The other substance formed in this chemical reaction is a colourless gas. This can be shown to be neutral and not very reactive: it is nitrogen. The nitrogen also comes from the ammonia.

The word and symbol equations for this chemical reaction are:

ammonia + copper(II) $\xrightarrow{\text{heat}}$ copper + water + nitrogen
             oxide

$2NH_3(g) + 3CuO(s) \longrightarrow 3Cu(s) + 3H_2O(l) + N_2(g)$

The copper(II) oxide is reduced by ammonia to copper. The oxygen given up by the copper(II) oxide is used to break up the ammonia into water and nitrogen. So ammonia is the reducing agent in this reaction and copper(II) oxide is the oxidizing agent.

Making nitric acid in industry

Another way of oxidizing ammonia, this time using the oxygen in the air, is shown in Figure 65.5. Air is drawn through the apparatus by the tap pump. After it has passed through the ammonia solution, it contains some ammonia gas. The hot platinized mineral wool acts as a catalyst (page 288) for the chemical reaction between ammonia and oxygen.

A brown gas (nitrogen dioxide) collects in the flask on the right-hand side of the apparatus.

ammonia + oxygen $\xrightarrow[\substack{\text{Pt} \\ \text{catalyst}}]{\text{heat}}$ nitrogen dioxide + water

$4NH_3(g) + 7O_2(g) \longrightarrow 4NO_2(g) + 6H_2O(l)$

This is the first stage in the manufacture of nitric acid from ammonia. In industry, the chemical reaction is carried out using a catalyst of platinum–rhodium gauze sheets. These are expensive but they are not used up during the chemical reaction.

In the second stage of the process, nitrogen dioxide is further oxidized by oxygen in the presence of water to form nitric acid.

nitrogen dioxide + oxygen + water $\longrightarrow$ nitric acid

$4NO_2(g) + O_2(g) + 2H_2O(l) \longrightarrow 4HNO_3(aq)$

This chemical reaction is the reverse of the breakdown of concentrated nitric acid by heat (page 62).

Synthesizing ammonia

It is difficult to synthesize ammonia from its elements in the school laboratory, but in industry this chemical reaction is carried out on a large scale using the *Haber process*. Nitrogen and hydrogen are mixed and then heated under pressure in the presence of an iron catalyst. The word and symbol

Figure 65.4

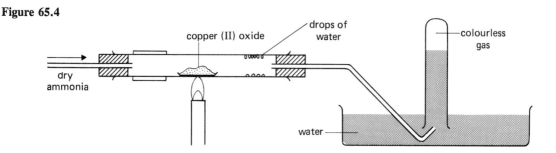

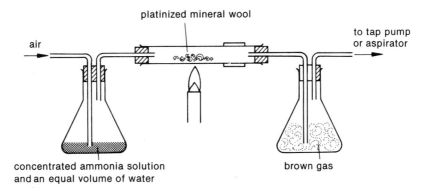

Figure 65.5

equations for this chemical reaction are:

nitrogen + hydrogen $\xrightarrow[\text{Fe catalyst}]{\text{heat}}$ ammonia

$N_2(g) + 3H_2(g) \longrightarrow 2NH_3(g)$

Figure 65.6 shows how diagrams of molecules can be used to explain the synthesis of ammonia.

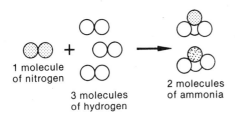

Figure 65.6

As Figure 65.1 (page 270) shows, the nitrogen for the Haber process comes from the air and the hydrogen from either natural gas (methane) or petroleum (crude oil). It is also possible to obtain the hydrogen by passing steam over hot coke (carbon).

The importance of nitrogen compounds to life

Nitrogen, in the form of various compounds or as the element itself, gets circulated between the air, soils and living things. Like carbon (page 281), nitrogen moves in a cycle which is vital to life. The *nitrogen cycle* is discussed in Figure 26.4, page 98.

If it were not for man's intensive use of the land for growing crops, a balance would be kept between the amount of nitrogen removed from the soil by plants and the amount of nitrogen returned to it by the fixation processes and by the decay of animal and plant proteins. Intensive cultivation of crops requires far more fixed nitrogen than natural processes can supply. For this reason, the large scale production of food crops from the land is only made possible by adding nitrogen-rich fertilizers to the land.

Experiment 4 Making a fertilizer from ammonia

Add about 20 cm³ of dilute sulphuric acid to an evaporating dish. Stirring all the time, add to this some dilute ammonia solution until the liquid in the dish smells slightly of ammonia.

Then begin to evaporate the solution. From time to time, remove a small sample of the solution on the end of a glass rod and let it cool. If crystals form, then stop the evaporation and leave the partially evaporated solution to crystallize.

The white crystals that form are of ammonium sulphate ('sulphate of ammonia'), a nitrogenous fertilizer.

ammonia + sulphuric acid ⟶ ammonium sulphate
(dilute)

$2NH_3(aq) + H_2SO_4(aq) \longrightarrow (NH_4)_2SO_4(aq)$

Ammonium sulphate is a salt and this chemical reaction involves the neutralization of an acid by an alkali (ammonia solution). What is special about this salt (and all other ammonium salts) is that it does not contain a metal. Instead, the cation is NH_4^+.

Questions

1. Explain why the Haber process is so important to modern food production.

2. Egg contains a lot of proteins. Which gas or gases would be given off when egg powder is heated with (*a*) copper(II) oxide, and (*b*) soda lime? Explain your answers.

3. Dry ammonia is passed over heated lead(II) oxide and a chemical reaction takes place.

(a) Draw a diagram that shows the apparatus for the preparation of dry ammonia connected to the apparatus for the reaction between ammonia and heated lead(II) oxide and then to the apparatus that could be used to collect the insoluble gas produced by the chemical reaction.

(b) Write the word and symbol equations for this chemical reaction.

(c) Explain this chemical reaction in terms of oxidation and reduction. Say which of the reactants acts as the oxidizing agent and which is the reducing agent.

4. Here are the names and formulas of some fertilizers.

(a) Which element is common to all these fertilizers?

Ammonium sulphate	$(NH_4)_2SO_4$
Ammonium nitrate	NH_4NO_3
Calcium nitrate	$Ca(NO_3)_2$
Carbamide (urea)	$CO(NH_2)_2$
Ammonia	NH_3

(b) Why is this particular element essential to plant growth?

(c) Which of the fertilizers are salts?

(d) Ammonium sulphate solution has a pH value of about 5.5. Why is this a disadvantage for a substance that is used as a fertilizer?

(e) All these fertilizers are soluble in water. Why is this property essential for a fertilizer?

Chemical Energy

66 ENERGY IN CHEMISTRY

Heat energy and chemical reactions

Suppose that a chemical reaction is described as being *vigorous*. This means that either or both the following statements about the reaction is true:

a A lot of heat (and perhaps light) energy is given out

b The reaction goes on very quickly (at a high rate).

In this chapter, we will concentrate on reactions of type **a**.

All chemicals can be thought of as storing chemical potential energy (page 159). It is like the energy that a man at the top of a hill has because of the height. The energy is only apparent once something happens—the man begins to run downhill or the chemicals are mixed to produce a chemical reaction. In a chemical reaction, some of the stored chemical energy is often released as heat and light energy (see Figures 66.1 and 66.2).

It is important to realize the difference between the heat energy given out (or taken in) while the chemical reaction is going on and the heat energy sometimes needed to start the chemical reaction. Charcoal and coal need a lot of heat energy to start them burning but, once begun, a lot of heat energy is given out to the surroundings. So the burning of charcoal or coal is an *exothermic* reaction.

There are many examples of exothermic reactions given in this book. But sometimes the stored chemical energy in substances is *increased* during a chemical reaction by heat energy being taken in from the surroundings (the air and the containers). These chemical reactions are said to be *endothermic*.

Experiment 1 Two endothermic reactions

In each of these chemical reactions, a hydrogencarbonate is reacting with an acid to give off carbon dioxide.

a Find the temperature of about 25 cm³ of dilute hydrochloric acid in a small beaker. Add to this about four spatula measures of potassium hydrogencarbonate.

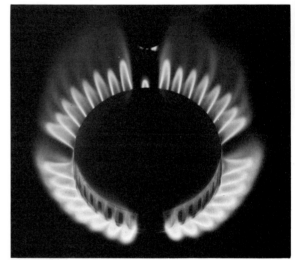

Figure 66.1

Figure 66.2

- What happens to the temperature shown on the thermometer?

b Mix together roughly equal amounts of sodium hydrogencarbonate ('bicarbonate of soda') and citric acid crystals (the acid contained in citrus fruits). Find the temperature of about 25 cm³ of water in a small beaker. Then add the solid mixture to the water.

- What happens to the temperature shown on the thermometer?

Heat energy may also be given out or taken in during changes that are not chemical reactions. It is given out when concentrated sulphuric acid is added to water to make dilute sulphuric acid, and also when water vapour condenses on a cool surface. It is taken in when a volatile liquid evaporates and when certain solids, such as potassium iodide, are dissolved in water.

Light energy and chemical reactions

Flames from burning fuels (Figure 66.1) or from other chemical reactions (Figure 66.2) provide both heat and light energy. For a burning fuel, the flame is the region where the gas from the fuel is reacting with the oxygen in the air. The gases glow brightly because they are hot. In the case of a burning candle, there is an extra brightness caused by the white hot grains of *solid* carbon that are produced by the reaction between candle wax and oxygen.

A few chemical reactions give out much of their energy in the form of light rather than heat. These include the chemical reactions that take place in a fire-fly or in the tail of a glow-worm.

Light can also be used like heat energy to start off a chemical reaction. Not many chemical reactions are *photosensitive*, but one of them is the vitally important natural process called *photosynthesis* (page 86).

Experiment 2 A photosensitive chemical reaction

Mix equal volumes of silver nitrate solution and potassium bromide solution in a boiling tube.

- What is the name of the precipitate that is formed (page 69)? (HINT: bromides have similar solubilities to chlorides.)

Filter the mixture and spread out the filter paper containing the precipitate on a bench by a window.

- What happens?

If you are not sure about the answer to this question, make a fresh mixture and compare the colour of the precipitate in that with the colour of the precipitate on the filter paper.

Electrical energy and chemical reactions

Electrical energy is changed into stored chemical energy during electrolysis (page 58). For example, the electrolysis of molten sodium chloride gives sodium and chlorine. These two elements contain more stored chemical energy than sodium chloride, and this extra chemical energy is released as heat and light energy when sodium burns in chlorine.

Chemical energy can also be changed directly into electrical energy. A torch battery (a cell) can be used to light the torch bulb because some of the stored chemical energy makes electrical energy in a chemical reaction taking place in the battery. This electrical energy is then changed into light energy, and also into some heat energy.

The chemical reaction in the torch cell is not an easy one to investigate, but many other chemical reactions can make electrical energy under the right conditions.

Experiment 3 Making electricity in a simple cell

Set up the apparatus shown in Figure 66.3a.

- How do you know that a chemical reaction is taking place in the beaker?
- What is the name of the gas being given off?
- What happens to the bulb?

The apparatus used in this experiment is a single cell. The first battery, which is made of several cells joined together, was invented in 1800 by Volta. It is called the *Voltaic pile* in his honour. Volta found that electrical energy is always made when two different metals are joined by a wire and then put in a solution of an electrolyte. An electric current is a flow of electrons. So in a cell there must be a way of causing electrons to flow through the connections between the two metal foils. It has been found that these always flow from the more reactive metal to the less reactive metal.

In the cell shown in Figure 66.3a, electrons flow through the wire from the magnesium to the copper, and in the other cell (Figure 66.3b), they flow from the zinc to the copper through the place where the foils are touching one another. In each case, the reactive metal loses electrons to form its cations.

zinc atoms $\longrightarrow$ zinc cations + electrons
(in foil)   (in electrolyte)   (in connections)

The cations move into the electrolyte. As the reaction continues, the more reactive metal slowly dissolves away. The cell stops producing electricity when there is no reactive metal left.

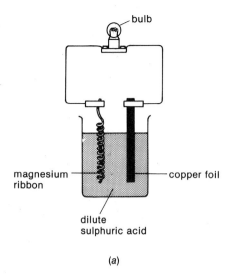
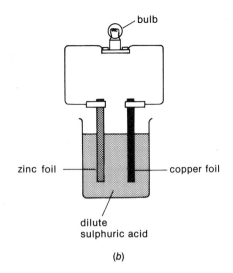

Figure 66.3

The torch cell (the so-called 'dry cell', Figure 66.4) is more complicated than the cells shown in Figure 66.3. Instead of two metal foils, there is a zinc case and a central carbon rod surrounded by a paste of manganese(IV) oxide, carbon and ammonium chloride. The electrolyte is a jelly-like paste of ammonium chloride. But it still works by the zinc atoms changing into zinc cations, so releasing electrons to flow through the connections to the carbon rod.

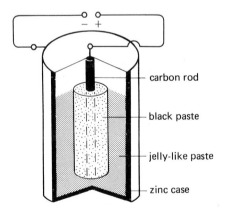

Figure 66.4

Questions

1. Explain the difference between an exothermic and an endothermic chemical reaction, and give an example of each.

2. Look at the electric lemon in Figure 66.5.
   (a) Why does the bulb glow?
   (b) Suppose that instead of the zinc foil you used another copper foil. Would the bulb still glow? Explain your answer.

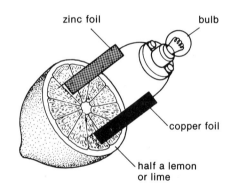

Figure 66.5

3. When a strip of zinc foil is dipped into dilute sulphuric acid, only a little hydrogen is given off (Figure 66.6a). If a strip of copper foil is placed in the same acid alongside the zinc foil, nothing else happens (Figure 66.6b). But if the copper foil is moved so that it touches the zinc foil, a lot of hydrogen is given off the *copper* foil (Figure 66.6c).

If the reaction shown in (c) is allowed to carry on for a few minutes, the zinc foil loses mass while the copper foil stays the same mass.

(i) In view of its position in the reactivity series, what is surprising about the chemical reaction of zinc foil with the acid in parts (a) and (b) of the experiment?

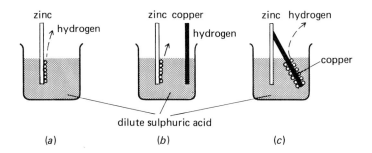

Figure 66.6

(*ii*) In part (*c*) of the experiment, which metal must be reacting so that hydrogen ions from the acid form hydrogen gas? Give the evidence for your answer.

(*iii*) The arrangement in part (*c*) is a cell. Are the electrons moving from the zinc foil into the copper foil, or from the copper foil into the zinc foil?

(*iv*) Why is the zinc foil losing mass in part (*c*) of the experiment?

67 INVESTIGATING FUELS

What makes a good fuel?

The energy released when chemical fuels are burned is the major source of all the energy used by today's civilizations.

Fuels can be solids (like coal, peat, charcoal and wood), liquids (like petroleum, fuel oil and kerosene) or gases (like natural gas). Particular fuels are chosen for particular jobs.

Suppose we want a fuel to propel a rocket-powered vehicle at 40 000 km/h away from the earth's surface. The fuel must produce a huge thrust. To do this a great deal of heat energy must be produced in a very short time so that the gases produced will expand rapidly and move away from the rocket very quickly. In addition, the volume and mass of these gases should be as high as possible.

Different properties are needed for fuels used in cooking and for keeping us warm. Quick burning is not so important as in space rockets. But it is very important for the fuel to have a high *heating value* ('calorific value'). This is the heat energy given out per unit mass, or sometimes per unit volume if the fuel is a gas. The heating values for some fuels are given in Table 67.1.

Table 67.1

Fuel	Heating value in kJ/g
SOLIDS	
Charcoal	33
Coal	25–33
Wood	17
LIQUIDS	
Ethanol	30
Fuel oil	45
Paraffin	48
GASES	
Methane (natural gas)	55
Propane (bottled gas)	50
Butane (bottled gas)	50

Besides having a high heating value, a fuel used in the home, the laboratory and industry should be as cheap as possible. What you need to know is how much it costs for each fuel to give out the same amount of energy. Follow these steps for two common fuels used in your own area.

1. Find the cost of the fuel per kilogram. (Even bottled gas is sold by mass rather than volume because it is always under pressure in its container.)
2. Use Table 67.1 to obtain the heating value of the fuel in kJ/g (kilojoules per gram). Change this to a value in kJ/kg (kilojoules per kilogram).
3. Divide your answer in **1** by your answer in **2** to work out the cost per kilojoule of energy given out.
4. Decide which is the cheaper of the two fuels. If this fuel is not the more widely used fuel, suggest some reasons for this.

What are fuels made of?

Fuels burn in air to form oxides. If these oxides can be identified, they provide clues about the elements that the fuels contain.

For example, the only product of the burning of charcoal is carbon dioxide (Figure 67.1). The molecules of this compound are made of the atoms

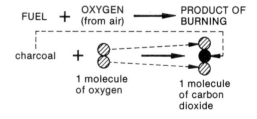

Figure 67.1

of the elements carbon and oxygen. At least some of the oxygen atoms must come from the air and all the carbon atoms must come from the charcoal. So it is clear that charcoal contains carbon. (This result does not prove for certain that charcoal is made only of carbon because it is possible that the oxygen comes from the charcoal as well as from the air. But in this case, it so happens that all the oxygen does come from the air.)

Experiment 1 Investigating a burning candle

Put a large beaker over a burning candle (Figure 67.2). When the flame begins to go out, lift up one side of the beaker to let some more air into the space around the candle. After 1–2 minutes, lift up the beaker from the bench and immediately place some lime water under it.

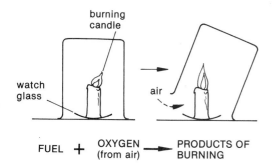

Figure 67.2

- What happens?
- Which element must be present in candle wax to give this result?

Do the experiment again with a completely dry beaker. When enough moisture has collected on the inside, lift up the beaker. Rub the inside of the beaker with a piece of cotton wool sprinkled with white copper sulphate.

- What happens?
- What is the moisture formed inside the beaker?
- Which element must be in candle wax for this substance to be a product of burning?

Candle wax burns to form two different oxides. Figure 67.3 shows that this must mean candle wax is made of at least two elements joined together. In fact, candle wax does not contain any oxygen and it is a compound of carbon and hydrogen only.

Many fuels are *hydrocarbons* (compounds of carbon and hydrogen only) or mixtures of hydrocarbons. Examples are paraffin, natural gas and bottled gas. Natural gas is often found in rocks along with crude oil (petroleum) and the other hydrocarbon fuels come from the fractional distillation (page 6) of crude oil.

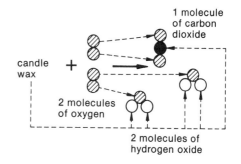

Figure 67.3

The carbon cycle

The part of the carbon cycle that involves fuels is shown in Figure 67.4. In the form of various compounds, carbon (like nitrogen, page 98) is circulated between the air, the earth's surface and the rocks beneath the surface. On the surface, many of these compounds are in living things. Coal, crude oil and natural gas are formed by the death, partial decay and burial of some of these living things.

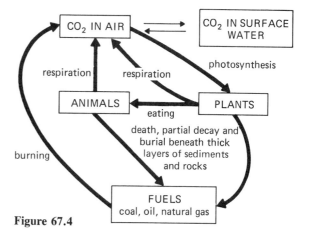

Figure 67.4

Fuels and air pollution

In a good supply of oxygen, a hydrocarbon fuel burns to form carbon dioxide and water vapour: these gases are already present in the air.

methane + oxygen ⟶ carbon + water
(natural gas)   dioxide   vapour
$CH_4(g) + 2O_2(g) \longrightarrow CO_2(g) + 2H_2O(g)$

But if the supply of oxygen is limited, other products can be formed. In Experiment 1, soot (carbon) can be seen on the inside of the upturned beaker when there is not enough air entering from outside it. Other fuels can also burn to form

Figure 67.5

carbon, and soot in the air is a pollutant (Figure 67.5).

methane + oxygen ⟶ carbon + water vapour
$CH_4(g) + O_2(g) \longrightarrow C(s) + 2H_2O(g)$

There is a half-way stage between the extremes of complete burning of a hydrocarbon fuel (which forms carbon dioxide) and very incomplete burning (which forms carbon). Here there is enough oxygen to oxidize the carbon, but not all the way to carbon dioxide. The products of this burning are carbon monoxide and water vapour.

methane + oxygen ⟶ carbon + water
                             monoxide vapour
$2CH_4(g) + 3O_2(g) \longrightarrow 2CO(g) + 4H_2O(g)$

Carbon monoxide is a very poisonous gas. People can be killed within half an hour if they breathe air containing a very small proportion of this gas. It acts by lowering the blood's ability to carry oxygen round our bodies.

When petrol (mostly octane, C_8H_{18}) burns in a car engine, some carbon monoxide is formed and is passed into the air through the exhaust pipe. This kind of air pollution can be a serious problem in busy city centres. In the worst areas, police directing traffic can only work for short periods before they are affected by it.

Questions

1. Wood and charcoal can both be used as fuels. Write down the advantages and disadvantages of each under the following headings: heating value, speed of burning, cleanliness of burning (that is, how much ash they produce) and availability.

2. When a beaker containing cold water is held over a Bunsen flame, moisture quickly forms on the outside. After a short time, this disappears and the outside of the beaker gets dry again. Explain these changes.

3. Describe an experiment you can carry out to show that bottled gas or natural gas contains carbon. Explain how the results of your experiment would prove this.

4. Depending on the conditions, hydrocarbon fuels can burn to form carbon, carbon monoxide or carbon dioxide.

(a) Explain why this is so.
(b) Which two of these products are undesirable? What effects can each of them have on our environment?
(c) The third of these products is already present in the air and is not a pollutant. Which one is this?

68 FOSSIL FUELS

How fossil fuels are formed

The way that coal, crude oil and natural gas were formed is shown in the carbon cycle (Figure 67.4, page 281). They are all *fossil fuels*. The word *fossil* (page 29) shows that the fuels were formed from once-living material.

Coal forms thin layers in between the layers of sedimentary rocks (page 27). Oil and gas are also found in sedimentary rocks, but unlike coal, they do not stay where they were formed. Instead, they move upwards through the spaces between the grains of rock.

Experiment 1 Permeable and impermeable rocks

Use the apparatus shown in Figure 68.1. The rock is a friable (easily crumbled) sandstone. Fill the tube with water from a dropper.

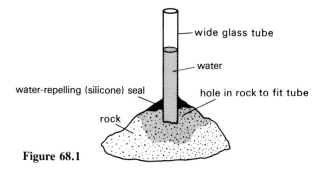

Figure 68.1

- How long does it take for the water to move out of sight down the tube?

Do the experiment again with other sedimentary rocks, such as limestone and shale, and also with slate.

- Which rocks let the water seep in quickly?
- Which rocks let the water seep in very slowly, if at all?

Rocks that are easily crumbled between your fingers have plenty of spaces between their grains. So long as the spaces join up into a network, oil and gas can pass through them quite easily. Such rocks are said to be *permeable*.

Rocks that do not have a network of joined-up spaces are usually *impermeable* to oil and gas. They become permeable only if they are broken up by large cracks through which oil and gas can move easily. Some limestones are like this and some cracks can be seen in Figure 62.6 (page 259).

If you were to drill a hole from the surface down through the layers of rocks, you would find that some layers are permeable and others are impermeable. These control what happens to the oil or gas as it rises towards the surface.

Experiment 2 The effect of rock layers on a moving liquid

Put a small lump of a friable sandstone in some coloured water in a Petri dish. Rest another lump of the same rock on top (Figure 68.2). Watch the water as it seeps upwards.

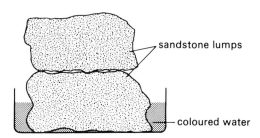

Figure 68.2

- What happens when the water gets to the 'join' between the two lumps?

Now set up the arrangement shown in Figure 68.3 with a thin piece of shale or slate between two more lumps of the sandstone.

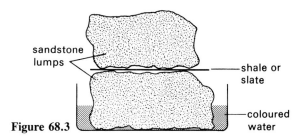

Figure 68.3

- What happens when the water gets to the shale or slate?

The movement of oil and gas up to the surface can be blocked by layers of impermeable rocks. Of course, the liquid can still move sideways beneath the shale or slate.

Suppose that a permeable rock is 'sandwiched' between sloping layers of impermeable rocks (Figure 68.4). Oil or gas could keep on moving

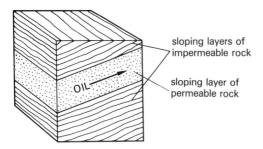

Figure 68.4

upwards through the permeable layer, even though it could not enter the impermeable layer above. But if the 'sandwich' is folded into a 'hill' of rock layers (Figure 68.5), the oil or gas would then be trapped in a place where it could not move

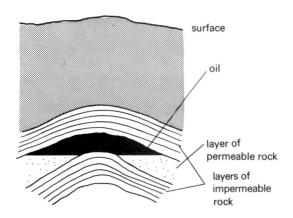

Figure 68.5

upwards or sideways. The oil or gas would also be trapped if the permeable rock was cut out by an impermeable barrier (Figure 68.6).

The 'hill' of rock under the ground is called an *anticline*, and the sudden break in the rocks (Figure 68.6) is called a *fault*. These are two of the most common kinds of oil and gas traps but there are other kinds as well.

Finding oil and gas

The search for fossil fuels is being carried out all over the world, both under land and under the sea. Early oil and gas fields were all under land, but recent advances in technology have given rise to the development of many offshore fields such as the one under Britain's North Sea.

The first task of the oil prospectors is to locate certain arrangements of rock layers that might act as oil or gas traps (see Figures 68.5 and 68.6). One way of doing this is to use a *seismic survey*.

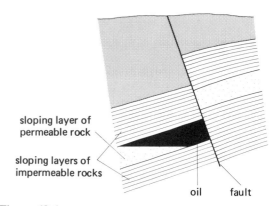

Figure 68.6

Small explosions are set off at the surface of the land or the sea and the shock waves from these move through the rocks. One kind of wave is like a sound wave travelling through the air: a 'pulse of vibration' moves through the rock. Detectors 'listen' for the waves that arrive back at the surface after having been reflected by some of the rock layers. Figure 68.7 shows the method that is used to determine the underground rock formation.

Once some promising arrangements of rock layers have been found, the next task is to put down some boreholes. A diamond core head or drill bit is used at the end of a 10 m length of pipe. When the 10 m of pipe have disappeared into the rock, another length is screwed on to the end (Figure 68.8) and drilling begins again. The head or bit gets very hot, and specially prepared 'mud' has to be pumped down the drill pipe to cool it.

If a drill bit is used for the drilling, this mud carries rock chippings up to the surface where they are examined. If the core head is used, whole cores of rock are lifted up to the surface. Whichever method is used, scientists aim to discover the order of the rocks in the borehole and whether

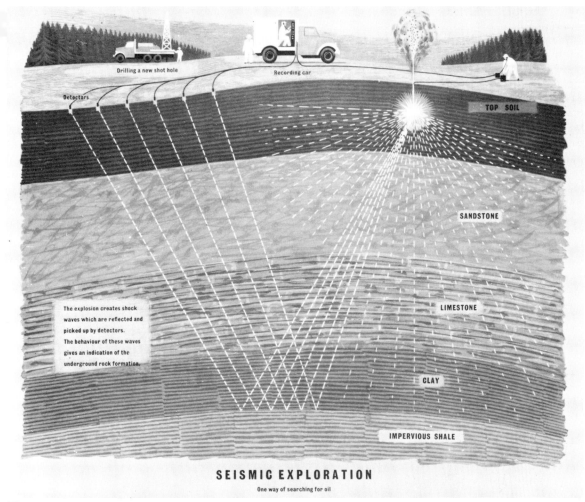

Figure 68.7

Figure 68.8

285

any of them contain fossil fuels or other useful materials.

Boreholes for oil and gas may go down to a depth of several thousand metres. In offshore drilling, the main problem is to keep the drilling rig completely steady, despite the movements of the wind and the waves.

Most boreholes produce no oil or gas at all. But if a strike of oil or gas is made, the next step is to drill several more boreholes around it to see how large the field is. It is also important to find out how easy it will be to get the oil or gas out. Finally, a decision is made about whether or not to go into full production.

Looking for fossil fuels of all three kinds is now very big business. The sums of money that need to be spent on carrying out seismic surveys and on drilling boreholes are large. But the rewards for success are much larger still.

Questions

1. Crude oil is much more likely to be found in an easily crumbled sandstone than in a tougher, well-cemented sandstone. It is never found in granite. Explain these facts using the idea of rock permeability.

2. 'Fuels such as oil, natural gas and our food have stored-up energy from the sun which is released when we burn the fuels in air or oxidize the food in our bodies.'

Write a short essay explaining this statement.

3. (*a*) If your country has reserves of coal, crude oil or natural gas, use reference books to complete a table with these headings.

Kind of fossil fuel	Places where fossil fuel comes from	Places where prospecting for fossil fuel is going on

(*b*) If your country imports fossil fuels, find out which countries are the main suppliers.

69 RATES OF CHEMICAL REACTIONS

Different reactions—different speeds

When you add water to a mixture of sodium hydrogencarbonate and citric acid (page 276), the mixture froths as the carbon dioxide is given off. Though there is a drop in temperature (the reaction is endothermic), this is still a fairly vigorous chemical reaction because it goes quite quickly.

The speed (rate) of a chemical reaction is a different thing altogether from the energy given out or taken in by the reaction mixture. This is true even though there are many chemical reactions which have a high rate and, at the same time, give out a lot of heat energy to the surroundings: the chemical reaction of zinc and sulphur is an example (see Figure 66.2, page 276).

Different chemical reactions go at different rates. The rusting of iron or steel is a fairly slow chemical reaction (though too fast for many car owners!) and the breaking up of the stonework in Figure 64.3 (page 268) by acidic gases in the air only happens at the rate of a few millimetres every hundred years. On the other hand, many chemical reactions you will have carried out in the laboratory are very fast indeed.

The fastest chemical reactions seem to be instantaneous because they happen as soon as the reactants are mixed. Many chemical reactions in solution are very fast, particularly precipitation reactions in which cations combine with anions to form an insoluble salt (page 243).

The precipitation reaction shown in Figure 69.1 does not only involve the joining together of cations and anions, and it is slow enough to be studied in the laboratory.

Changing the rate of a chemical reaction

Magnesium reacts with dilute hydrochloric acid to give off hydrogen.

magnesium + hydrochloric ⟶ magnesium + hydrogen
                acid   chloride

$Mg(s) + 2HCl(aq) \longrightarrow MgCl_2(aq) + H_2(g)$

A rough idea of the rate of this chemical reaction can be obtained by timing how long it takes for a strip of magnesium ribbon to 'disappear' in the acid. A long time means a low rate and a short time means a high rate of reaction.

In the next experiment, you can find out how the chemical reaction between magnesium and dilute hydrochloric acid can be speeded up or slowed down by changing the conditions.

Figure 69.1 The reaction between sodium thiosulphate and an acid takes about a minute. The cross below the beaker fades as more and more of the sulphur precipitate is formed.

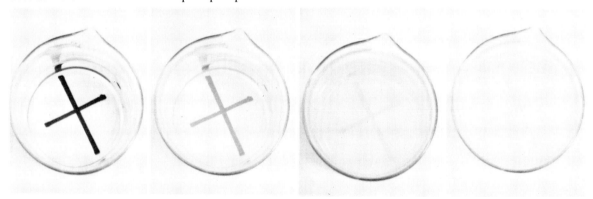

Table 69.1

Part	Temperature	Amount of magnesium	Volume of dilute hydrochloric acid* in cm³	Volume of water in cm³
1	Room temperature	30 mm strips of clean ribbon	(a) 50 (b) 40 (c) 30 (d) 20	0 10 20 30
2	(a) 30 °C (b) 40 °C (c) 50 °C (d) 50 °C	30 mm strips of clean ribbon	30	20
3	Room temperature	(a) 0·03 g of powder (b) 0·03 g of turnings (c) 30 mm of clean ribbon (d) 9 mm of clean wire	50	0

* The acid should be of normal 'bench' concentration (i.e. 2 M).

Experiment 1 Changing the rate of the chemical reaction between magnesium and dilute hydrochloric acid

For each separate experiment shown in Table 69.1, add the magnesium to the acid and use a stopclock to find the time taken for the metal to dissolve completely.

In part 1, the acid and the water should be thoroughly mixed before the magnesium is added. In part 2, the acid can be heated to a temperature of 2 or 3 °C above the value you want so as to allow for some cooling during the chemical reaction. Do not heat the acid after the magnesium has been added. In part 3, the different samples of magnesium all have about the same mass.

- What is being varied in each part of the experiment?
- What effect does each change have on the rate?

The rate of a chemical reaction can be increased by:

(i) raising the temperature
(ii) using any solid in as fine-grained a form as possible
(iii) increasing the concentration of any solution.

It can also be increased in some cases by another method (see below).

Catalysts

Adding certain substances (which are not reactants) to some reaction mixtures can increase the rate a lot. These substances are called *catalysts*. Catalysts are widely used in industry to speed up chemical reactions which otherwise would be uneconomically slow. Some examples of industrial catalysts are mentioned in this book (pages 267 and 273).

A catalysed reaction is often used to prepare oxygen in the laboratory. The decomposition of hydrogen peroxide solution gives oxygen but the reaction is slow at room temperature unless a catalyst (manganese(IV) oxide) is used.

$$\text{hydrogen peroxide} \xrightarrow{\text{MnO}_2 \text{ catalyst}} \text{water} + \text{oxygen}$$
$$2H_2O_2 \text{(aq)} \longrightarrow 2H_2O(l) + O_2(g)$$

A rough idea of the rate of this chemical reaction is given by the amount of 'bubbling' in the solution that is caused by the oxygen being given off.

Experiment 2 Changing the rate of a catalysed reaction

For this experiment, prepare a dilute solution of hydrogen peroxide by adding about 5 cm³ of '20 volume' hydrogen peroxide solution to about 45 cm³ of water.

a Add about half a spatula measure of granular manganese (IV) oxide to a few cm³ of the hydrogen peroxide solution in a boiling tube. Then add about the same amount of powdered manganese (IV) oxide to the same volume of the hydrogen peroxide solution in another boiling tube.

• Which catalyst gives the higher rate of reaction?

b Repeat **a**, this time using just enough powdered manganese (IV) oxide to cover the tip of a spatula in the first boiling tube, and about half a spatula measure of the same catalyst in the second boiling tube.

• Which amount of catalyst gives the higher rate of reaction?

The rate of this chemical reaction can be increased by:

(i) using the catalyst in as fine-grained a form as possible;
(ii) increasing the mass of the catalyst.

Even so, unlike a true reactant, a catalyst is not used up during the chemical reaction: it always has the same mass at the end of the chemical reaction as it had at the start.

A catalyst takes an active part in a chemical reaction, but is reformed at the end so that its mass stays the same. It reacts with one or more of the reactants to form a new compound which then rapidly reacts to form the products along with the original catalyst (Figure 69.2).

Questions

1. Make a list of chemical reactions that occur naturally in everyday life. Classify these as fast reactions (take place over a few minutes), slow reactions (take place over a few days, or even longer), and in-between reactions (take place over a few hours).

2. Calcium carbonate reacts with dilute hydrochloric acid to give off carbon dioxide. Which of the following changes in the conditions would speed up this chemical reaction?

(a) increasing the temperature of the reaction mixture
(b) using the acid after it has been further diluted with water
(c) using powdered calcium carbonate instead of limestone chippings (marble chips).

3. (a) Describe an experiment you could carry out to decide how effective each of the following solids is in catalysing the decomposition of hydrogen peroxide solution: copper(II) oxide (CuO), lead (IV) oxide (PbO_2) and manganese (IV) oxide (MnO_2).

(b) The following results are obtained in a series of experiments using equal volumes of hydrogen peroxide solution of the same concentration.

Solid	Mass of solid added in grams	Volume of oxygen given off in the first minute of the reaction in cm³
Powdered MnO_2	0.1	24
Powdered MnO_2	0.2	56
Granular MnO_2	0.1	3
Powdered PbO_2	0.1	90
Powdered CuO	0.1	0

(i) Which of the solids in the above table does not act as a catalyst in the decomposition of hydrogen peroxide solution?
(ii) Explain the differences between the results for the three kinds of manganese (IV) oxide?
(iii) Which of the solids is the most effective catalyst for this reaction?

4. Find an example of:

(a) a solid catalyst used to speed up a chemical reaction between two gases,
(b) a solid catalyst used to speed up the preparation of a common gas from a solution,
(c) a very fast reaction between two solutions that is finished as soon as the solutions are mixed.

Figure 69.2

THE UNCATALYSED REACTION

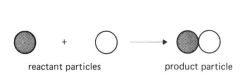

reactant particles → product particle

A POSSIBLE ROUTE FOR THE CATALYSED REACTION

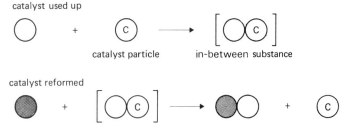

catalyst used up
catalyst particle — in-between substance
catalyst reformed

The Biology of Man

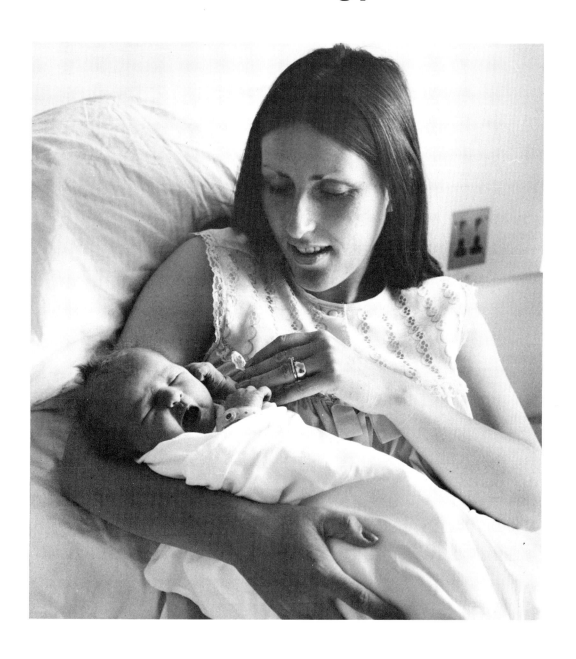

70 FOOD AND DIET

THE NEED FOR FOOD

All living organisms need food. An important difference between plants and animals is that plants can make the food in their leaves (page 86) while animals have to take it in 'ready made' by eating plants or the bodies of other animals. In both plants and animals food is used as follows.

(*a*) *For growth.* It provides the substances needed for making new cells and tissues.

(*b*) *For energy.* Food is the source of energy for the chemical reactions which take place in living organisms. When food is broken down during *respiration* (see page 80), energy is made available for driving other reactions. Movement, heart beat and nerve impulses are examples of animal activities which need energy.

(*c*) *Replacement of worn and damaged tissues.* The substances provided by food are needed to replace, for example, the millions of red blood cells that break down each day; to replace the skin which is worn away and to repair wounds.

CLASSES OF FOOD

The three classes of food are carbohydrates, proteins and fats.

Carbohydrates

Sugar and starch are important carbohydrates in our diet. Starch is abundant in potatoes, maize, wheat, rice and other cereals. Sugar appears in our diet mainly as sucrose (table sugar) which is added to drinks and many prepared foods such as biscuits and cakes.

Although all foods provide us with energy, carbohydrates are the cheapest and most readily available energy source. They contain the elements carbon, hydrogen and oxygen (e.g. glucose $C_6H_{12}O_6$).

When oxidized to provide energy in respiration, they are broken down to carbon dioxide and water, (see page 80). If we eat more carbohydrates than we need for our immediate energy requirements, the excess is converted in the liver to either *glycogen* (see page 298) or fat. The glycogen is stored in the liver and muscles; the fat is stored in fat depots round the kidneys or under the skin.

The *cellulose* in the cell walls of all plant tissues is an abundant carbohydrate. Humans do not have enzymes for digesting cellulose but it is important as 'fibre' or 'roughage' in our diets.

Proteins

Lean meat, fish, eggs, milk and cheese are important sources of animal protein. All plants contain some protein but nuts, beans and cereals are the best sources.

Proteins, when digested, provide the chemical substances needed to build cells and tissues, e.g. skin, muscles, blood and bones. Neither carbohydrates nor fats can do this so it is essential to include some protein in the diet.

When proteins are digested, they are broken down to substances called *amino acids*. The amino

A–B–C–D–E–F–G–H–I–J–K

(a) representation of a small protein molecule. The letters are the amino acids

(b) the protein is digested and the amino acids are set free

B–J–K–D–A–F
|
E–C–G–H–I

(c) the same amino acids are built up into a different protein

Figure 70.1

acids are absorbed into the blood stream and used to build up a variety of different proteins which contribute to the cytoplasm of cells and, eventually, all tissues. Figure 70.1 illustrates this process.

The amino acids which are not used for making new tissues cannot be stored, but the body converts them to glycogen which can then be stored or oxidized to provide energy.

Chemically, proteins differ from both carbohydrates and fats because they contain nitrogen and sulphur as well as carbon, hydrogen and oxygen.

Fats

Animal fats are found in meat, milk, cheese, butter and egg yolk. Plant fats occur as oils in fruits (palm oil) and seeds (sunflower seeds) and are used for cooking and making margarine.

Fats are used in the body to form parts of the cell structure such as the cell membrane. When oxidized in respiration, they give twice as much energy as carbohydrates and they provide long-term storage in fat depots as mentioned above.

DIET

A balanced diet must provide the requirements listed below.

(*a*) *Sufficient energy*. The quantity of food taken in each day must be sufficient to provide the energy needed to stay alive and be active. Even during sleep the body needs energy to keep the blood circulating and the lungs filling and emptying. The energy needed during the waking hours depends on the age and activity of the person. A manual labourer needs more energy than a bus driver, for example. If the diet contains insufficient energy, reserves of glycogen and fat are used. When these run out, the person cannot keep up his rate of activity and finally, to stay alive, he has to oxidize the protein of his own tissues to provide the necessary energy.

(*b*) *The correct proportions of proteins, fats and carbohydrates*. The diet must include some protein for making and replacing the structures of the body. Pregnant women and growing children have particular need of protein because of the new tissues they are producing. A shortage of protein in young children not only reduces growth, but leads to protein deficiency disease and permanent mental retardation due to the failure to make sufficient brain cells.

Insufficient carbohydrates and fats in the diet will result in a shortage of energy, while an excess of these substances may lead to obesity when the surplus is stored in the fat depots.

(*c*) *Mineral salts*. The red cells of our blood need *iron* for their haemoglobin (page 301); our bones (page 321) need *calcium phosphate*; the thyroid gland requires *iodine* and all cells need *sodium* and *potassium*. These mineral elements are normally obtained from the diet, though they can be supplied artificially if necessary. Lack of iron leads to anaemia; shortage of iodine causes goitre, a swelling of the thyroid gland in the neck; inadequate calcium and phosphorus leads to weak bones and teeth.

Most mixed diets will contain adequate quantities of these mineral elements. Red meat, for example, contains iron; milk contains calcium and phosphorus.

(*d*) *Vitamins*. These are chemical substances which play a part in essential chemical reactions in the body but cannot be used for energy or built into cell structures. They often function in association with enzymes (page 79), but whereas we can build up our enzymes from amino acids, we cannot build up vitamins and have to obtain them ready made in our diet. A mixed diet will usually contain all the vitamins we need, but a diet consisting mainly of one food, such as rice, may be lacking in vitamins. A lack of any one vitamin leads to an illness which can be remedied only by provision of the missing vitamin.

Table 70.1 gives the sources and importance of three vitamins (fifteen or more are known).

(*e*) *Water*. About 70 per cent of most tissue consists of water; it is an essential constituent of cytoplasm. Substances are carried round the body as a watery solution in the blood; the process of digestion takes place in water. Since we lose water by evaporation, sweating, urinating and breathing, it is necessary to make good this loss by taking in water with the diet.

(*f*) *Fibre* (roughage). The cellulose in plant cell walls is not digested by our own enzymes, but there are bacteria in the intestine which digest part of it and produce some useful substances. This cellulose, or *dietary fibre*, as it is called, retains water and provides a solid mass for the muscles of the intestine to work against as they move the food along. Plenty of dietary fibre helps to prevent constipation and other, more serious, intestinal disorders.

Table 70.1

Name and source of vitamin	Diseases and symptoms caused by lack of vitamin	Notes
Vitamin A Milk, butter, cheese, liver, cod-liver oil, fresh green vegetables	Reduced resistance to diseases, especially those affecting the skin. Poor night vision	Green vegetables contain a yellow substance called *carotene* which is turned into vitamin A by the body
Vitamin C Oranges, lemons, black currants, tomatoes, potatoes, green vegetables	Bleeding under the skin at the joints. Swollen, bleeding gums; poor wound healing. All symptoms of scurvy	Scurvy is only likely to occur when fresh food is not available. Milk is deficient in vitamin C so babies need additional sources
Vitamin D Fish-liver oil, butter, milk, cheese, egg yolk, liver	Children's bones do not harden properly, leading to deformities	Fats in the skin are changed to vitamin D by sunlight

Food tests

Test for starch: Shake a little starch powder in a test-tube with some cold water and then boil to make a clear solution. When the solution is cold, add 3 or 4 drops of iodine solution*. The dark blue colour that results is characteristic of the reaction between starch and iodine.

Test for glucose: Set up a water-bath as shown in Figure 24.2 (page 88). Add a little glucose to some Benedict's solution* in a test-tube and place the test-tube in the boiling water. The solution changes from blue to opaque green, yellow and finally to a red precipitate of copper(I) oxide.

Sucrose is recognized by its failure to react with Benedict's solution until after it has been boiled with dilute hydrochloric acid and neutralized with sodium hydrogencarbonate. For this reason, sucrose is called a non-reducing sugar.

Test for protein (Biuret test): Add $5\,cm^3$ dilute sodium hydroxide (CARE: this solution is caustic) and $5\,cm^3$ of a 1 per cent copper sulphate solution to a 1 per cent solution of albumen. A purple colour indicates protein.

Test for fat: Shake two drops of cooking oil thoroughly with about $5\,cm^3$ ethanol in a dry test-tube until the fat dissolves. Then pour the alcoholic solution into a test-tube containing a few cm^3 of water. A cloudy white emulsion is formed.

* Instructions for preparing these reagents are given on page 84.

Application of the food tests

If possible, you should apply these tests to samples of food such as milk, raisins, potato, onion, beans, egg yolk, ground almonds and find what food materials are present in each. Crush the solid samples in a mortar and shake with warm water to extract the soluble products. Use the tests on both the solution and suspension of crushed food. To test for fats, the food must first be crushed with ethanol (not water) and filtered.

Questions

1. Write a list of the requirements for a fully balanced diet.

2. What sources of protein are available to a vegetarian who (*a*) will eat animal products but not meat itself, (*b*) will eat only plants and their products?

3. In theory, at least, could you survive on a diet which consisted only of meat? Explain your answer. What might be the disadvantages of this diet?

4. Why must all diets contain some protein?

5. In what sense can the fats in your diet be said to contribute to 'keeping you warm'?

6. How do vitamins differ from proteins in the use the body makes of them?

7. How does the 'protein' in Figure 70.1c (page 292) differ from the 'protein' in Figure 70.1a?

8. Discuss whether a man engaged in heavy physical work needs (*a*) more carbohydrate, (*b*) more protein, than a person in an office job.

71 DIGESTION, ABSORPTION AND USE OF FOOD

To be of use to the body, the food we take in through the mouth must enter the blood stream and be distributed to all the living regions.

Digestion is the process by which insoluble food, consisting of large molecules, is broken down into soluble compounds having smaller molecules. These smaller molecules, in solution, pass through the walls of the intestine and enter the blood stream. Digestion and absorption take place in the *alimentary canal* (Figure 71.1). Digestion is brought about by means of *enzymes* (see page 79). The digestive enzymes speed up the rate at which the food is digested. In a few hours, enzymes can digest proteins. Without enzymes, this process could take years.

The *alimentary canal* is a muscular tube, with an internal glandular lining, running from mouth to anus. Some regions have particular functions and, accordingly, different structures. Juices are poured into the alimentary canal from glands in its lining or through ducts from digestive glands outside it. As the food passes through the alimentary canal it is broken down in stages until all the digestible material has been broken down and absorbed. The indigestible residue is expelled through the anus.

MOVEMENT OF FOOD THROUGH THE ALIMENTARY CANAL

Ingestion. This is the act of taking food into the alimentary canal through the mouth.

Swallowing (see Figures 71.2 and 3). In swallowing, the following actions take place: (*i*) the tongue

Figure 71.1

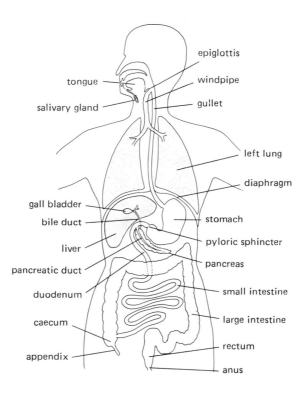

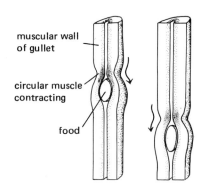

Figure 71.2

presses upwards and back against the roof of the mouth, forcing the food to the back of the mouth; (*ii*) the *soft palate* closes the nasal cavity at the back; (*iii*) the *laryngeal cartilage* round the top of the windpipe is pulled upwards by muscles so that the opening of the windpipe lies under the back of the tongue; (*iv*) this opening is reduced by the contraction of a ring of muscle; and (*v*) the *epiglottis*, a flap of cartilage, directs food over the opening of the windpipe. In this way, food is able to pass over the windpipe without entering it.

The food is then forced down the gullet by peristalsis.

Peristalsis (Figure 71.2). The walls of the alimentary canal contain muscle fibres. By contracting and relaxing alternately the muscles urge the food, in a wave-like motion, through all regions of the alimentary canal.

Egestion. This is the expulsion from the alimentary canal of the undigested remains of the food.

DIGESTION

Mouth

In the mouth, the food is chewed and mixed with saliva. *Saliva* is a digestive juice produced by three pairs of glands whose ducts lead into the mouth (Figure 71.3). It helps to lubricate the food and make the particles stick together. Saliva contains one enzyme, *salivary amylase*, which acts on cooked starch and begins to break it down into a soluble sugar.

Stomach

The stomach has flexible walls and extends as the food accumulates. The *pyloric sphincter* at the end of the stomach stops solid particles of food from passing through. The food from a meal is thus stored for some time and released slowly to the rest of the alimentary canal.

Glands in the lining of the stomach produce *gastric juice* containing the enzyme *pepsin*. Pepsin acts on proteins and breaks them down into soluble compounds called *peptides*. The stomach wall also produces *hydrochloric acid* which makes a weak solution in the gastric juices. The acid provides the optimal degree of acidity for pepsin to work in and probably kills many of the bacteria taken in with the food.

Figure 71.3

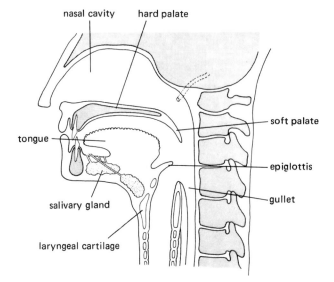

The rhythmic, peristaltic movements of the stomach, about once every 20 seconds, mix up the food and gastric juice to a creamy liquid. How long food remains in the stomach depends on its chemical nature. Water may pass through in a few minutes, a meal of carbohydrate such as porridge may be retained less than an hour but a mixed meal containing protein and fat may be in the stomach for 1 or 2 hours. Each wave of peristalsis down the stomach, forces a little of the partly digested liquid through the narrow pyloric sphincter into the first part of the small intestine called the *duodenum*.

Small intestine

An alkaline juice from the pancreas, and bile from the liver, are poured into the duodenum. The *pancreas* is a cream-coloured gland lying below the stomach (Figure 71.1). It makes several enzymes which act on all classes of food: one breaks down proteins and peptides to amino acids; another attacks starch and converts it to sugar; and a third digests fats to fatty acids and glycerol. Pancreatic juice also contains sodium hydrogencarbonate which partly neutralizes the acid fluid from the stomach, and so creates a suitable pH (see page 53) for the enzymes of the pancreas and intestine to work in.

Bile is a green, watery fluid made in the liver, stored in the gall bladder and conducted to the duodenum by the bile duct. It contains no enzymes but acts like a detergent on fats, breaking them into a suspension of tiny droplets which allows more rapid digestion.

All the digestible material is now reduced to soluble compounds which can pass through the lining of the intestine and into the blood stream. The final products of digestion are:

starch ⟶ glucose
proteins ⟶ amino acids
fats ⟶ fatty acids and glycerol

Caecum and appendix

In man, these are small structures which do not seem to do any useful job. In herbivores like the rabbit and the horse, however, they are much larger and it is here that most of the cellulose digestion takes place, largely as a result of bacterial activity.

ABSORPTION

Small intestine

Nearly all the absorption of digested food takes place in the small intestine which is efficient at this because:

a It is fairly long and presents a large absorbing surface to the digested food.
b Its internal surface is greatly increased by thousands of tiny, finger-like projections called *villi* (singular = villus) (Figure 71.4).
c The lining epithelium is very thin and the fluids can pass through it quickly.
d There is a dense network of blood capillaries in each villus (Figure 71.5).

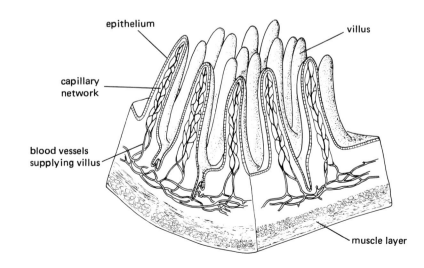

Figure 71.4

Figure 71.5

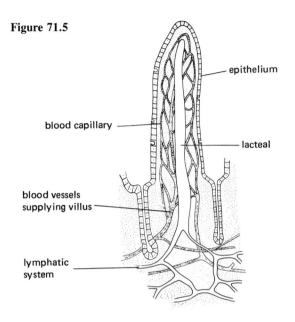

The small molecules of glucose and amino acids pass through the epithelium and the capillary walls and enter the blood stream. They are then carried away in the capillaries which join up to form veins; eventually these unite to form one large vein, the *hepatic portal vein*. This vein carries all the blood from the intestine to the liver, which may store or alter any of the digestion products. When the digested food is released from the liver it enters the general circulation.

Some of the fatty acids and glycerol from the digestion of fats, enter the blood capillaries of the villi but a large proportion may be recombined in the intestinal lining to form fats once again and then these fats pass into the *lacteals* (Figure 71.5). The fluid in the lacteals enters the *lymphatic system* which forms a network all over the body and eventually empties its contents into the blood stream (page 304).

The large intestine

The material passing into the large intestine consists of water with undigested matter, largely cellulose and vegetable fibres (*roughage*), bacteria, mucus and dead cells from the lining of the alimentary canal. The large intestine absorbs much of the water from the undigested remains. This semisolid waste, the faeces, is passed into the rectum by peristalsis and is expelled at intervals through the anus.

USE OF DIGESTED FOOD

The products of digestion are carried round the body in the blood. From the blood, cells absorb and use glucose, fats and amino acids.

(*a*) *Glucose*. During respiration in the cells, glucose is oxidized to carbon dioxide and water (see page 80). This reaction provides energy to drive the many chemical processes in the cells which result in, for example, the building up of proteins, contraction of muscles or electrical changes in nerves.

(*b*) *Fats*. These are built into cell membranes and other cell structures. Any fats not used for growth or maintenance in this way are oxidized to carbon dioxide and water, providing energy for the vital processes of the cell.

(*c*) *Amino acids*. These are absorbed by the cells and built up into proteins. These proteins may form structures such as the cell membrane or they may become enzymes which control and co-ordinate the chemical activity within the cell.

STORAGE OF DIGESTED FOOD

If the quantity of food taken in exceeds the energy requirements of the body or the demand for structural materials, it is stored in one of the following ways.

(*a*) *Glucose*. The sugar not required immediately for the energy supply in the cells is converted in the liver and in the muscles to *glycogen*. The glycogen molecule is built up by combining many glucose molecules into a long chain similar to the starch molecule. Some of this insoluble glycogen is stored in the liver and the rest in the muscles. When the blood sugar falls below a certain level, the liver converts its glycogen back to glucose and releases it into the circulation. The muscle glycogen is not returned to the circulation but is used by muscle as a source of energy during muscular activity.

The glycogen in the liver is a 'short-term' store, sufficient for about only 6 hours. Excess glucose not stored as glycogen is converted to fat and stored in the fat depots.

(*b*) *Fats*. These are stored in special cells which form fatty tissue under the skin or in the abdomen. Unlike glycogen, there is no limit to the amount of fat stored and because of its high energy value, it is an important long-term reserve of energy-giving food.

(c) *Amino acids.* Amino acids are not stored in the body. Those not used in protein formation are deaminated (see below). The protein of the liver and tissues can act as a kind of protein store to maintain the protein level in the blood but absence of protein in the diet soon leads to serious disorders.

Body weight

The rate at which glucose is oxidized or changed into glycogen and fat is controlled by hormones. When intake of carbohydrate and fat exceeds the energy requirements of the body, the surplus will be stored mainly as fat. Some people never seem to get fat no matter how much they eat, while others start to lay down fat when their intake only just exceeds their needs. Putting on weight is certainly the result of eating more food than the body needs but it is not clear why people should differ so much in this respect. The explanation probably lies in the balance of hormones which, to some extent, is determined by heredity. A slimming diet designed to reduce food intake must, nevertheless, always include the essential proteins, mineral salts and vitamins.

THE LIVER

The liver has been mentioned several times in connection with the digestion, use and storage of food. This is only one aspect of its many important functions, some of which are summarized below. It is a large, reddish-brown organ which lies just beneath the diaphragm and partly overlaps the stomach. All the blood from the alimentary canal passes through the liver. In the liver the composition of the blood is adjusted before it is released into the general circulation.

(a) *Regulation of blood sugar.* After a meal, the liver removes excess glucose from the blood and stores it as glycogen. In the periods between meals, when the glucose concentration in the blood starts to fall, the liver converts some of its stored glycogen into glucose and releases it into the bloodstream. In this way, the concentration of sugar in the blood is kept at a fairly steady level.

(b) *Production of bile.* Cells in the liver make bile all the time. This is stored in the gall bladder until needed to assist digestion of fats in the small intestine. The bile breaks up the liquid fats into small droplets (an *emulsion*) which makes them easier to digest and absorb.

(c) *Deamination.* The amino acids not needed for making proteins are converted into glycogen in the liver. During this process, the nitrogen-containing amino part ($-NH_2$) of the amino acid is removed and changed to *urea* which is later excreted by the kidneys.

(d) *Storage of iron.* Millions of red blood cells break down every day. The iron from their haemoglobin (see page 301) is stored in the liver.

(e) *Detoxication.* Most poisonous compounds which are absorbed in the intestine are made harmless when the blood passes through the liver on its way to the general circulation.

EXPERIMENTS WITH DIGESTIVE ENZYMES

Experiment 1 The action of saliva on starch

The mouth must be rinsed with water to remove traces of food. Collect some saliva in two test-tubes, labelled A and B, to a depth of about 15 mm. Heat the saliva in tube B over a small Bunsen flame until it boils for about 30 seconds. Then cool the tube under the tap. To each tube add about 2 cm³ of a 2 per cent starch solution, shake and leave for 5 minutes.

Divide the contents of tube A between two test-tubes. One test-tube should have iodine solution added to it and the other should be boiled with Benedict's solution as described on page 294. Test the contents of tube B in the same way.

Results. The contents of tube A fail to give a blue colour with iodine, showing that the starch has gone. The other half of the contents, however, gives a red precipitate with Benedict's solution showing that sugar is present.

The contents of tube B still give a blue colour with iodine but do not form a red precipitate on boiling with Benedict's solution.

Interpretation. The results with tube A suggest that something in saliva has converted starch into sugar. The fact that the boiled saliva in tube B fails to do this, confirms that it was an enzyme in saliva which brought about the change (see page 80), because enzymes, being proteins, are destroyed by boiling.

Experiment 2 The action of pepsin on egg-white protein

A cloudy suspension of egg-white can be made by stirring the white of one egg into 500 cm³ tap-water, heating it to boiling point and filtering it

Figure 71.6

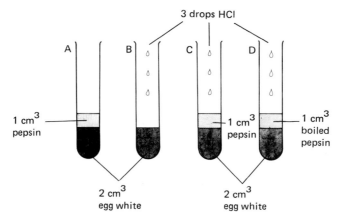

through glass wool to remove the larger particles.

Label four test-tubes A to D and place 2 cm³ egg-white suspension in each of them. Then add pepsin solution and/or dilute hydrochloric acid to the tubes as follows:

A Egg-white suspension + 1 cm³ pepsin solution (1%)
B Egg-white suspension + 3 drops dilute hydrochloric acid
C Egg-white suspension + 1 cm³ pepsin + 3 drops hydrochloric acid
D Egg-white suspension + boiled pepsin + 3 drops hydrochloric acid

Place all four tubes in a beaker of warm water at 35 °C for 10–15 minutes. Notice that the contents of tube C go clear. The rest remain cloudy.

Interpretation. The change from a cloudy suspension to a clear solution shows that the solid particles of egg protein have been digested to soluble products. The failure of the other three tubes to give a clear solution shows that:

a pepsin will work only in acid solution
b it is the pepsin and not the hydrochloric acid which does the digestion
c pepsin is an enzyme, because its activity is destroyed by boiling.

Questions

1. Why is it necessary for an animal's food to be digested? Why do plants not need a digestive system? (see page 86).

2. Write down the menu for your breakfast and lunch (or supper). Indicate the principal food substances present in each component of the meal. State the final digestion product of each and the use your body is likely to make of it.

3. Explain what happens to a protein from the time it is swallowed, to the time its products are built up into the cytoplasm of a muscle cell.

4. Study the classes and characteristics of enzymes on pages 79–80. Suggest a more logical name for *pepsin*. In what ways does *pepsin* exhibit the characteristics of an enzyme?

5. List the ways in which the body can store an excess of carbohydrates taken in with the diet.

6. If you were deprived of food for several days, how would your body meet the demands for energy by, for example, your heart?

7. Explain how the liver exercises control over the substances coming from the intestine and entering the general circulation.

72 THE BLOOD CIRCULATORY SYSTEM

COMPOSITION

Blood is a suspension of red and white cells in a liquid called plasma. There are 5–6 litres of blood in the body of an adult.

Red cells

These are tiny, disc-like cells (Figures 72.1a and 72.2) without nuclei. They consist of spongy cytoplasm enclosed in an elastic cell membrane. In their cytoplasm is the red pigment, *haemoglobin,* which is a protein combined with iron. Haemoglobin readily combines with oxygen when the oxygen concentration is high. The reaction between haemoglobin and oxygen forms an unstable compound, *oxyhaemoglobin.* In conditions of low oxygen concentration, oxyhaemoglobin breaks down and releases its oxygen. This property makes haemoglobin very efficient in transporting oxygen from the lungs to the tissues. Each red cell lasts for about 4 months, after which it breaks down and is disintegrated in the liver. About 200,000 million are formed and destroyed each day.

White cells

There are several different kinds of white cell (Figures 72.1b and 72.2), all of which have a nucleus. Some of them move about by a flowing action of their cytoplasm and can escape from blood capillaries into the tissues by squeezing between the cells of the capillary walls. This type accumulates at the site of an infection (e.g. round a cut), engulfing and digesting harmful bacteria (Figure 72.1c). In this way they prevent the spread of infection through the body. There is one white cell to every six hundred red cells.

Platelets

These are tiny cell fragments without nuclei. They help to clot the blood at the sites of wounds. Red cells, white cells and platelets are all made by the red bone marrow of the short bones such as the ribs, vertebrae and breast bone.

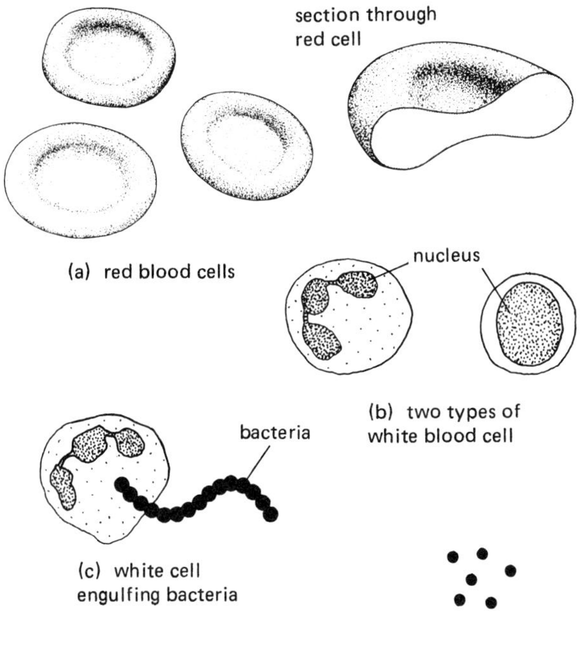

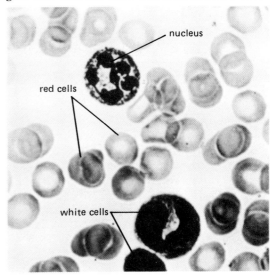

Figure 72.2 Red and white cells from human blood.

Figure 72.1

(a) red blood cells
(b) two types of white blood cell
(c) white cell engulfing bacteria
(d) blood platelets

Plasma

The liquid part of the blood is called plasma. It consists of water with a large number of substances dissolved in it, for example, the products of digestion such as amino acids, glucose and salts. There are also nitrogen-containing waste products such as urea (page 299), hormones like adrenalin and the plasma proteins. The plasma proteins have important functions in the blood but many of the other substances are simply being carried from one place to another, e.g. from the intestine to the muscles.

THE HEART

The heart pumps blood through the circulatory system all round the body. Figure 72.3 shows its appearance from the outside while Figure 72.4 is a diagrammatic, vertical section to show its internal structure. Since the heart is seen as if in a dissection of a person facing you, the left side is drawn on the right.

Study of Figure 72.4 shows that there are four chambers. The upper, thin-walled chambers are the *atria* (singular = atrium) and each of these opens into a thick-walled *ventricle* below.

Blood enters the atria from veins. The *pulmonary vein* brings oxygenated blood from the lungs into the left atrium. The *vena cava* brings deoxygenated blood from the body tissues into the right atrium. The blood passes from each atrium to its corresponding ventricle and the ventricle pumps it out into arteries. The artery carrying oxygenated blood to the body from the left ventricle is the *aorta*. The *pulmonary artery* carries deoxygenated blood from the right ventricle to the lungs.

In pumping the blood, the muscle in the walls of the atria and ventricles contracts and relaxes (Figure 72.5). The atrial walls contract first and force blood into the two ventricles. Then the ventricles contract and expel blood into the arteries. The blood is prevented from flowing backwards by two sets of valves. When the ventricles contract, blood

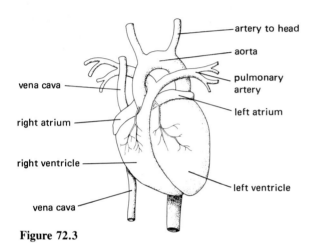

Figure 72.3

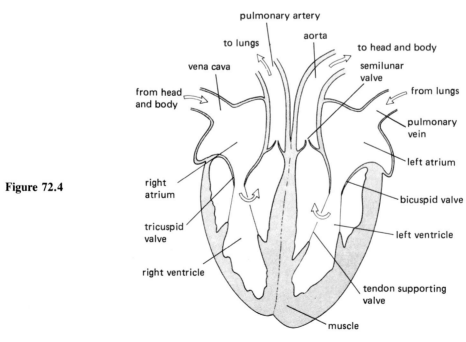

Figure 72.4

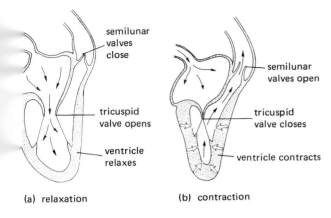

Figure 72.5

pressure closes the *bicuspid* and *tricuspid valves* and these prevent blood returning to the atria. When the ventricles relax, the blood pressure in the arteries closes the *semi-lunar valves* so preventing the return of blood to the ventricles. The way the valves work is shown in Figure 72.6. The valves between the atria and ventricles have string-like tendons to stop them being turned inside out.

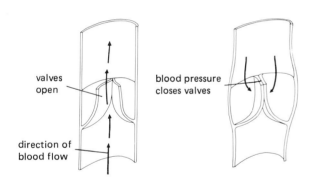

Figure 72.6

The heart contracts and relaxes about sixty to eighty times a minute. During exercise, this rate increases to over a hundred, so increasing the supply of oxygen and food to the tissues.

THE CIRCULATION

Pumped by the heart, the blood travels all round the body in blood vessels. It leaves the heart in arteries and returns in veins. Figure 72.7 shows the route of the circulation in diagrammatic form. The blood passes twice through the heart during one complete circuit; once on its way to the body and again on its way to the lungs. On average a red cell would complete the circulation of the body in 45 seconds.

Arteries

These are fairly wide vessels (Figure 72.8*a*) which carry blood from the heart to the limbs and organs of the body. They are thick-walled, muscular and elastic, and must stand up to the surges of high pressure caused by the heart beat. The arteries divide into smaller vessels called arterioles which themselves divide repeatedly until they form a dense network of microscopic vessels penetrating between the cells of every living tissue. These final branches are called capillaries.

Figure 72.7

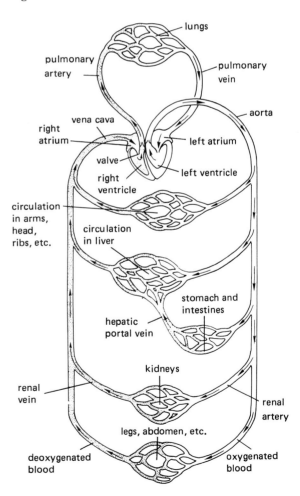

Capillaries

These are tiny vessels with walls often only one cell thick (Figures 72.8c and 72.9). Although the blood is confined in the capillary, the capillary walls are permeable, allowing water and dissolved substances to pass in and out. The blood cells and the blood proteins cannot pass through the capillary walls. The liquid which does escape is called *tissue fluid* and it supplies the cells of the body with the dissolved food and salts which they need. Oxygen diffuses out of the capillaries into the body cells and the carbon dioxide produced by these cells diffuses back into the capillaries. Some of the tissue fluid is absorbed back into the capillaries and some of it enters the lymphatic system (below) (Figure 72.10).

The capillary network is so dense that no living cell is far from a supply of oxygen and food.

Eventually, the capillaries unite into larger vessels *venules*, which join to form veins and these return blood to the heart.

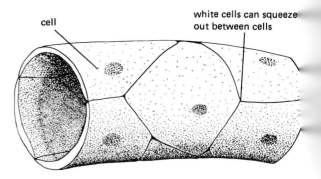

Figure 72.10

Veins

Veins return blood from the tissues to the heart. The blood pressure in them is steady and is less than in the arteries. They are wider and have thinner walls than the arteries (Figure 72.8b). They also have valves in them which prevent blood flowing away from the heart (Figure 72.6).

The blood in most veins will contain less oxygen and food, and more nitrogenous waste and carbon dioxide, while most arterial blood has a higher concentration of oxygen and dissolved food.

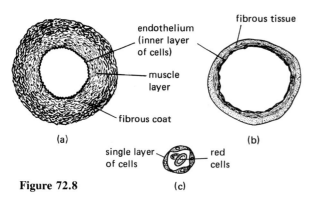

Figure 72.8

Lymphatics

Not all the tissue fluid returns to the capillaries. Some of it enters the blindly-ending, thin-walled vessels called lymphatics (Figure 72.9). These vessels join up with those from the lacteals of the

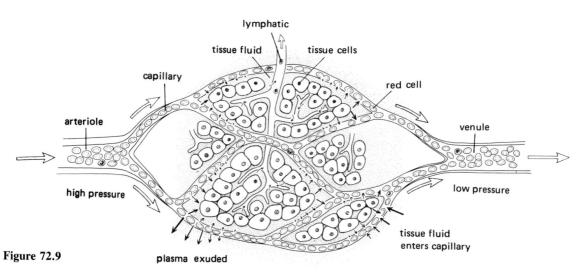

Figure 72.9

small intestine (page 298) and empty their contents into the circulatory system near the heart. The fluid in the lymphatics is similar to plasma and tissue fluid and is called *lymph*.

FUNCTIONS OF THE BLOOD

All the living tissues of the body are bathed in the fluid which escapes from the blood capillaries. This tissue fluid supplies the cells with food and oxygen and removes their waste products. The blood circulating round the body replenishes this tissue fluid and so acts as a transport system.

(*a*) *Transport of oxygen from the lungs to tissues.* In the lungs, the concentration of oxygen is high and so the oxygen combines with the haemoglobin in the red cells, forming oxyhaemoglobin. When this oxygenated blood reaches tissues where oxygen is being used up, the oxyhaemoglobin breaks down and releases its oxygen to the tissues. Oxygenated blood is a bright red colour; deoxygenated blood is dark red.

(*b*) *Transport of carbon dioxide from the tissues to the lungs.* The blood picks up carbon dioxide from actively respiring cells and carries it to the lungs. In the lungs, the carbon dioxide escapes from the blood and is breathed out (see page 308).

(*c*) *Transport of digested food from the intestine to the tissues.* The soluble products of digestion pass into capillaries of the villi lining the small intestine (page 297). They are carried in solution by the plasma and after passing through the liver enter the general circulation. Glucose and amino acids diffuse out of the capillaries and into the cells of the body. Glucose may be oxidized in a muscle, for example, and provide the energy for contraction; amino acids will be built up into new proteins and make new cells and tissues.

(*d*) *Transport of urea from the liver to the kidneys.* When amino acids break down in the liver, they produce a nitrogenous waste product called *urea* (see pages 299 and 310). The urea is carried in the blood all round the body. When it reaches the kidneys, they remove the urea from the blood and get rid of it in the urine.

(*e*) *Transport of hormones.* Hormones are chemicals made by certain glands in the body. The blood carries these chemicals from the glands to the organs where they affect the rate of activity. For example, a hormone, insulin, made in the pancreas is carried by the blood to the liver where it controls the rate of storage of glucose (see pages 298 and 299).

(*f*) *Transport of heat.* The limbs and head lose heat to the surrounding air but contraction of muscles and chemical activity in the liver produce heat. The blood carries the heat from warmer to colder places and so helps to keep an even temperature in all regions. Also, by opening or closing blood vessels in the skin, the blood system helps to regulate the body temperature (see page 313).

Questions

1. In what ways are white cells different from red cells in (*a*) their structure, (*b*) their function?

2. Where, in the body, would you expect haemoglobin to be combining with oxygen to form oxyhaemoglobin? In what parts of the body would you expect oxyhaemoglobin to be decomposing to oxygen and haemoglobin?

3. Why is it incorrect to say that arteries carry oxygenated blood and veins carry deoxygenated blood?

4. Why is it important for oxyhaemoglobin to be an *unstable* compound, i.e. easily decomposed?

5. Why do you think that (*a*) the walls of the ventricles are more muscular than the walls of the atria and (*b*) the muscle of the left ventricle is thicker than that of the right ventricle? (Consult Figure 72.7, page 303.)

6. Which important veins are not shown in Figure 72.3 (page 302)?

7. Why is a person whose heart valves are damaged by disease unable to take part in active sport?

8. How do capillaries differ from other blood vessels in (*a*) their structure and (*b*) their function?

9. What substances would the blood (*a*) gain, (*b*) lose, on passing through (*i*) the kidneys, (*ii*) the lungs, (*iii*) an active muscle? Remember that respiration (page 80) is taking place in all these organs.

10. State in detail the course taken by (*a*) a glucose molecule from the time of its absorption in the intestine, (*b*) a molecule of oxygen in the lungs, to the time when they both meet in a muscle cell of the leg.

73 BREATHING

The various processes carried out by the body, e.g. movement, growth and reproduction, require energy. In animals this energy can be obtained only from the food they eat. Before the energy can be used by the cells of the body it must be set free from the chemicals of the food. This process involves the use of oxygen and the production of carbon dioxide (see page 80).

In man and other mammals, the oxygen is obtained from the air by means of the lungs. In the lungs, the oxygen dissolves in the blood and is distributed to the tissues by the circulatory system.

Lung structure

The lungs are enclosed in the *thorax* (chest region), (see Figure 73.5). They have a spongy texture and can be expanded or compressed by movements of the thorax. This causes air to be taken in and expelled. The lungs are connected to the back of the mouth by the windpipe or *trachea* (Figure 73.1). The trachea divides into two tubes which enter the lungs and divide into smaller branches called *bronchi*. These fine branches end up in a mass of little, thin-walled, pouch-like air sacs called *alveoli* (Figure 73.2).

Rings of gristle (*cartilage*) keep the trachea and bronchi open when the pressure inside them falls during breathing in. The *epiglottis* at the top of the trachea prevents food and drink entering the air passages during swallowing (see page 295).

The alveoli have thin elastic walls of a single cell layer or *epithelium*. Beneath the epithelium is a dense network of capillaries (Figure 73.2b)

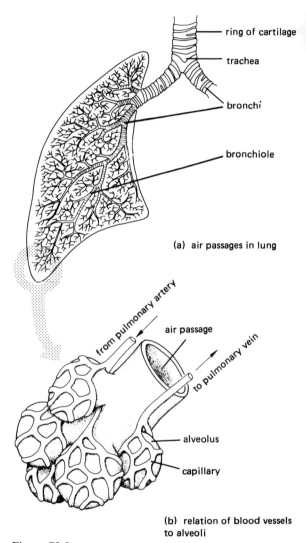

(a) air passages in lung

(b) relation of blood vessels to alveoli

Figure 73.2

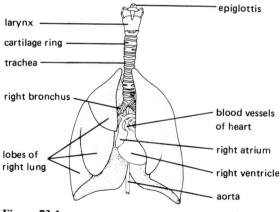

Figure 73.1

supplied with deoxygenated blood (page 305) pumped from the right ventricle through the pulmonary artery (see Figure 72.4, page 302). In man, there are about 350 million alveoli with a total absorbing surface of about 90 m².

The *diaphragm* is a sheet of muscular tissue which separates the thorax from the abdomen (see Figure 71.1, page 295). When relaxed, it is domed slightly upwards. The ribs are moved by the *intercostal muscles* which run from one rib to the next (Figure 73.3). Figure 73.4 shows how the contraction of the intercostal muscles makes the ribs move upwards.

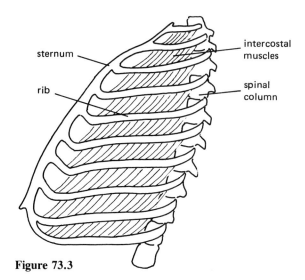

Figure 73.3

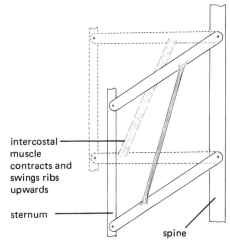

Figure 73.4

Ventilation of the lungs

The movement of air into and out of the lungs, called ventilation, serves to renew the oxygen supply in the lungs and remove the surplus carbon dioxide from them. The lungs contain no muscles and are made to expand and contract by movements of the ribs and diaphragm.

Inhaling

a The diaphragm muscles contract and pull it down (Figure 73.5a).
b The intercostal muscles contract and pull the rib

Figure 73.5

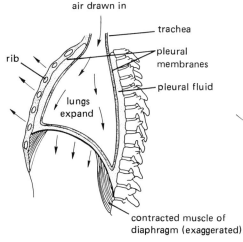

(a) inhaling: diaphragm depressed, ribs raised

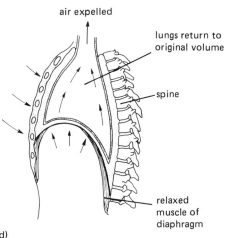

(b) exhaling: diaphragm relaxes and returns to its domed shape, ribs return to original position

307

Figure 73.6

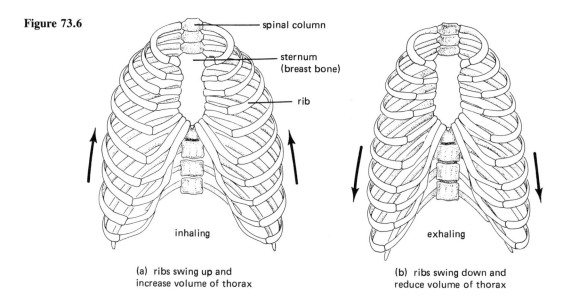

(a) ribs swing up and increase volume of thorax

(b) ribs swing down and reduce volume of thorax

cage upwards (Figure 73.6a). These two movements make the cavity of the thorax bigger, so forcing the lungs to expand and draw air in through the nose and trachea.

Exhaling

a The diaphragm muscles relax allowing the diaphragm to return to its domed shape (Figure 73.5b).
b The intercostal muscles relax, allowing the ribs to move downwards under their own weight (Figure 73.6b). The lungs are elastic and shrink back to their relaxed size, forcing air out again.

Lining the inside of the thorax and the outside of the lungs is a layer of tissue called the *pleural membrane* (Figure 73.5a). This membrane produces a thin layer of *pleural fluid* between the lungs and the chest wall. The pleural fluid lubricates the lungs and stops them rubbing against the chest wall during the breathing movements.

Gaseous exchange

Even in the relaxed state, the lungs contain 1½ litres of air which cannot be expelled no matter how hard you try to breathe out. As a result, the air in the alveoli is not exchanged during ventilation and the oxygen has to reach the blood capillaries by the slower process of diffusion. Figure 73.7 shows how oxygen reaches the red blood cells and how carbon dioxide escapes from the blood. The capillaries carrying oxygenated blood from the alveoli join up to form the pulmonary vein (see Figure 72.7, page 303) which returns to the left atrium of the heart. From here it enters the left ventricle and is pumped all round the body, so supplying the tissues with oxygen.

The process of gaseous exchange in the alveoli does not remove all the oxygen from the air. The air breathed in contains about 21 per cent of oxygen; the air breathed out still contains 17 per cent of oxygen.

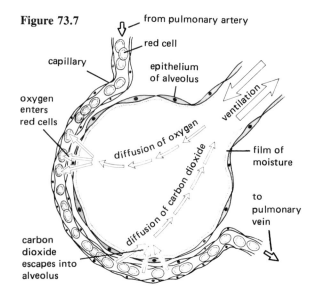

Figure 73.7

Questions

1. Try to make a clear distinction between *respiration* (page 80), *gaseous exchange* and *ventilation*. Say how one depends on the other.

2. Describe the forces acting on a molecule of oxygen and the path it takes from the time it is breathed in to the time it enters the heart in some oxygenated blood.

3. One function of the small intestine is to absorb food (page 297). One function of the lungs it to absorb oxygen. Point out the basic similarities in these two structures which help to speed up the process of absorption.

4. Place the following in the correct order: lungs expand, ribs rise, air enters lungs, intercostal muscles contract, thorax expands.

5. Figure 73.7 shows oxygen and carbon dioxide diffusing across an alveolus. What causes them to diffuse in opposite directions? (See page 100.)

6. During vigorous exercise, what two changes in your breathing help to increase the oxygen supply to the blood? What other change has to occur if the extra oxygen is to reach your muscles quickly enough?

7. In 'mouth to mouth' resuscitation, air is breathed from the rescuer's lungs into the lungs of the person who has stopped breathing. How can this 'used' air help to restore the person?

74 EXCRETION

An enormous number of chemical reactions take place inside the cells of a living body in order to keep it alive. The breakdown products of some of these reactions are poisonous and must be removed from the body. For example, the breakdown of glucose during respiration (page 80) produces carbon dioxide. This is removed by the blood and eliminated in the lungs. Excess amino acids are deaminated to form glycogen and urea as explained on page 299. The urea is poisonous and is removed from the tissues by the blood and eliminated in the kidneys.

In the course of feeding, more water and salts are taken in than are needed by the body and so these excess substances must be removed as fast as they accumulate.

Excretion is the name given to the elimination from the body of the waste products of its chemical reactions and the excess water and salts taken in with the diet. The process also includes the removal of drugs and other foreign substances taken into the alimentary canal and absorbed by the blood. The term, excretion, is not usually applied to the elimination of faeces (page 298). This is because most of the substances in the faeces, such as undigested cellulose, have not taken part in reactions within the body tissues.

Figure 74.1

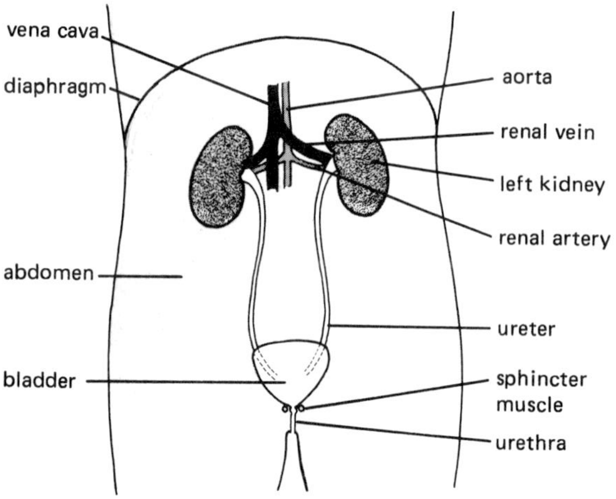

EXCRETORY ORGANS

The lungs which supply the tissues with oxygen are also excretory organs because they get rid of excess carbon dioxide. They also, incidentally, lose a great deal of water vapour, but this process is not used to regulate the water content of the body.

The kidneys eliminate urea and other nitrogen-containing substances such as uric acid from the blood. They also expel excess water, salts, hormones and drugs.

THE KIDNEYS

Structure

The two kidneys are fairly solid, oval structures. They are reddish-brown, enclosed in a transparent membrane and attached to the back of the abdominal cavity (Figure 74.1). The *renal artery* branching from the aorta, brings oxygenated blood to them, and the *renal vein* takes deoxygenated blood away to the vena cava (see Figure 72.7, page 303). A tube, the *ureter*, runs from each kidney to the base of the *bladder* in the lower part of the abdomen.

The kidney tissue consists of many capillaries and tiny tubes, called *renal tubules*, held together with connective tissue. If the kidney is cut down its length (sectioned), it is shown to have a darker, outer region called the *cortex* and a lighter inner zone, the *medulla*. Where the ureter leaves the kidney is a space called the *pelvis* (Figure 74.2).

The renal artery divides up into a great many arterioles and capillaries (Figure 74.3), mostly in the cortex. Each arteriole leads to a *glomerulus*, which is a capillary repeatedly divided and coiled, making a little knot of vessels (Figure 74.4). The glomerulus is almost entirely surrounded by a cup-shaped organ called a *Bowman's capsule*, which leads to a coiled renal tubule. This tubule, after a series of coils and loops, joins other tubules and passes through the medulla to open into the pelvis.

Figure 74.2

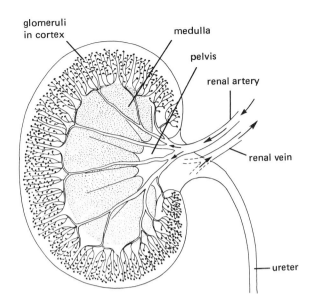

Function

The blood pressure in a glomerulus causes blood fluid to leak through the capillary walls. The blood cells and the plasma proteins cannot pass out of the capillary, so the fluid that does filter through, is plasma without the proteins (see page 302). The fluid thus consists mainly of water with dissolved salts, glucose, amino acids, urea and uric acid. This *filtrate* from the glomerulus collects in the Bowman's capsule and trickles down the renal tubule (Figures 74.3 and 74.4). As it does so, the capillaries which surround the tubule absorb back into the blood those substances which the body needs.

Firstly, all the amino acids and glucose are reabsorbed with some of the water. Then some of the salts are taken back to maintain their correct concentration in the blood. Salts in excess of these needs plus urea and uric acid are left to pass on down the kidney tubule into the pelvis of the kidney. From here the fluid, now called *urine*, passes down the ureter to the bladder.

In an average person, the bladder can expand to hold about 400 cm³ of urine. When the sphincter muscle (Figure 74.1) at the base of the bladder relaxes, the muscular walls of the bladder expel the urine through the *urethra*.

Figure 74.3

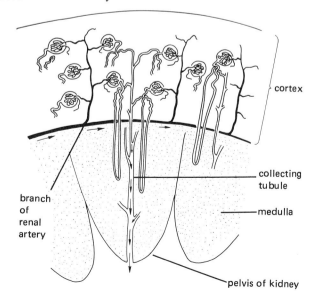

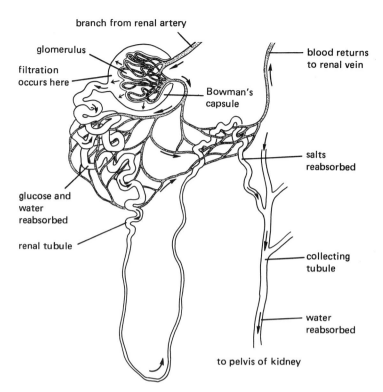

Figure 74.4

The kidneys also control the amount of water in the blood. If the blood is too dilute, less water is reabsorbed from the renal tubules, leaving more to enter the bladder. If the blood is too concentrated, more water is absorbed back into the blood, so that less passes to the bladder. This regulatory process maintains the blood at a constant concentration and is called *osmo-regulation*.

Questions

1. In what ways is water lost from the body? Which of these would you consider to be excretion? Justify your answers.

2. How does the composition of urine differ from the composition of blood plasma?

3. Where, in the urinary system, do the following take place, (answer as precisely as possible): filtration, selective reabsorption, storage of urine, transport of urine, osmo-regulation?

4. In hot weather when you sweat a great deal, urination is infrequent and the urine is a dark colour. In cold weather when you sweat little, urination occurs frequently and the urine is pale in colour. Use your knowledge of kidney function to explain these observations.

5. Trace the path taken by a molecule of urea from the time it is produced in the liver, to the time it leaves the body in the urine. (See also pages 299 and 303.)

75 TEMPERATURE REGULATION

HEAT BALANCE

Heat gain

(*a*) *Internal.* A great many of the chemical reactions in cells release heat. The chief heat producers are the contracting cells of active muscle and the cells of the liver, where so many chemical changes are taking place. Any increase in muscular activity or chemical reactions in the liver will result in more heat being produced in the body.

(*b*) *External.* Direct heat from the sun will be absorbed by the body. Hot food and drink are further sources of heat.

Heat loss

Heat is lost to the air from the exposed surfaces of the body by conduction, convection and radiation (see page 180). Evaporation from the skin takes place continually and is a cause of heat loss. The cold air breathed into the lungs and cold food or drink taken into the stomach all absorb heat from the body.

Birds and mammals are 'warm-blooded'. That is, their body temperatures are, in most instances, higher than that of their surroundings. These animals have a covering of feathers or fur which traps a layer of stationary air round the body and so reduces convection and conduction. We wear clothing to achieve the same effect.

To a large extent, in these animals, the heat lost from the body is balanced by the heat absorbed or produced. However, changes in the external temperature or in the rate of activity may cause a temporary imbalance which, in man, is regulated largely by processes in the skin.

Figure 75.1 Section through the skin.

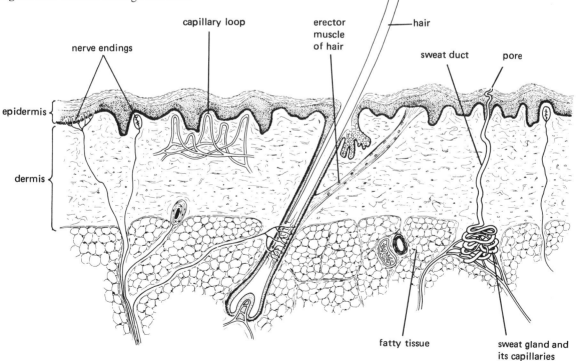

Temperature regulation by the skin

Skin structure

The structures in the skin which play a part in temperature control are the small blood vessels, the sweat glands and, in furry animals, the hairs (Figure 75.1).

The arterioles and capillaries near the surface of the skin can increase or decrease their diameter and therefore increase or reduce the amount of blood flowing through them. An increase in diameter of the vessel is called *vaso-dilation*; a reduction in diameter is called *vaso-constriction*.

The sweat glands are inactive when the loss and gain of heat are in balance. As soon as the body temperature starts to rise, the coiled tubular parts of the glands lying in the dermis of the skin extract water and salts from the blood capillaries which supply them and expel this solution, *sweat*, through the sweat ducts and pores on to the skin surface.

In cold weather, the hairs in furry mammals can be pulled into a more upstanding position by contraction of the hair muscles. The fur will thus trap a thicker layer of air and help to reduce the heat losses. In man, however, the contraction of the hair muscles produces only 'goose pimples'.

Over-heating

If the body gains or produces heat faster than it is losing it, the following processes occur.

(*a*) *Vaso-dilation.* The widening of the blood vessels in the skin allows more warm blood to flow near the surface and so lose more heat (Figure 75.2*a*).

(*b*) *Sweating.* The sweat glands pour sweat on to the skin surface. When this layer of liquid evaporates it takes latent heat (page 184) from the body and so cools it down.

Over-cooling

If the body begins to lose heat faster than it can produce it, the following changes occur:

(*a*) *Sweat production stops.* This reduces the heat lost by evaporation.

(*b*) *Vaso-constriction.* The blood vessels in the skin constrict, which reduces the amount of warm blood flowing near the surface (Figure 75.2*b*).

(*c*) *Shivering.* Uncontrollable bursts of rapid muscular contraction in the limbs release heat as a result of the chemical changes in the muscles.

Nearly all chemical reactions are speeded up by a rise in temperature so the 'warm-blooded' creatures have the advantage that the reactions in their cells take place as rapidly as possible in a living system. By regulating their temperatures to keep them more or less constant, they also become independent of changes in their surroundings. On a cold day, a lizard will cool down and may become very sluggish in its actions but a mammal's temperature will not change and so its body chemistry goes on at the same rate as before.

Questions

1. Draw up a balance sheet showing all the possible ways the human body can gain and lose heat.

2. (*a*) Which *structural* features of the skin of a mammal help to reduce heat loss?
   (*b*) What physiological changes take place in the skin of man to reduce heat loss?

3. What is vaso-dilation? How does it help to lower body temperature?

4. Why do you think it is more difficult to keep cool in hot humid conditions than in hot dry conditions?

5. What conscious actions does man take in order to reduce heat losses from his body?

Figure 75.2

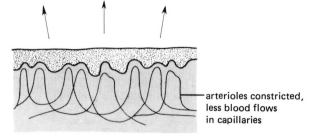

(a)   (b)

76 SEXUAL REPRODUCTION

Sexual reproduction involves the joining or fusing together of two cells. One of these reproductive cells comes from a male animal and the other from a female. The male reproductive cell is called a *sperm* and the female cell is the *ovum* (plural = *ova*) (Figure 76.1). Reproductive cells, both male and female are called *gametes*.

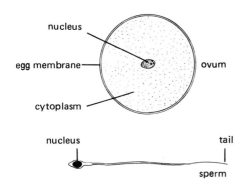

Figure 76.1 Sperm and ovum of a mammal (not to scale).

Fertilization is the fusion of these two cells to form a composite cell which can grow into a new individual.

REPRODUCTIVE ORGANS

Female

The female reproductive organs (Figure 76.2) are the *ovaries*, two whitish oval bodies, 3–4 cm long. They lie in the lower half of the abdomen, one on each side of the uterus. Close to each ovary is the expanded, funnel-shaped opening of the *oviduct*, the tube down which the ova pass when released from the ovary.

The oviducts are narrow tubes that open into a wider tube, the *uterus* or *womb*, lower down in the abdomen. When there is no embryo developing in it, the uterus is only about 80 mm long. It communicates with the outside through a muscular tube, the *vagina*. The *cervix* is a ring of muscle closing the lower end of the uterus where it joins the vagina. There is normally only a very small aperture connecting these two organs at this point, The *urethra*, from the bladder, opens into the *vulva* just in front of the vagina (see Figure 76.10).

Male

The male reproductive organs (Figure 76.3) are the *testes* (singular = *testis*) which lie outside the abdominal cavity in man, in a special sac called the

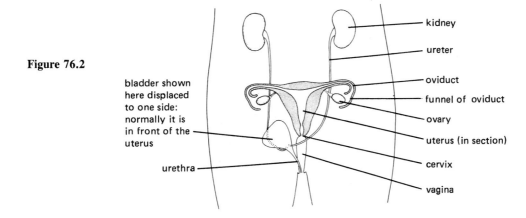

Figure 76.2

315

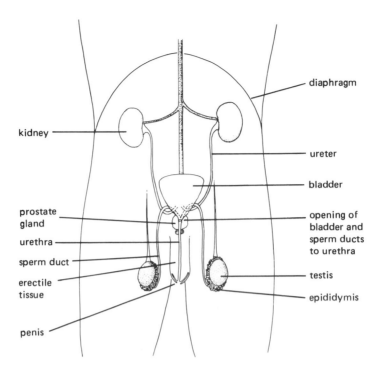

Figure 76.3

scrotum. In this position they remain at a temperature rather below that of the rest of the body, which is favourable to sperm production. The testes consist of a mass of sperm-producing tubes. These tubes join to form ducts leading to the *epididymis*, a coiled tube about 6 metres long on the outside of each testis. The epididymis, in turn, leads into a muscular *sperm duct*. The two sperm ducts, one from each testis, open into the top of the urethra just after it leaves the bladder. A short, coiled tube, the *seminal vesicle*, branches from each sperm duct just before the latter enters the *prostate gland*, which surrounds the urethra at this point. The urethra conducts, at different times, either urine or sperms and is prolonged into a *penis,* which consists of connective tissue with numerous small blood spaces in it.

PRODUCTION OF GAMETES

Ovulation

Ovulation is the release of an ovum from the ovary. The immature egg cells are present in the ovary from the time of birth. No more are formed during the life time but between the ages of about 11–16 years, (*puberty*), the egg cells start to ripen and are released, one at a time about every 4 weeks. During the process of maturation, the cells round the ovum divide rapidly and eventually produce a fluid-filled sac. This sac is called a *follicle* and when mature, it projects from the surface of the ovary like a small blister (Figure 76.4). Finally, the follicle bursts and releases the ovum with its coating of cells into the funnel of the oviduct. From here it is wafted down the oviduct by the action of cilia in the lining of the tube. If it meets sperm cells in the oviduct, it may be fertilized by one of them.

Sperm production

The testes produce sperm cells which collect in the epididymis. During copulation, contractions of the epididymis and sperm ducts force sperms out through the urethra. The prostate gland and seminal vesicle add fluid to the sperm. This fluid plus the sperms it contains is called *semen*, and the ejection of sperms through the penis is called *ejaculation*.

COPULATION AND FERTILIZATION

Copulation

As a result of sexual stimulation, the male's penis becomes erect. This is due to an inflow of blood to the erectile tissue surrounding the urethra. In the female, the lining of the vagina produces a lubricating mucus which makes it possible for the penis to enter. The sensory stimulus produced by copula-

Figure 76.4 Section through an ovary (as seen under the low power of a microscope).

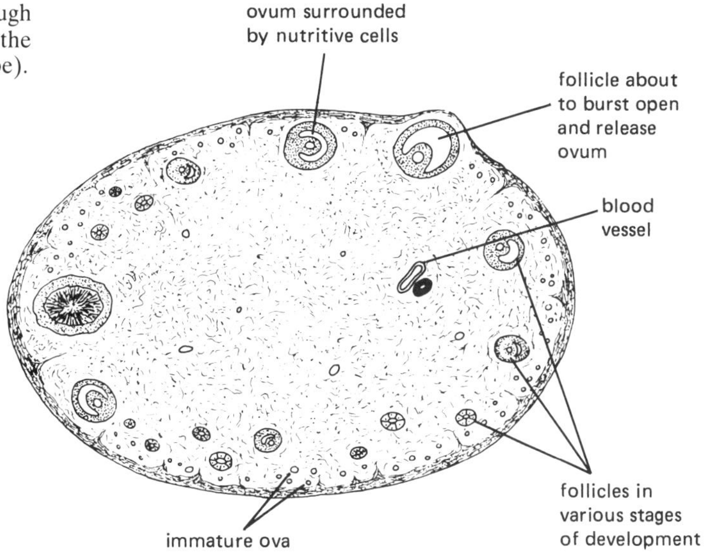

tion causes a reflex action (page 334) in the male resulting in ejaculation of semen into the top of the vagina.

Fertilization

The sperms pass through the cervix and uterus and enter the oviduct. This movement is achieved partly by the wriggling of the sperm tails but how the sperms travel through the uterus and oviduct is not known. If the sperms meet an ovum in the oviduct, one of the sperms may collide with it and stick to its surface. The male nucleus from the sperm head then enters the cytoplasm of the ovum and fuses with the female nucleus. This is the moment of fertilization and is shown in more detail in Figure 76.5. Although a single ejaculation may contain 2 or 3 hundred million sperms, only one

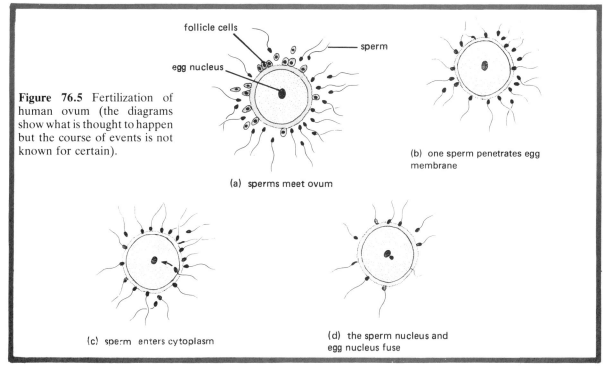

Figure 76.5 Fertilization of human ovum (the diagrams show what is thought to happen but the course of events is not known for certain).

(a) sperms meet ovum
(b) one sperm penetrates egg membrane
(c) sperm enters cytoplasm
(d) the sperm nucleus and egg nucleus fuse

will actually fertilize the ovum. The function of the others is not understood but it is likely that a great many do not even complete the journey from the vagina to the oviduct.

The released ovum is thought to survive for up to 24 hours, while the sperms might be able to fertilize an ovum for 2–3 days. There is thus only a relatively short period of 3–4 days each month when fertilization might occur.

PREGNANCY AND DEVELOPMENT

The fertilized ovum firstly divides into two cells. Each of these divides again so producing four cells. The cells continue to divide in this way to produce a solid ball of cells which, by now, has travelled down the oviduct and reached the uterus. The ball of cells sinks into the lining of the uterus where it continues to grow and produce new cells, thus beginning to form the *embryo* (Figure 76.6a).

As the embryo grows, the uterus enlarges to contain it. Inside the uterus the embryo becomes enclosed in a fluid-filled sac called the *amnion* or *water sac* which protects it from damage and prevents unequal pressures from acting on it (Figure 76.6b and 76.7). The oxygen and food needed to keep the embryo alive and growing are obtained from the mother's blood by means of a structure called the *placenta*.

Placenta

Soon after the ball of cells reaches the uterus, some of the cells, instead of contributing to the tissues of the embryo, grow into a disc-like structure, the placenta. The placenta becomes closely attached to the lining of the uterus and communicates with the embryo through a tube called the *umbilical cord* (Figure 76.6c). After a few weeks, the

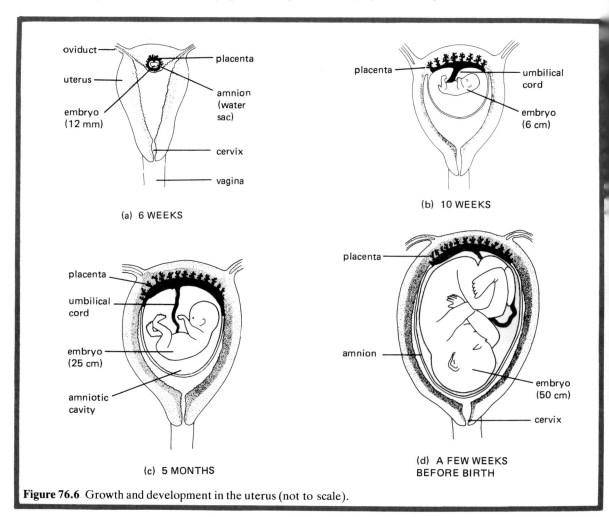

Figure 76.6 Growth and development in the uterus (not to scale).

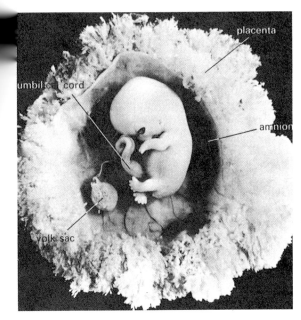

Figure 76.7 Human embryo, 7 weeks after conception.

embryo's heart has developed and is circulating blood through the placenta as well as through its own tissues. The blood vessels in the placenta are very close to the blood vessels in the uterus so that oxygen, glucose, amino acids and salts can pass from the mother's blood to that of the embryo (Figure 76.8). Thus, the blood returning in the *umbilical vein* to the embryo from the placenta carries food and oxygen to be used by the living, growing tissues of the embryo. Similarly the carbon dioxide and urea in the embryo's blood, escape from the vessels in the placenta and are carried away by the mother's blood in the uterus (Figure 76.9). In this way the embryo gets rid of its excretory products. There is no direct communication between the mother's blood system and the embryo's. The exchange of substances takes place

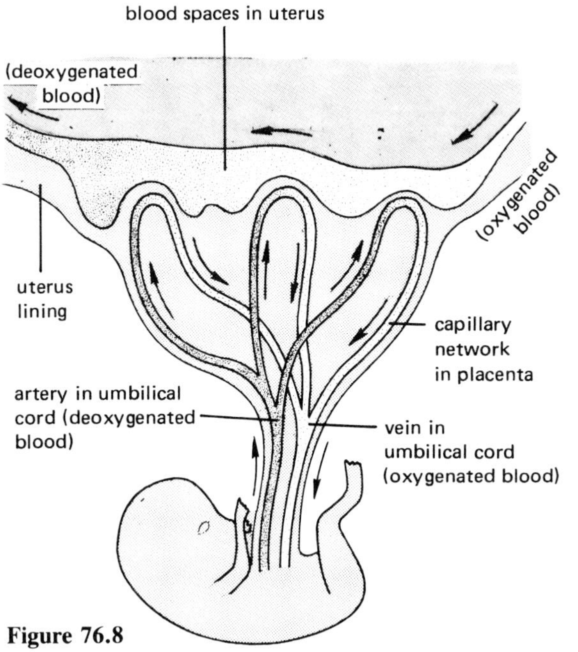

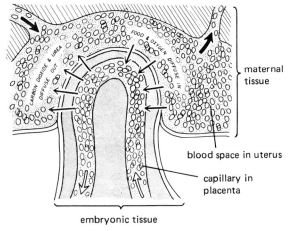

Figure 76.9

across the thin walls of the blood vessels. In this way, the mother's blood pressure cannot damage the delicate vessels of the embryo and it is possible for the placenta to exercise some selection over the substances allowed to pass into the embryo's blood.

BIRTH

From fertilization to birth takes about 9 months in humans. A few weeks before birth, the embryo has come to lie head downwards in the uterus, with its head just above the cervix (Figures 76.6d and 76.10). When the birth starts, the uterus begins to contract rhythmically. This is the onset of what is called 'labour'. These rhythmic contractions become stronger and more frequent. The opening of the cervix gradually widens enough to let the child's head pass through. The contractions of the uterus are reinforced by muscular contractions of the abdomen. The water sac breaks at some stage in labour and the fluid escapes through the vagina. Finally, the muscular contractions of the uterus and abdomen expel the child head first through the widened cervix and vagina. The umbilical cord

Figure 76.8

which still connects the child to the placenta is tied and cut. Later, the placenta breaks away from the uterus and is expelled separately as the 'afterbirth'.

The sudden fall in temperature experienced by the newly born baby stimulates it to take its first breath, usually accompanied by crying. In a few days the remains of the umbilical cord attached to the baby's abdomen shrivel and fall away, leaving a scar in the abdominal wall, called the *navel*.

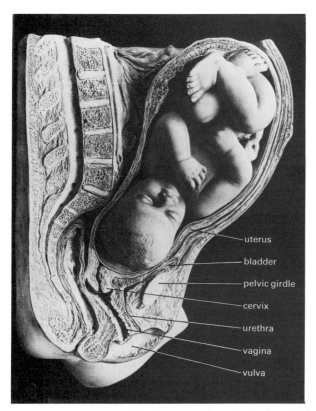

Figure 76.10 Model of human embryo just before birth.

Questions

1. How do sperms differ from ova in (*a*) their structure, (*b*) the numbers produced and (*c*) their activities?

2. Place the following structures in the correct order for the production and ejaculation of sperms: prostate gland, sperm duct, epididymis, urethra, testis.

3. List, in the correct order, the parts of the female reproductive system through which sperms must pass before reaching and fertilizing an ovum.

4. State exactly what happens at the moment of fertilization.

5. Why is there only a short period each month when copulation is likely to result in fertilization?

6. What changes will take place in the composition of the mother's blood as it passes through the placenta?

77 THE SKELETON AND LOCOMOTION

Figure 77.1

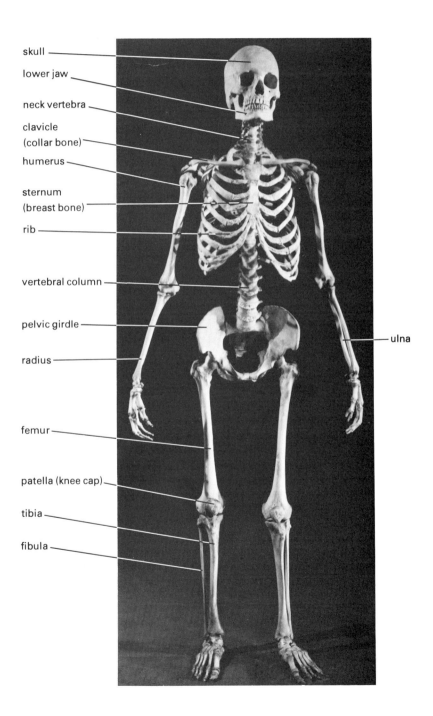

- skull
- lower jaw
- neck vertebra
- clavicle (collar bone)
- humerus
- sternum (breast bone)
- rib
- vertebral column
- pelvic girdle
- radius
- ulna
- femur
- patella (knee cap)
- tibia
- fibula

FUNCTIONS OF THE SKELETON

Support: The skeleton raises the body off the ground and maintains its shape even during locomotion when muscles are contracting vigorously.

Protection: The brain is protected from injury by being enclosed in the skull; the heart, lungs and liver are protected by the rib cage, and the spinal cord is enclosed inside the vertebral column.

Movement: Many bones of the skeleton act as levers (page 137). When muscles pull on these bones, they produce movements such as the raising of the ribs during breathing (see page 307) or the chewing action of the jaws. For a muscle to produce effective movement, both ends need to have a firm attachment. The skeleton provides a suitable anchorage for the ends of the muscles.

STRUCTURE OF THE SKELETON

The *skull* and *vertebral column* form the axis of the skeleton. The legs are connected to the vertebral column by the *pelvic girdle* (*pelvis*) and the arms are connected by means of muscles, through the shoulder blade (*scapula*) and collar bone (*clavicle*) (Figure 77.1).

The vertebral column (Figure 77.2a) is made up of about thirty separate bones called *vertebrae* (singular = *vertebra*) one of which is illustrated in Figure 77.2b. Each consists of a solid *centrum* and a *neural arch* through which runs the spinal cord. The bony processes sticking out from the vertebrae serve to lock the vertebrae together and limit their movement (see Figure 77.2c), or provide for the attachment of muscles which flex or straighten the vertebral column.

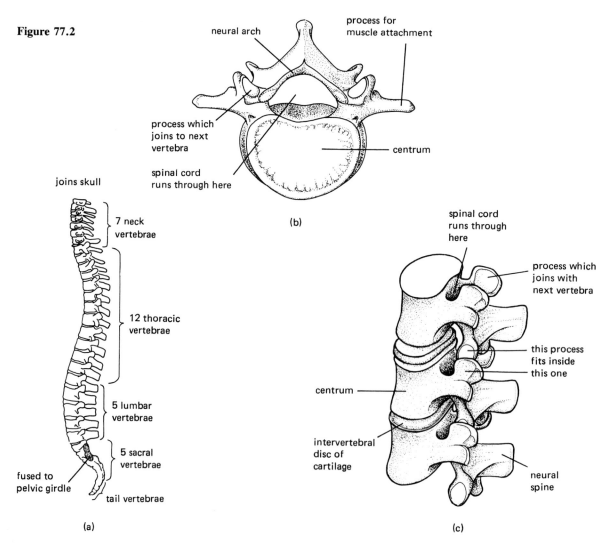

Figure 77.2

In the various regions of the vertebral column the vertebrae have slightly different functions and therefore variations in their structure. The *thoracic vertebrae* have processes for rib attachment and the five *sacral vertebrae* are fused together to form a rigid structure for the attachment of the pelvic girdle.

Separating the vertebrae are discs of cartilage and connective tissue which allow the vertebrae to move slightly and thus give flexibility to the vertebral column as a whole. These *inter-vertebral discs* also act as shock absorbers.

THE LIMBS

Arm

The upper arm bone is the *humerus*. It is attached by a ball joint to the shoulder blade; at the elbow it is attached by a hinge joint to the lower arm bones, the *radius* and *ulna* (Figure 77.3). These two bones make a joint with a group of small wrist bones which in turn, join to a series of five hand and finger bones. The ulna and radius can partly rotate round each other so that the hand can be held palm up or palm down.

Figure 77.3

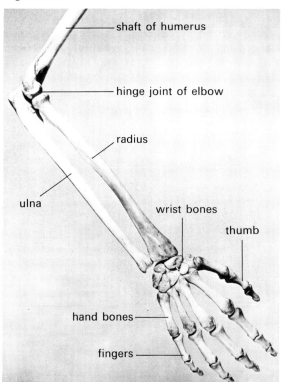

Leg

At the hip, the thigh bone or *femur* is attached to the pelvic girdle by a ball joint and at the knee it makes a hinge joint with the *tibia*. The *fibula* runs parallel to the tibia but does not contribute to the knee joint. The ankle, foot and toe bones are similar to those of the wrist, hand and fingers.

Joints

Where two bones meet they form a joint. It may be a fixed joint as in the junction of the pelvic girdle and the sacral vertebrae, or a moveable joint as in the knee. Two important types of moveable joint already mentioned are the *ball and socket joints* of the hip or the shoulder, and the *hinge joints* of the elbow and knee. The ball and socket joint allows

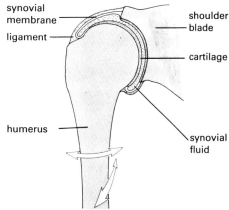

Figure 77.4 Ball-and-socket joint of shoulder

movement forwards, backwards and sideways, while the hinge joint permits movement in only one plane.

Where the surfaces of the bones in a joint rub over each other, they are covered with smooth cartilage which reduces the friction between them. Friction is further reduced by a lubricating liquid called *synovial fluid* (Figure 77.4). Where the bones form a joint they are held in place by tough, fibrous bands of tissue called *ligaments*.

MOVEMENT AND LOCOMOTION

Muscles and movement

Muscles are bundles of long cells which are able to contract and so shorten the muscles as a whole. The ends of the limb muscles are drawn out into inextensible *tendons* which attach the muscle at each end to the skeleton.

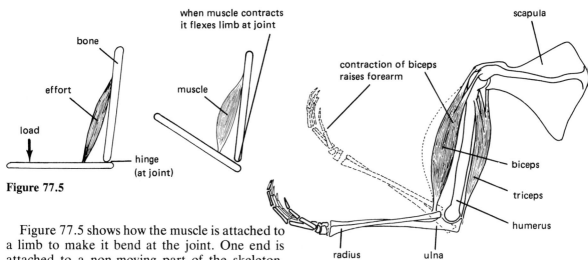

Figure 77.5

Figure 77.6

Figure 77.5 shows how the muscle is attached to a limb to make it bend at the joint. One end is attached to a non-moving part of the skeleton, while the other end is attached to the moveable bone close to the joint. The position of this attachment ensures that a small contraction of the muscle will produce a large movement at the end of the limb. Figure 77.6 shows how contraction of the *biceps* muscle bends (or flexes) the arm at the elbow, while the *triceps* straightens (or extends) the arm. Limb muscles like this are usually arranged in pairs having opposite effects because (*i*) muscles can only shorten or relax, they cannot elongate, so the triceps is needed to pull the relaxed biceps back to its extended shape after contraction, and (*ii*) to hold the limbs steady, as both muscles need to maintain tension at the same time.

Locomotion

Locomotion is brought about by the limb muscles contracting and relaxing in an orderly (co-ordinated) manner. Figure 77.7 shows how some of the muscles of the leg act on the limb bones to produce a forward movement of the body. When muscle A contracts, it pulls the femur backwards. Contraction of muscle B straightens the leg at the knee. Muscle C contracts and pulls the foot down at the ankle. The body is thrust forwards by these three actions which transmit their forces through the pelvic girdle to the vertebral column.

Contraction of limb muscles occurs when they receive nerve impulses. The brain sends out impulses in the nerves so that the muscular contractions are *co-ordinated* to produce locomotion. For example, only one leg at a time must be lifted forwards, and when the muscle which bends the limb is contracting it is essential that its opposite muscle is relaxed.

Figure 77.7

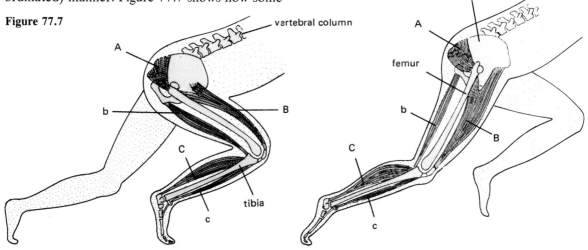

Questions

1. Where do you consider (*a*) ball and socket joints, (*b*) hinge joints occur in the body? Name some joints which are neither of these.

2. What is the difference between a ligament and a tendon?

3. What structures and events are needed to produce effective movement at a joint?

4. In Figure 77.7, (*a*) to what bones is the non-moving end of muscle B attached? (*b*) Which muscle has the opposite effect to B? (*c*) Which muscle has not had its opposite partner drawn in? (*d*) What movement would you expect this missing muscle to produce when it contracted?

5. In Figure 77.2*c* (page 322), to what part of the vertebrae would you expect those muscles to be attached which would bend the vertebral column backwards?

6. What is the principal action of (*a*) your calf muscle, (*b*) the muscle in the front of your thigh, (*c*) the muscles in your forearm? If you don't already know the answer, try making the muscles contract and feel where the tendons are pulling.

7. By making measurements on Figure 77.6 calculate the force that the biceps would have to exert for the hand to lift a 3 kg mass (see page 137).

78 THE SENSES

The senses make us aware of changes in our surroundings and, to some extent, in our own bodies. We have sense cells which respond to touch, light, temperature, vibrations (sound) and chemicals. Structures which respond to stimuli such as these are called *receptors*. Some of these receptors are distributed throughout the skin while others are concentrated into special organs such as the eye and the ear.

SKIN SENSES

There are a great many sensory nerve endings in the skin which respond to touch, pressure, heat and cold, and some which give rise to the sensation of pain. These sensory nerve endings are very small; they can be seen only in sections of the skin when studied under the microscope and some have not been identified with certainty.

When the nerve ending is stimulated, it sends a nerve impulse to the brain which makes us aware of the sensation. It is generally assumed that each type of nerve ending responds only to one kind of stimulus. For example, a heat receptor would fire off a nerve impulse if its temperature was raised but not if it were simply touched. Figure 78.1 shows some of the skin receptors and indicates the stimulus to which each is thought to respond.

TASTE AND SMELL

In the lining of the nasal cavity and on the tongue are groups of sensory cells which respond to chemicals. On the tongue, these groups of cells are called *taste buds* and they lie mostly in the grooves round the bases of the little projections on the upper surface of the tongue (Figure 78.2). The receptor cells recognize only four classes of chemicals; those which give rise to the taste sensations of either sweet, sour, salt or bitter. Generally, the taste cells are particularly sensitive to only one or two of these classes. For a substance to produce a sensation of taste it must be able to dissolve in the film of water covering the tongue. The smell receptors in the nasal cavity, however, can recognize a much wider variety of air-borne chemicals which do not have to be soluble in water. The sensation we call *flavour* (as distinct from 'taste') is the result of the vapours from the food reaching the sensory cells in the nasal cavity from the back of the mouth (see Figure 71.3, page 296).

Figure 78.1

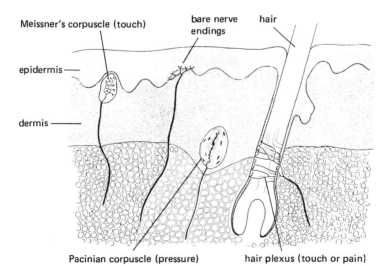

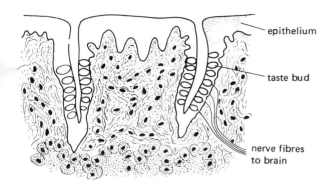

Figure 78.2

SIGHT

The eye

The structure of the eye is shown in Figure 78.3. The *sclera* is the tough, outer coating which, in the front becomes transparent to form the *cornea*. The eye contains clear fluid whose outward pressure on the sclera maintains the spherical shape of the eye. The fluid, *vitreous humour*, in the back chamber of the eye is jelly-like, while the *aqueous humour* in the front chamber is watery. The *lens* is a transparent structure, held in place by the *suspensory ligament*. Unlike the lens of a camera or telescope, the eye lens is flexible and can change its shape. In front of the lens is a disc of tissue called the *iris*. It is the iris we refer to when we describe the colour of the eye as blue or brown. There is a hole in the centre of the iris, the *pupil*, which admits light to the rest of the eye. The pupil looks black because all the light entering the eye is absorbed and none is reflected. Muscles in the iris can increase or decrease the size of the pupil in response to changes in light intensity.

The internal lining at the back of the eye is the *retina* and consists of thousands of cells which respond to light. When light falls on these cells they fire off nervous impulses which travel in nerve fibres through the optic nerve to the brain and so give rise to the sensation of sight.

Vision

Figure 78.5a explains how light from an object produces a focused image on the retina. The curved surfaces of the cornea and lens both 'bend' or refract (page 202) the light rays in such a way that each 'point of light' from the object is reproduced as a 'point of light' on the retina, so forming an image, upside down and smaller than the object.

The pattern of sensory cells stimulated by the image will affect the pattern of nerve impulses reaching the brain. The brain interprets this pattern using its past experience and learning, and so forms an impression of the real size, distance and upright nature of the object.

There are no sensory cells in the retina at the point where the optic nerve leaves it and so no information reaches the brain about the part of the image which falls on this '*blind spot*' (see Figure 78.4).

Figure 78.3 Horizontal section through left eye.

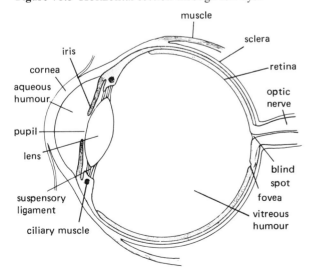

Figure 78.4 The blind spot. Hold the book about 60 cm away. Close the left eye and concentrate on the cross with the right eye. Slowly bring the book closer to the face. When the image of the dot falls on the blind spot, it will seem to disappear.

The millions of light-sensitive cells in the retina are of two kinds, the *rods* and the *cones* (according to shape). The cones enable us to distinguish *colours*. There are thought to be three types of cone cell. One type responds to yellow light, one to green and one to blue. If all three types are equally stimulated we get the sensation of white. The cone cells are concentrated in a central part of the retina, called the *fovea* (Figure 78.3), and when you study an object closely, you are making its image fall on the fovea.

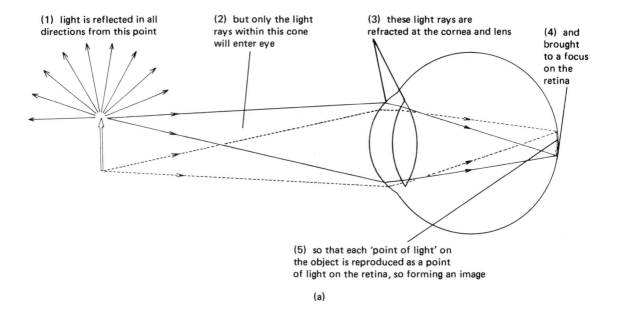

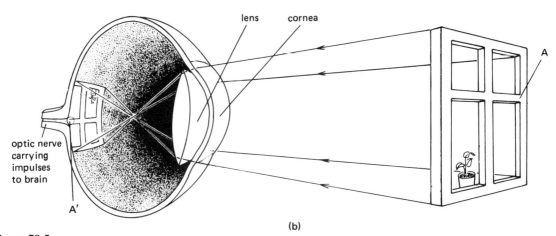

Figure 78.5

Accommodation

The eye can produce a focused image of either a near object or a distant object. To do this, the lens changes its shape, becoming thinner for distant objects and fatter for near objects. This change in shape is brought about by contraction or relaxation of the *ciliary muscle* (Figure 78.3) which runs in a circle round the lens (Figure 78.7). When the ciliary muscle is relaxed, the outward pressure of the humours on the sclera pulls on the suspensory ligament and stretches the lens to its thin shape, i.e. accommodated for distant objects (Figures 78.6a and 78.7a). To focus a near object the ciliary muscle contracts, so reducing its diameter and taking the tension out of the suspensory ligament (Figures 78.6b and 78.7b). The lens being elastic, is thus allowed to change to its fatter shape. This shape is better able to refract the strongly divergent light rays from a close object.

As we become older, the lens loses some of its elasticity. Thus, it is more difficult to accommodate for close objects and we have to use spectacles for reading and other close work.

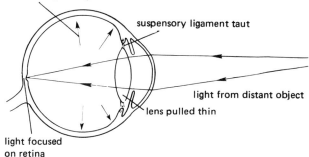
(a) eye relaxed (distant accommodation)

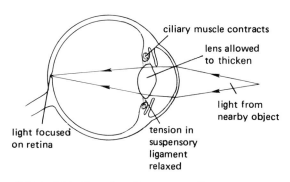
(b) eye focused on near object

distorts shape of eye (greatly exaggerated)

Figure 78.6

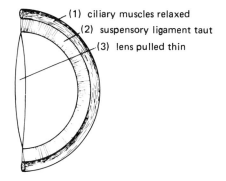

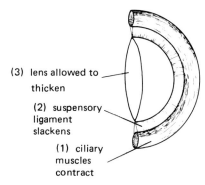

Figure 78.7 (a) DISTANT ADAPTED (b) NEAR ADAPTED

Long and short sight

In some people the size of the eyeball does not exactly match the power of the lens. For example, in short-sighted people the eyeball may be too long or the lens too powerful, so that the image of a distant object comes to a focus in front of the retina and the image on the retina itself is blurred. The causes of and corrections for long and short sight are explained in Figure 78.8.

Figure 78.8

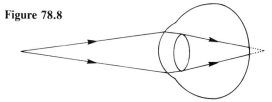

Long sight is caused by short eyeballs or 'weak' lenses. Light from a distant object is brought to a focus on the retina but from a close object its focus is behind the retina

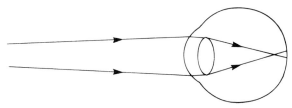

Short sight is usually caused by long eyeballs. Light from a distant object is focused in front of the retina

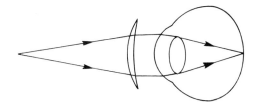

long sight can be corrected by wearing converging lenses

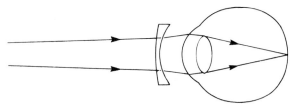

short sight can be corrected by wearing diverging lenses

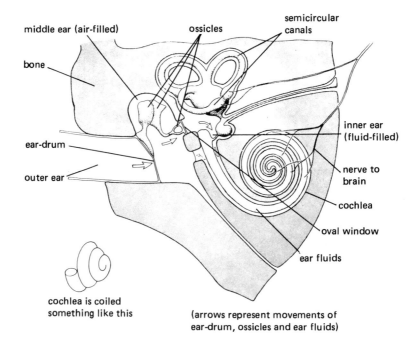

Figure 78.9

(arrows represent movements of ear-drum, ossicles and ear fluids)

HEARING

The hearing apparatus is housed in the skull on each side of the head and communicates with the atmosphere by a wide tube. Figure 78.9 shows the structure of the ear.

Outer ear

Sound is the name we give to sensations resulting from vibrations in the air. These vibrations, in the form of pulses of compressed air (see page 220), enter the tube of the outer ear and hit the *ear drum*, a thin membrane sealing off the inner end of the tube. The air vibrations cause the ear drum to vibrate as well. If there are two hundred pulses of compression every second, the ear drum will move backwards and forwards at the same rate.

Middle ear

This is an air-filled cavity containing a chain of three tiny bones or *ossicles*. The outermost of these ossicles is attached to the ear drum and the inner one fits into a small hole in the skull called the *oval window*. When the ear drum vibrates back and forth, it forces the ossicles to vibrate in the same way, so that the innermost ossicle moves rapidly backwards and forwards like a tiny piston in the oval window.

Inner ear

This is where the vibrations are converted into nerve impulses. The inner ear contains fluid and the vibrations of the ossicles are transmitted to this fluid. The sensitive part of the inner ear is the *cochlea*, a coiled tube containing sensory nerve endings. When the fluid in the cochlea is made to vibrate, the nerve endings fire off impulses to the brain. Very approximately, the nerve endings in the last part of the cochlea are sensitive to low frequency vibrations (low notes) and those in the first part are particularly sensitive to high frequency vibrations (high notes). Thus, if the brain receives nerve impulses coming from the first part of the cochlea, it interprets this as a high-pitched noise or a high musical note (see page 221).

BALANCE

Semi-circular canals

These are in the inner ear, (see Figure 78.9) but do not play a part in hearing. There are two vertical canals at right angles to each other and one horizontal canal, each canal having a swelling at one end called an *ampulla* (Figure 78.10). When the head rotates, the fluid in the canal lags slightly behind and stimulates sensory nerve endings in the ampullae. The resulting nerve impulses to the brain inform it of the direction and speed of rotation. The horizontal canal responds best to rotations in the horizontal plane, e.g. twisting the body round while in an upright position, and the vertical canals respond particularly to tilting movements of the body forwards, backwards or sideways. The information from the semi-circular canals and other sense organs in the middle ear enables us to maintain our posture when standing still, and keep our balance while moving about.

Questions

1. If a piece of ice is pressed on to the skin, which receptors are likely to send impulses to the brain?

2. Apart from the cells which detect chemicals, what other types of receptor must be present in the tongue?

3. What is the difference between taste and smell?

4. In Figure 78.5*b* (page 328) what structures of the eye are *not* shown in the diagram?

5. In Figure 78.5*a*, explain what the broken lines are meant to represent.

6. An eye defect known as 'cataract' results in the lens becoming opaque. To relieve the condition, the lens can be removed. Make a diagram to show how an eye without a lens could, with the aid of spectacles, form an image on the retina. What disadvantages would result from such an operation?

7. Many people over the age of 50 have to wear spectacles for reading. What sort of lenses will these spectacles need? Explain your answer.

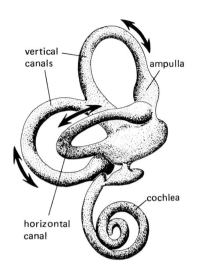

Figure 78.10

8. In what functional way does a sensory cell in the retina differ from a sensory cell in the cochlea?

9. How do you think you tell the difference between (*a*) a high and a low note, (*b*) a loud and a quiet sound?

10. Which semi-circular canals are likely to be maximally stimulated while you are turning a somersault? What other sensory information will be reaching the brain while you are doing this?

79 THE NERVOUS SYSTEM

Figure 79.1 is a diagram of the human nervous system. The brain and spinal cord together form the *central nervous system*. Nerves carry electrical impulses from the central nervous system to all parts of the body, making muscles contract or glands produce enzymes or hormones. The nerves also carry impulses back to the central nervous system from the sense organs of the body. These impulses from the eyes, ears, skin, etc., make us aware of changes in our surroundings or in ourselves. Nerve impulses from the sense organs to the central nervous system are called *sensory impulses*; those from the central nervous system to the muscles, resulting in action of some kind, are called *motor impulses*.

Figure 79.1

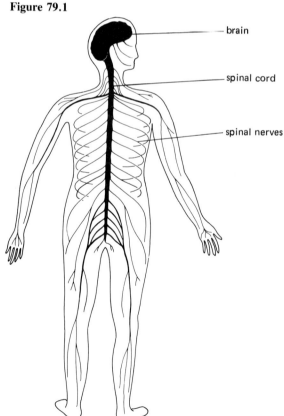

NERVOUS SYSTEM

Nerve cells (neurones)

The central nervous system and the nerves are made up of nerve cells, called *neurones*. Figure 79.3 shows three types of neurone. The *motor neurone* carries motor impulses from the central nervous system to muscles and glands. The *sensory neurone* carries sensory impulses from the sense organs to the central nervous system. The *multipolar neurone* is neither sensory nor motor but makes connections to other neurones inside the central nervous system.

All the neurones have a *cell body* consisting of a nucleus surrounded by a little cytoplasm. Branching processes, called *dendrites*, from the cell body of a multi-polar or motor neurone make contact with other neurones.

From the cell body of the sensory and motor neurones there extends a long filament of cytoplasm surrounded by an insulating sheath. This filament is called a *nerve fibre*. The cell bodies of the neurones are mostly located in the brain or spinal cord and it is the fibres which run in the nerves (see Figure 79.8, page 336). A nerve, therefore, which is an easily visible, white, tough, stringy structure, consists of hundreds of microscopic fibres bundled together. Figure 79.2 shows this structure in a simplified form. Most nerves will contain a mixture of sensory and motor fibres. The insulating sheath round the fibres prevents the impulses crossing from one fibre to another.

Some of the nerve fibres are very long. The

Figure 79.2

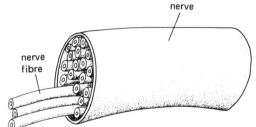

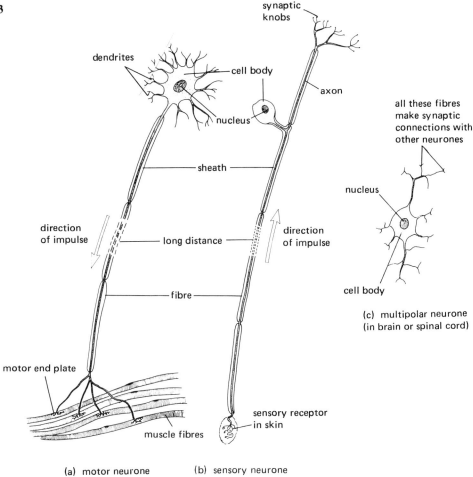

Figure 79.3

(a) motor neurone (b) sensory neurone (c) multipolar neurone (in brain or spinal cord)

nerve fibres to the foot will have their cell bodies in the spinal cord and the fibres will run in the nerves, without a break, down to the skin of the toes or the muscles of the foot. Thus, a single nerve cell may have a fibre about one metre long.

Synapses

Although nerve fibres are insulated, it is necessary for impulses to pass from one neurone to another. An impulse from the finger tips will have to pass through at least three neurones before reaching the brain and so produce a conscious sensation. The regions where impulses are able to cross from one neurone to the next are called *synapses*.

At these points, branches at the end of one fibre are in close contact with the cell body or dendrites of another neurone (Figure 79.4). When an impulse arrives at the synapse it releases a chemical which sets off an impulse in the next neurone. Sometimes several impulses have to arrive at the synapse before an impulse is fired off in the next neurone.

The nerve impulse

The nerve fibres do not carry sensations like pain or cold. These sensations occur only when a nerve impulse reaches the brain. The impulse itself is a series of electrical pulses which travel down the fibre rather like morse code. All nerve impulses are similar; there is no difference between nerve impulses from the eyes, ears or hands. We are able to tell the difference only because the impulses are sent to different parts of the brain.

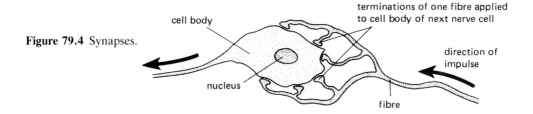

Figure 79.4 Synapses.

The reflex arc

One of the simplest situations where impulses are relayed across synapses to produce action is in the reflex arc. A reflex action is an automatic response which cannot be consciously controlled. When a particle of dust touches the cornea of your eye, you will blink; you cannot prevent yourself from blinking. A particle of food touching the lining of the windpipe will set off a coughing reflex you cannot suppress. When a bright light shines in your eye, the pupil contracts (see page 327). We cannot stop this reflex and are not usually aware that it is happening.

The nervous pathway for such reflexes is called a *reflex arc*. Figure 79.5 shows such a nervous pathway from a receptor in the hand to a muscle in the arm. The organ which responds to the motor impulse is sometimes called the *effector*. In this case, the biceps muscle is the effector. Figure 79.6 shows the reflex arc in more detail. Here the spinal cord is drawn in transverse section. The spinal nerve divides into two 'roots' at the point where it joins the spinal cord. All the sensory fibres enter through the dorsal root and the motor fibres all leave through the ventral root, but they both travel in the same spinal nerve. The cell bodies of all the sensory fibres are situated in the dorsal root and they make a bulge called a *ganglion* (Figure 79.7).

A well-known reflex is the 'knee jerk'. One leg is crossed over the other and the muscles totally relaxed. If the tendon just below the knee cap of the upper leg is tapped sharply, a reflex arc makes the thigh muscle contract and the lower part of the leg swings forward. Figure 79.6 traces the pathway of this reflex arc. Striking the tendon causes the muscle's stretch receptor to fire off impulses in the sensory fibre. These impulses travel in the nerve to the spinal cord. In the central region of the spinal cord, the sensory fibre makes a synapse with the motor neurone which conducts the impulse down the fibre, back to the thigh muscle. The arrival of the impulse at the muscle makes it contract and jerk the lower part of the limb forward. You are aware that this is happening (which means that sensory impulses must be reaching the brain) but there is nothing you can do to stop it.

In even the simplest reflex action, many more nerve fibres, synapses and muscles are involved than are described here. Figure 79.8 shows the reflex arc which would result in someone's hand being removed from a painful stimulus. On the right side of the spinal cord, some of the incoming sensory fibres are shown making synapses with neurones sending fibres to the brain, thus keeping the brain informed about events in the body. Also, nerve fibres from the brain make synapses with motor neurones in the spinal cord so that 'commands' from the brain can be transmitted to muscles of the body.

The reflex just described is a *spinal reflex*. The brain, theoretically, is not needed for it to happen. The reflex arcs for reactions in the head such as blinking, coughing and iris contraction are located in the brain.

Figure 79.5

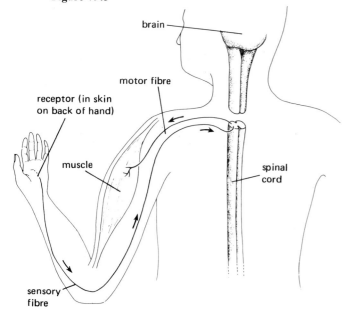

Figure 79.6

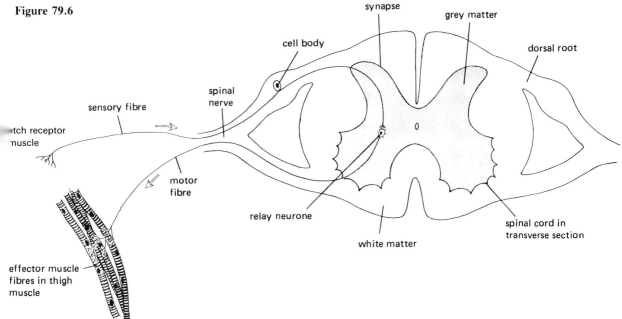

CENTRAL NERVOUS SYSTEM

Spinal cord

Like all parts of the nervous system, the spinal cord consists of thousands of nerve cells. Figure 79.8 shows its structure diagrammatically. Figure 79.6 shows it in transverse section and Figure 79.9 is a photograph of such a section. None of these diagrams show how it is protected by being enclosed in the vertebral column (see page 322).

All the cell bodies, apart from those in the dorsal root ganglion, are concentrated in the central region called the *grey matter*. The *white matter* consists of nerve fibres. Some of these will be passing from the grey matter to the spinal nerves and others will be running along the spinal cord, connecting the spinal nerve fibres to the brain. The spinal cord is thus concerned with (*i*) reflex actions involving body structures below the neck, (*ii*) conducting sensory impulses from the skin and muscles to the brain, and (*iii*) carrying motor impulses from the brain to the muscles of the limbs and trunk.

Figure 79.7

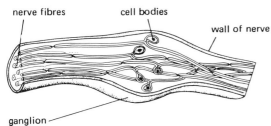

Voluntary actions

A voluntary action starts in the brain. It may be the result of external events, e.g. seeing a book on the floor which needs picking up, but the resulting action is entirely voluntary; it does not have to happen. The brain sends motor impulses down the spinal cord in the nerve fibres. These make synapses with motor fibres which enter spinal nerves and make connections to the sets of muscles needed to produce effective action. Many sets of muscles in the arms, legs and trunk would be brought into play in order to stoop and pick up the book, and impulses passing between the eye, brain and arm would direct the hand to the right place and 'tell' the fingers when to close on the book.

The brain

The brain may be regarded as the expanded front end of the spinal cord with certain areas greatly enlarged to deal with all the complicated information arriving from the ears, eyes, tongue, nose and semi-circular canals. Figure 79.9 gives a simplified diagram of the main regions of the brain as seen in vertical section. The *medulla* is concerned

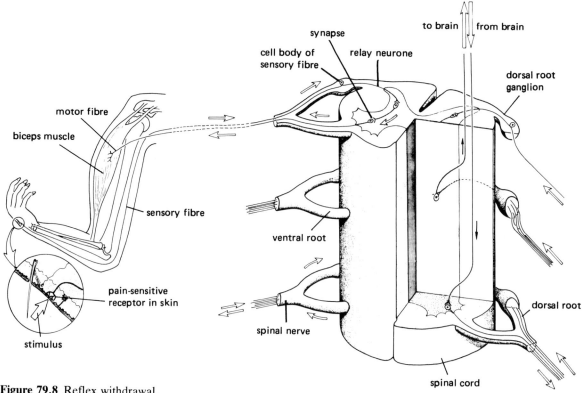

Figure 79.8 Reflex withdrawal.

with regulation of the heart beat, body temperature and breathing rate. The *cerebellum* controls balance and movement. The *mid-brain* deals with reflexes involving the eye. The major parts of the brain, however, are the *cerebral hemispheres*. These are very large and highly developed in mammals, especially man, and are thought to be the regions concerned with intelligence, memory, reasoning ability and acquired skills.

In the cerebral hemispheres and the cerebellum, there is an outer layer of grey matter, the *cortex*, with hundreds of thousands of multi-polar neurones (Figure 79.3c, page 333) which make possible an enormous number of synaptic connections between dendrites.

CO-ORDINATION

From this account, you can see that the nervous system links together, or *co-ordinates*, all bodily activities. Without co-ordination, the functions of different sets of organs would be quite unrelated. The nervous system co-ordinates impulses from the eye with motor impulses to the arm muscles so that we can pick up objects; it co-ordinates the heart with bodily activity so that, when the tissues need more oxygen during exercise the heart beats faster and supplies this need. The movements of the limbs are co-ordinated by the nervous system so that, in walking, the legs are

Figure 79.9

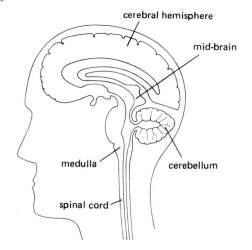

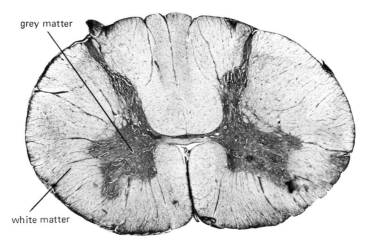

Figure 79.10 Transverse section through spinal cord.

moved alternately and not both together. The action of the lips, epiglottis and tongue must be co-ordinated by the nervous system to make sure that when food is swallowed, it enters the gullet and not the windpipe (see page 295).

Without co-ordination, the bodily activities would be thrown into chaos and disorder; food would pass undigested through the alimentary canal because enzymes were not secreted at the right time; a runner would collapse after a few metres from lack of oxygen when his heart failed to speed up to meet the increased demands, and so on.

Questions

1. In what ways are sensory neurones and motor neurones similar (*a*) in structure, (*b*) in function? How do they differ?

2. Can (*a*) a nerve fibre, (*b*) a nerve, carry both sensory and motor impulses? Explain your answer.

3. What might be the longest distance a nerve impulse could travel in a nerve fibre without having to cross a synapse?

4. Explain why the tongue may be considered to be both a receptor and an effector organ.

5. If you could intercept and 'listen' to the nerve impulses travelling in the spinal cord, could you tell which ones came from pain receptors and which ones from cold receptors? Explain your answer.

6. Discuss whether coughing is a voluntary or a reflex action.

7. Draw a *simple* diagram of the reflex arc which might produce a blinking response to a particle blown into the eye. (Treat the brain as simply part of the spinal cord for this purpose.)

8. In general terms, how might the nervous system co-ordinate the rate and depth of breathing, with the body's need for oxygen during exercise?

9. Describe the biological events involved when you hear a sound and turn your head towards it.

Magnetism and Electricity

80 MAGNETS AND MAGNETIC FIELDS

Permanent magnets

Permanent magnets do not readily lose their magnetism with normal treatment. The first permanent magnets were made of steel (an alloy of iron); modern magnets are much stronger and are of two types.

a *Alloy magnets* contain metals, e.g. iron, nickel, copper, cobalt, aluminium. They have trade names such as Alnico and Alcomax.

b *Ceramic magnets* are made from compounds called ferrites which consist of iron oxide and barium oxide. They are brittle. One has the trade name Magnadur.

Magnets are used in cycle dynamos, electric motors, loudspeakers and telephones. Ferrite powder can be bonded with plastic and rubber to give a flexible magnet or one of any shape. Very fine powder, each particle of which can be magnetized, is used to coat tapes for audio and video recorders, and tapes and discs for computer memories. Figure 80.1 shows a 'floppy' disc which stores data magnetically in small computers.

Properties of magnets

(*a*) *Magnetic materials.* Magnets only attract strongly certain materials such as iron, steel, nickel, cobalt and ferrites, which are called ferromagnetics.

(*b*) *Magnetic poles.* These are the places in a magnet to which magnetic materials are attracted, e.g. iron filings. They are near the ends of a bar magnet and occur in pairs of equal strength.

(*c*) *North and south poles.* If a magnet is supported so that it can swing in a horizontal plane it comes to rest with one pole, the north-seeking or N pole, always pointing roughly towards the earth's N pole. A magnet can therefore be used as a compass.

(*d*) *Law of magnetic poles.* If the N pole of a magnet is brought near the N pole of a suspended magnet repulsion occurs (Figure 80.2*a*). Two S poles also repel. By contrast, N and S poles always attract (Figure 80.2*b*). The law of magnetic poles summarizes these facts and states:

Like poles repel, unlike poles attract.

Figure 80.1

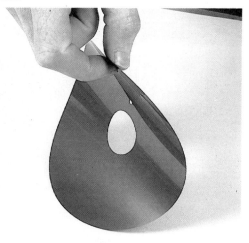

Figure 80.2

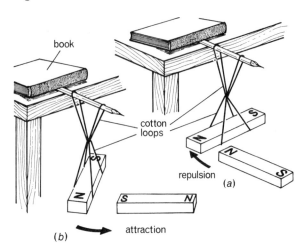

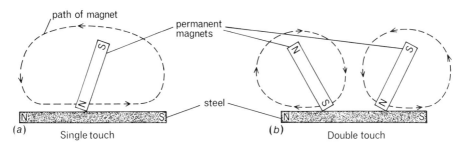

Figure 80.3

Test for a magnet

A permanent magnet causes repulsion with one pole when both poles are, in turn, brought near to a suspended magnet. An unmagnetized magnetic material would give attraction with both poles of the suspended magnet.

Repulsion is the only sure test for a magnet.

Making a magnet

(*a*) *By stroking.* The methods of single and double touch are shown in Figure 80.3*a* and *b*; steel knitting needles, hair grips or pieces of clock spring can be magnetized. In single touch, the steel is stroked from end to end about twenty times in the same direction by the same pole of a permanent magnet. In the better method of double touch, stroking is done from the centre outwards with unlike poles of two magnets at the same time. The magnets must be lifted high above the steel at the end of each stroke in both methods.

The pole produced at the end of the steel where the stroke ends is of the opposite kind to that of the stroking pole.

(*b*) *Electrically.* The magnetic material is placed inside a cylindrical coil called a solenoid (page 362), having several hundred turns of insulated copper wire, which is connected to a 6–12 V *direct current* (d.c.) supply (Figure 80.4*a*). If the current is switched on for a second and then off, the material is found to be magnetized when removed from the solenoid.

The polarity of the magnet depends on the direction of the current and is given by the *right-hand grip rule.* It states that if the fingers of the right hand grip the solenoid in the direction of the current (i.e. from the positive of the supply), the thumb points to the N pole (Figure 80.4*b*).

In practice magnets are made electrically using a very large current for a fraction of a second. Figure 80.5 shows loudspeakers being magnetized.

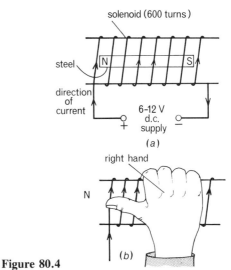

Figure 80.4

Figure 80.5

Demagnetizing a magnet

The magnet is placed inside a solenoid through which *alternating current* (a.c.) is flowing (Figure 80.6). With the current still passing, the magnet is slowly removed to a distance from the solenoid.

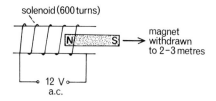

Figure 80.6

Induced magnetism

When a piece of unmagnetized magnetic material touches or is brought near to the pole of a permanent magnet, it becomes a magnet itself. The material is said to have magnetism *induced* in it. Figure 80.7 shows that a N pole induces a N pole in the far end.

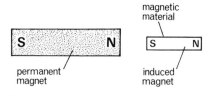

Figure 80.7

This can be checked by hanging two iron nails from the N pole of a magnet. Their lower ends repel each other (Figure 80.8a) and both are repelled by the N pole of another magnet (Figure 80.8b).

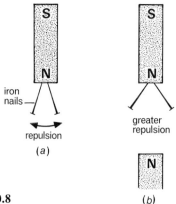

Figure 80.8

Magnetic properties of iron and steel

Chains of small iron paper clips and steel pen nibs can be hung from a magnet (Figure 80.9). Each clip or nib magnetizes the one below it by induction and the unlike poles so formed attract.

If the iron chain is removed by pulling the top clip away from the magnet, the chain collapses,

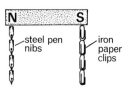

Figure 80.9

showing that *magnetism induced in iron is temporary*. When the same is done with the steel chain, it does not collapse; *magnetism induced in steel is permanent*.

Magnetic materials like iron which magnetize easily but do not keep their magnetism, are said to be 'soft'. Those like steel which are harder to magnetize than iron but stay magnetized, are 'hard'. Both types have their uses; very hard ones are used to make permanent magnets.

Theory of magnetism

If a magnetized piece of steel clock spring or thin rod is cut into smaller pieces, each piece is a magnet with a N and a S pole (Figure 80.10). It is therefore reasonable to suppose that a magnet is made up of lots of 'tiny' magnets all lined up with their N poles pointing in the same direction (Figure 80.11a). At the ends, the 'free' poles of the 'tiny' magnets repel each other and fan out so that the poles of the magnet are *round* the ends.

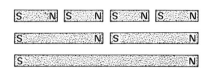

Figure 80.10

In an unmagnetized bar we can imagine the 'tiny' magnets pointing in all directions, the N pole of one being neutralized by the S pole of another. Their magnetic effects cancel out and there are no 'free' poles near the ends (Figure 80.11b).

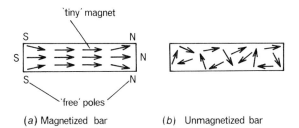

(a) Magnetized bar   (b) Unmagnetized bar

Figure 80.11

As well as explaining the breaking of a magnet, this theory accounts for the following.

a *Magnetic saturation.* There is a limit to the strength of a magnet. It occurs when all the 'tiny' magnets are lined up.

b *Demagnetization by heating or hammering.* Both processes cause the atoms of the magnet to vibrate more vigorously and disturb the alignment of the 'tiny' magnets.

There is evidence to show that the 'tiny' magnets are groups of millions of atoms, called *domains*. In a ferromagnetic material each atom is a magnet and the magnetic effect of every atom in a particular domain acts in the same direction.

Storing magnets

A magnet tends to become weaker with time due to the 'free' poles near the ends repelling each other and upsetting the alignment of the domains. To prevent this, bar magnets are stored in pairs with unlike poles opposite and pieces of soft iron, called *keepers*, across the ends (Figure 80.12).

The keepers become induced magnets and their poles neutralize the poles of the bar magnets. The domains in both magnets and keepers form closed chains with no 'free' poles.

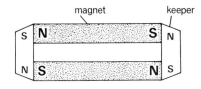

Figure 80.12

Magnetic fields

The space surrounding a magnet where it produces a magnetic force is called a *magnetic field*. The force around a bar magnet can be detected and shown to vary in direction using the apparatus in Figure 80.13. If the floating magnet is released near the N pole of the bar magnet, it is repelled to the S pole and moves along a curved path known as a *line of force* or a *field line*. It moves in the opposite direction if its S pole is uppermost.

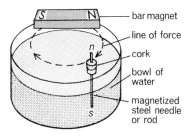

Figure 80.13

It is useful to consider that a magnetic field has a direction and to represent the field by lines of force. It has been decided that *the direction of the field at any point should be the direction of the force on a N pole*. To show the direction, arrows are put on the lines of force and point away from a N pole towards a S pole.

Experiment 1 Plotting lines of force

a *Plotting compass method.* A plotting compass is a small pivoted magnet in a glass case with brass walls (Figure 80.14a).

Lay a bar magnet NS on a sheet of paper (Figure 80.14b). Place the plotting compass at a point such

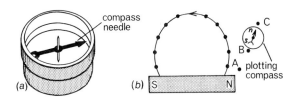

Figure 80.14

as A near one pole of the magnet. Mark the position of the poles *n*, *s* of the compass by pencil dots A, B. Move the compass so that pole *s* is exactly over B, mark the new position of *n* by dot C.

Continue this process until the S pole of the bar magnet is reached. Join the dots to give one line of force and show its direction by putting an arrow on it. Plot other lines by starting at different points round the magnet.

This method is slow but suitable for weak fields.

b *Iron filings method.* Place a sheet of paper *on top of* a bar magnet and sprinkle iron filings *thinly and evenly* on to the paper from a 'pepper pot'.

Tap the paper gently with a pencil and the filings should form patterns of the lines of force. Each filing is magnetized by induction and turns in the direction of the field when the paper is tapped.

The method is quick but no use for weak fields. Figure 80.15 shows typical patterns with two magnets.

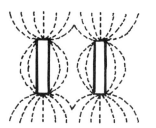
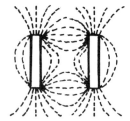

Figure 80.15

- Why are they different?

Earth's magnetic field

If lines of force are plotted on a sheet of paper with no magnets near, a set of parallel straight lines are obtained. They run roughly from S to N geographically (Figure 80.16) and represent a small part of the earth's magnetic field in a horizontal plane.

The *combined* field due to the earth and a bar magnet with its N pole pointing north is shown in

Figure 80.16

Figure 80.17a: it is obtained by the plotting compass method. At the points marked X, the fields of the earth and the magnet are equal and opposite and the resultant field is zero. They are called *neutral points* and no lines of force pass through them. At A the magnet's field is stronger than the earth's field; at B it is weaker.

When the N pole of the magnet points south, the neutral points Y are along the axis of the magnet (Figure 80.17b).

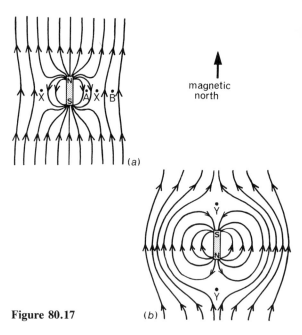

Figure 80.17

Declination and dip

(*a*) *Declination.* At most places on the earth's surface a magnetic compass points east or west of true north, i.e. the earth's geographical and magnetic N poles do not coincide. The angle between magnetic north and true north is called the *declination* (Figure 80.18). In London in 1982 it was 7°W of north and is decreasing. By about the year 2140 it should be 0°.

Figure 80.18

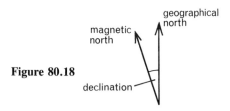

(*b*) *Dip.* A magnet pivoted at its centre of gravity so that it can rotate in a vertical plane is called a *dip needle* (Figure 80.19). When placed in the magnetic meridian, i.e. in the vertical plane containing

Figure 80.19

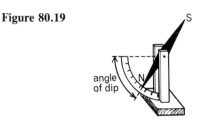

magnetic north and south, it comes to rest with its N pole pointing downwards at an angle, in London, of about 67°. The angle a dip needle makes with the horizontal in the magnetic meridian is called the *angle of dip* or *inclination*.

The angle of dip varies over the earth's surface from 0° at the earth's magnetic equator to 90° at its magnetic poles. This can be explained if we consider the earth behaves as though it had, at its centre, a strong bar magnet whose S pole points to magnetic north (Figure 80.20).

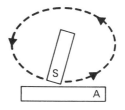

Figure 80.21

Figure 80.22

(b) Using this metal, what kind of pole would be produced at A?

3. In Figure 80.22, XY is an iron bar inside a coil of wire carrying a current in the direction shown. Which end of XY becomes a N pole?

4. Explain why needles hung from either end of a bar magnet held horizontally incline towards one another as shown in Figure 80.23.

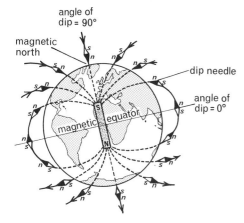

Figure 80.20

The cause of the earth's magnetism may be electric currents in the liquid core at its centre but there is no generally accepted theory.

Questions

1. Which one of the following does a magnet attract? **A** plastics, **B** any metal, **C** iron and steel, **D** aluminium, **E** carbon?

2. A metal bar is being stroked by the S pole of a magnet in Figure 80.21.

(a) What must the bar be made of to become a permanent magnet?

Figure 80.23

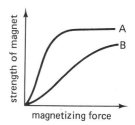

Figure 80.24

5. The graphs in Figure 80.24 are for two magnetic materials.

(a) Does material A or B become the stronger magnet?
(b) Which material is easier to magnetize?

6. In the three lines of force diagrams of Figure 80.25, say whether each of the poles A, B, C, D, E, F is a north or a south.

7. Explain with the help of diagrams the terms (a) neutral point, (b) dip, (c) declination.

Figure 80.25

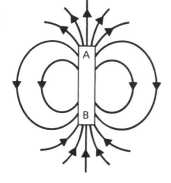 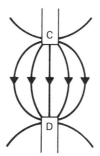 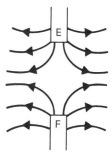

81 STATIC ELECTRICITY

A nylon garment often crackles when it is taken off. We say it has become 'charged with static electricity'; the crackles are caused by tiny electric sparks which can be seen in the dark. Pens and combs made of certain plastic become charged when rubbed on the sleeve and can then attract scraps of paper.

A flash of lightning is nature's most spectacular static electricity effect (Figure 81.1).

Figure 81.1

ELECTRICAL CHARGES

Positive and negative charges

When a strip of polythene (white) is rubbed with a cloth it becomes charged. If it is hung up and another rubbed polythene strip is brought near, repulsion occurs (Figure 81.2). Attraction occurs when a rubbed cellulose acetate (clear) strip approaches.

Figure 81.2

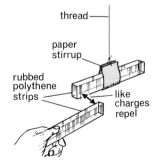

This shows there are two kinds of electric charge. That on cellulose acetate is taken as *positive* (+) and that on polythene is *negative* (−). It also shows that:

> *like charges* (+ and + or − and −) *repel*
> *unlike charges* (+ and −) *attract*.

Charges, electrons and atoms

There is evidence that we can picture an atom as made up of a small nucleus containing positively charged particles called *protons*, surrounded by an equal number of negatively charged *electrons* (see page 376). The charges on a proton and an electron are equal and so an atom as a whole is normally electrically neutral, i.e. has no net charge.

Hydrogen is the simplest atom (page 384) with one proton and one electron (Figure 81.3). A copper atom has 29 protons in the nucleus and 29 surrounding electrons. All nuclei (plural of nucleus) also contain uncharged particles called *neutrons*.

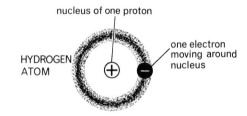

Figure 81.3

The production of charges by rubbing can be explained by supposing that electrons are transferred from one material to the other. For example, when cellulose acetate is rubbed, electrons go from the acetate to the cloth, leaving the acetate short of electrons, i.e. positively charged. The cloth now has more electrons than protons and becomes negatively charged. Note that it is electrons which move, the protons remain in the nucleus.

Experiment 1 Gold leaf electroscope

A gold leaf electroscope consists of a metal cap on a metal rod at the foot of which is a metal plate having a leaf of gold foil attached (Figure 81.4). The rod is held by an insulating plastic plug in a case with glass to protect the leaf from draughts.

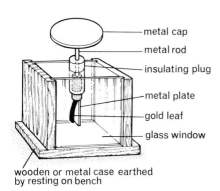

Figure 81.4

a *Detecting a charge.* Bring a charged polythene strip towards the cap: the leaf rises away from the plate. On removing the charged strip the leaf falls again. Repeat with a charged acetate strip.

b *Charging by contact.* Draw a charged polythene strip *firmly across the edge* of the cap. The leaf should rise and stay up when the strip is removed. If it does not, repeat the process but press harder. The electroscope has now become negatively charged by contact with the polythene strip.

c *Finding the sign of a charge.* Recharge the polythene strip and bring it near the cap of the negatively charged electroscope. The leaf should rise farther.

Remove the polythene strip and bring up a charged acetate strip. The leaf should fall a bit.

Discharge the electroscope by touching the cap with your hand. Its charge then passes through your body to the earth and the leaf collapses completely.

Charge the electroscope positively by contact with a charged acetate strip. Show that the leaf rises farther when a charged acetate strip approaches the cap but falls when a charged polythene strip approaches.

We can conclude that *if the leaf rises more, the sign of the charge being investigated is the same as the charge on the electroscope.*

If the leaf falls it does not always follow that a charge of the opposite sign has approached the cap. An uncharged body has the same effect, as you can show by holding your hand near the cap. The sure test is when the leaf rises farther, i.e. diverges.

d *Insulators and conductors.* Touch the cap of the charged electroscope with different things, e.g. a piece of paper, a wire, your finger, a comb, a cotton handkerchief, a piece of wood, a glass rod, a plastic pen, rubber tubing.

When the leaf falls, charge is passing to or from the earth through you and the material touching the cap. If the fall is rapid the material is a *good conductor*, if slow, it is a poor conductor and if the leaf does not alter, the material is a *good insulator*. Record your results.

Insulators and conductors

In an insulator all electrons are bound firmly to their atoms; in a conductor some electrons can move freely from atom to atom. An insulator can be charged by rubbing because the charge produced cannot move from where the rubbing occurs, i.e. the electric charge is static. A conductor will become charged only if it is held in an insulating handle; otherwise the charge passes to earth via our body.

Good insulators are plastics such as polythene, cellulose acetate, Perspex, nylon. All metals and carbon are good conductors. Inbetween are materials that are both poor conductors and (because they conduct to some extent) poor insulators. Examples are wood, paper, cotton, the human body, the earth. Water conducts and if it were not present in materials like wood and on the surface of, for example, glass, these would be good insulators. Dry air insulates well.

Questions

1. Explain in terms of electron movement, what happens when a polythene rod becomes negatively charged by being rubbed with a cloth.

2. An insulated metal ball is given an electric charge. It is then touched with a rod of material, one end of which is held in the hand. What happens to its charge if the rod is made from a material which is (*a*) a good insulator, (*b*) a poor insulator, (*c*) a good conductor?

82 ELECTRIC CURRENT

An electric current is a flow of moving electric charges. A battery produces a current if there is a *complete path* (i.e. a circuit) of *conductors* connected between its two terminals. One terminal is called the positive (+) and the other the negative (−) of the battery.

Wires of copper are used to connect batteries, lamps, etc., in a circuit because copper is a good electrical conductor. If wires are covered with insulation, e.g. plastic, this must be removed from the ends before they are connected.

In a metal the current consists of free electrons drifting slowly round the metal in the direction from the negative to the positive terminal of the battery.

Before the electron was discovered scientists agreed to think of current as positive charges moving round a circuit in the direction from positive to negative of a battery. This agreement is still kept—to change it would cause confusion.

Effects of current

A current has *three* effects, any one of which tells us when it is flowing in a circuit. All three effects may be shown with the apparatus shown in Figure 82.1.

Figure 82.1

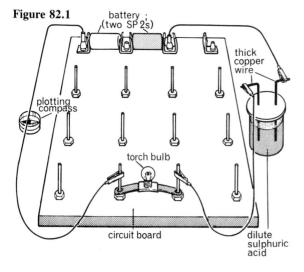

(*a*) *Heating and lighting effect.* The bulb lights due to the current making a small wire in it (called the filament) white hot.

(*b*) *Magnetic effect.* The plotting compass is deflected when it is placed in the magnetic field produced round the wire by the current.

(*c*) *Chemical effect.* Bubbles of gas are given off at the wires in the acid because of the chemical action caused by the current. See pages 58–60 for more detailed discussion.

The ampere and ammeters

The unit of current is the *ampere* (shortened to A). The current flowing in a small torch bulb is about $\frac{1}{4}$ A (0.25 A).

Current is measured by an *ammeter* (Figure 88.8, page 368). It works by using the magnetic effect of the current (see page 362). An ammeter should be treated carefully. One of its terminals is marked with a plus or positive sign (+) and is sometimes coloured red; the other terminal may be marked with a minus or negative sign (−) and is sometimes coloured black. The signs tell you how to connect the ammeter into a circuit (see the experiment that follows) to prevent damage.

Circuit diagrams

The signs or symbols used for various parts of an electrical circuit are shown in Figure 82.2. A battery consists of two or more electric cells joined together. A switch makes a gap in a circuit when it is 'off' and completes the circuit when it is 'on'.

Arrows on a circuit diagram show the direction in which positive charges would flow, that is, the direction of what is called the *conventional current*.

Experiment 1 Measuring current

a Connect the circuit of Figure 82.3*a* (on a circuit board if possible) ensuring that the + of the cell

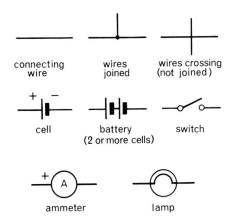

Figure 82.2

the metal stud on an SP2) goes to the + of the ammeter (marked red). Note the current.

b Connect the circuit of Figure 82.3*b*. The cells are *in series* (+ of one to − of other), as are the lamps. Record the current. Measure the current at B, C, and D by disconnecting the circuit at each point in turn and inserting the ammeter.

- What do you find?

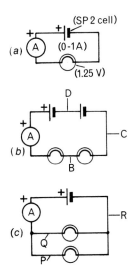

Figure 82.3

c Connect the circuit of Figure 82.3*c*. The lamps are *in parallel*. Read the ammeter. Also measure the currents at P, Q, and R.

- What is your conclusion?

Series and parallel circuits

In a *series* circuit the current is the same at all points; it is not used up.

In a *parallel* circuit the total current equals the sum of the currents in the separate branches. For example, in Figure 82.3*c*, if the current at P is 0·2 A and at Q 0·2 A then the current at R is $0·2 + 0·2 = 0·4$ A.

The coulomb

The unit of charge, the *coulomb* (C), is defined in terms of the ampere.

One coulomb is the charge passing any point in a circuit when a steady current of 1 ampere flows for 1 second.

A charge of 3 C would pass each point in 1 s if the current was 3 A. In 2 s, $3 \times 2 = 6$ C would pass. In general, if a steady current I (amperes) flows for time t (seconds) the charge Q (coulombs) passing any point is given by:

$$Q = I \times t$$

The charge on about 6 million million million (6×10^{18}) electrons equals 1 C.

Questions

1. Draw the circuit diagram for a circuit containing two batteries, two lamps and a switch so that all are in the same circuit and the lamps have their full, normal brightness when the switch is pressed.

2. The lamps and the batteries in all the circuits of Figure 82.4 are the same. If the lamp in circuit *a* has its full, normal brightness, what can you say about the brightness of the lamps in *b*, *c*, and *d*?

Figure 82.4

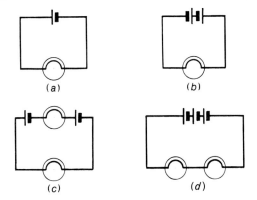

349

3. The batteries and the lamps in Figure 82.5 are exactly alike. If A reads 0·2 A, say whether the readings on A_1, A_2, A_3 and A_4 will be the same, greater or less.

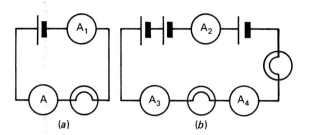

Figure 82.5

4. Draw two diagrams, (a) one to show how three lamps all in the same circuit could be fully lit by three batteries and (b) one to show how they could be fully lit by one battery. In which circuit would the lamps remain alight longest?

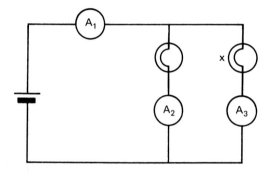

Figure 82.6

5. If the lamps are all the same in Figure 82.6 and if A_1 reads 0·5 A, what do A_2 and A_3 read?
If lamp X is removed what will be the reading on each of the three ammeters?

6. When the circuit of Figure 82.7 is set up it is found that the lamp is fully lit but the pointer of the ammeter not deflected over the scale. Why?

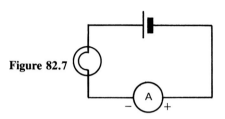

Figure 82.7

7. If the lamps are all the same in Figure 82.8 and if A_1 reads 0·5 A, what do A_2, A_3, A_4 and A_5 read?

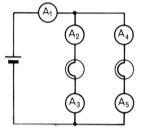

Figure 82.8

8. If the current through a floodlamp is 5 A, what charge passes in (a) 1 s, (b) 10 s, (c) 5 minutes?

9. What are the currents in a circuit if the charges passing each point are (a) 10 C in 2 s, (b) 20 C in 40 s, (c) 240 C in 2 minutes?

83 POTENTIAL DIFFERENCE

Heat flows naturally from a body at a higher temperature to one at a lower temperature. A liquid flows of its own accord from a higher pressure to a lower one. Temperature and pressure are the qualities which decide the direction of flow of heat and liquids, respectively.

With electricity we consider that electric charge flows from one point to another if there is an electrical pressure difference between the two points. Such a pressure difference is produced between the terminals of a battery as a result of the chemical action inside it creating a surplus of electrons at the negative terminal and a shortage at the positive terminal. The pressure difference is called the *potential difference* or *p.d.* in the electrical case and it causes electron flow in a circuit connected across the terminals of the battery.

Electrical p.d. is measured in volts (shortened to V) and we often refer to the p.d. of a supply as its *voltage*. The voltage of the domestic mains supply in Britain is 240 V, of a car battery 12 V and of one SP2 cell 1·5 V.

Cells and batteries

Greater p.ds are obtained when cells are joined in series, i.e. + of one to − of next, to form a battery. In Figure 83.1a the two 1·5 V cells give a p.d. of 3 V at the terminals A, B.

If two 1·5 V cells are connected in parallel (Figure 83.1b) the p.d. at terminals P, Q is still 1·5 V but the arrangement behaves like a larger cell and will last longer.

The cells in Figure 83.1c are in opposition and the p.d. at X, Y is zero.

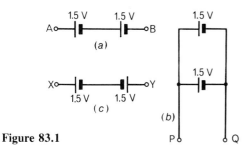

Figure 83.1

Experiment 1 Measuring p.d.

A *voltmeter* is an instrument for measuring p.d. It looks like an ammeter but has a scale marked in volts. Whereas an ammeter is inserted in *series* in a circuit to measure the current, a voltmeter is connected across that part of the circuit where the p.d. is required, i.e. *in parallel*. (We will see later, page 368, that a voltmeter should have a high resistance and an ammeter a low resistance.)

To prevent damage the + terminal (marked red) must be connected to the + of the battery.

a Connect the circuit of Figure 83.2a on a circuit board. The voltmeter gives the p.d. across the lamp. Read it.

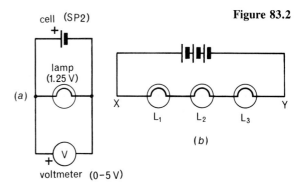

Figure 83.2

b Connect the circuit of Figure 83.2b. Measure:

(i) the p.d. V between X and Y
(ii) the p.d. V_1 across lamp L_1
(iii) the p.d. V_2 across lamp L_2
(iv) the p.d. V_3 across lamp L_3

- How does the value of V compare with $V_1 + V_2 + V_3$?

Energy changes and p.d.

In an electric circuit electrical energy is supplied from a source such as a battery and is changed into other forms of energy by devices in the circuit. A lamp produces heat and light.

If the circuits of Figure 83.3 are connected up, it

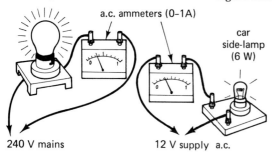

mains lamp (100 W)   a.c. ammeters (0–1A)   car side-lamp (6 W)
240 V mains   12 V supply a.c.

Figure 83.3

will be found from the ammeter readings that the current is about the same (0·4 A) in each lamp. However, the mains lamp with a p.d. of 240 V applied to it, gives much more light and heat than the car lamp with 12 V across it. In terms of energy, the mains lamp changes a great deal more electrical energy in a second than does the car lamp.

Evidently the p.d. across a device affects the rate at which it changes electrical energy. This gives us a way of defining the unit of p.d.—the volt.

Model of a circuit

It may help you to understand the definition of the volt, i.e. what a volt is, if you *imagine* that the current in a circuit is formed by 'drops' of electricity, each having a charge of 1 C and carrying equal-sized 'bundles' of electrical energy. In Figure 83.4, Mr Coulomb represents one such 'drop'. As a 'drop' moves round the circuit it gives up all its energy which is changed to other forms of energy. Note that electrical energy is 'used up', not charge or current.

In our imaginary representation, Mr Coulomb travels round the circuit and unloads energy as he goes, most of it in the lamp. We think of him receiving a fresh 'bundle' every time he passes through the battery, which suggests he must be travelling very fast. In fact, the electrons drift along quite slowly. As soon as the circuit is complete, energy is delivered at once to the lamp, not by electrons directly from the battery but from electrons that were in the connecting wires. The model is helpful but not an exact representation.

Figure 83.4

The volt

The demonstrations of Figure 83.3 show that the greater the p.d. at the terminals of a supply, the larger is the 'bundle' of electrical energy given to each coulomb and the greater is the rate at which light and heat are produced in a lamp.

The p.d. between two points in a circuit is 1 volt if 1 joule of electrical energy is changed into other forms of energy when 1 coulomb passes from one point to the other.

That is, 1 volt = 1 joule per coulomb (1 V = 1 J/C). If 2 J are given up by each coulomb, the p.d. is 2 V. If 6 J are changed when 2 C pass, the p.d. is 6/2 = 3 V.

Questions

1. The lamps and the cells in all the circuits of Figure 83.5 are the same. If the lamp in *a* has its full, normal brightness, what can you say about the brightness of the lamps in *b* and *c*?

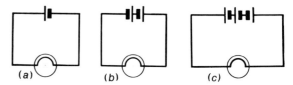

Figure 83.5

2. Three voltmeters V, V_1, V_2 are connected as in Figure 83.6.

   (a) If V reads 18 V and V_1 reads 12 V, what does V_2 read?

   (b) Copy Figure 83.6 and mark the positive terminals of the ammeter and voltmeters for correct connection.

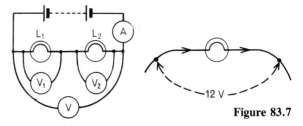

Figure 83.6

Figure 83.7

3. The p.d. across the lamp in Figure 83.7 is 12 V. How many joules of electrical energy are changed into light and heat when (a) a charge of 1 C passes through it, (b) a charge of 5 C passes through it?

4. Three 2 V cells are connected in series and used as the supply for a circuit.

   (a) What is the p.d. at the terminals of the supply?

   (b) How many joules of electrical energy does 1 C gain on passing through (i) one cell, (ii) all three cells?

84 RESISTANCE

When the same p.d. is applied across different conductors, different currents flow depending on how easily electrons can move in them. The opposition of a conductor to current flow is called its *resistance*. A long thin wire has more resistance than a short thick one of the same material because a given p.d. causes a smaller current in it.

A good conductor has a low resistance and a poor conductor a high resistance. Silver is the best conducting material, but copper, the next best, is much cheaper and is used for connecting wire and cables in electric circuits. Good insulators for static electricity are also good for current electricity.

RESISTORS AND THEIR PROPERTIES

Conductors intended to have resistance are called *resistors* (symbol ─▭─) and are made either from wires of special alloys or from carbon. Variable resistors or rheostats are useful in laboratory experiments and consist of a coil of constantan wire (an alloy of copper and nickel) wound on a tube with a sliding contact on a metal bar above the tube (Figure 84.1).

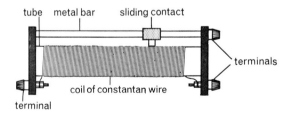

Figure 84.1

A rheostat is used to change the current in a circuit; only one end connection and the sliding contact are required. In Figure 84.2 moving the sliding contact to the left reduces the resistance and increases the current.

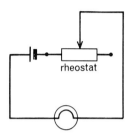

Figure 84.2

The ohm

If the current through a conductor is I when the p.d. across it is V (Figure 84.3) its resistance R is defined by:

$$R = \frac{V}{I}$$

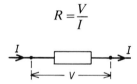

Figure 84.3

This is a reasonable way to measure resistance since the smaller I is for a given V, the greater is R. If V is in volts and I in amperes, R is in *ohms* (Ω: pronounced omega). For example, if $I = 2$ A when V is 12 V, then $R = 12/2 = 6\,\Omega$.

The ohm is the resistance of a conductor in which the current is 1 ampere when a p.d. of 1 volt is applied across it.

Experiment 1 Measuring resistance

The resistance R of a conductor can be found by measuring the current I through it when a p.d. V is applied across it and then using $R = V/I$. This is called the *ammeter–voltmeter method*.

Set up the circuit of Figure 84.4 in which the unknown resistance R is 1 m of S.W.G.34 constanton wire. Altering the rheostat changes both the p.d. V and the current I. Record in a table, with

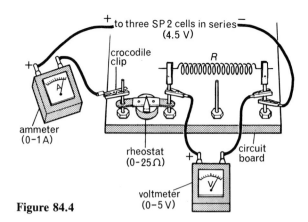

Figure 84.4

three columns, five values of I (e.g. 0·10, 0·15, 0·20, 0·25 and 0·30 A) and the corresponding values of V.

Work out R for each pair of readings.

Ohm's law

Experiments show that for metallic conductors, so long as their temperature does not change, V/I always has the same value when V is varied and the corresponding value of I found. That is, the resistance of a metallic conductor is constant whatever the p.d. applied. Hence V/I is a constant or $I \propto V$. This means, for example, that doubling V doubles I, and so on.

The current through a metallic conductor is directly proportional to the p.d. across its ends if the temperature and other physical conditions are constant.

This is *Ohm's law*. It applies to metals and some alloys, called ohmic conductors. Ohm's law can also be written as:

$$\frac{V}{I} = R \quad \text{or} \quad I = \frac{V}{R} \quad \text{or} \quad V = I \times R$$

The last one is easiest to remember; all are useful for calculations on ohmic conductors.

Figure 84.5

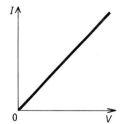

For an ohmic conductor a graph of I against V is a straight line through the origin since $I \propto V$ (Figure 84.5).

Resistors in series

The resistors in Figure 84.6 are in series. The *same current I flows through each and the total p.d. V across all three equals the separate p.d.s across*

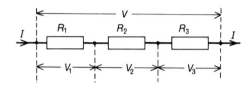

Figure 84.6

them, i.e.

$$V = V_1 + V_2 + V_3$$

But $V_1 = IR_1$, $V_2 = IR_2$ and $V_3 = IR_3$. Also, if R is the combined resistance, $V = IR$, and so.

$$IR = IR_1 + IR_2 + IR_3$$

Dividing both sides by I, then:

$$R = R_1 + R_2 + R_3$$

Resistors in parallel

The resistors in Figure 84.7 are in parallel. The *p.d. V between the ends of each is the same* and the total

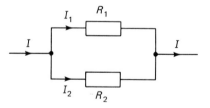

Figure 84.7

current I equals the sum of the currents in the separate branches, i.e.

$$I = I_1 + I_2$$

But $I_1 = V/R_1$ and $I_2 = V/R_2$. Also, if R is the combined resistance, $V = IR$, and so:

$$\frac{V}{R} = \frac{V}{R_1} + \frac{V}{R_2}$$

Dividing both sides by V, then:

$$\frac{1}{R} = \frac{1}{R_1} + \frac{1}{R_2}$$

Rearranging we get:

$$R = \frac{R_1 R_2}{R_1 + R_2} = \frac{\text{product of resistances}}{\text{sum of resistances}}$$

Worked example.

A p.d. of 24 V from a battery is applied to the network of resistors in Figure 84.8a.

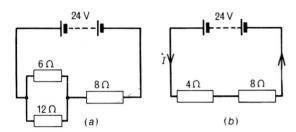

Figure 84.8

(a) What is the combined resistance of the 6 Ω and 12 Ω resistors in parallel?
(b) What is the current in the 8 Ω resistor?
(c) What is the p.d. across the parallel network?
(d) What is the current in the 6 Ω resistor?

(a) Let R_1 = resistance of 6 Ω and 12 Ω in parallel

$$\therefore \frac{1}{R_1} = \frac{1}{6} + \frac{1}{12} = \frac{2}{12} + \frac{1}{12} = \frac{3}{12}$$

$$\therefore R_1 = \frac{12}{3} = 4\,\Omega$$

(b) Let R = total resistance of circuit = $4 + 8 = 12\,\Omega$. The equivalent circuit is shown in Figure 84.8b and if I is the current in it then since $V = 24$ V

$$I = \frac{V}{R} = \frac{24}{12} = 2\,\text{A}$$

∴ Current in 8 Ω resistor = 2 A

(c) Let V_1 = p.d. across parallel network

$$\therefore V_1 = I \times R_1 = 2 \times 4 = 8\,\text{V}$$

(d) Let I_1 = current in 6 Ω resistor, then since $V_1 = 8$ V

$$I_1 = \frac{V_1}{6} = \frac{8}{6} = \frac{4}{3}\,\text{A}$$

Questions

1. What is the resistance of a lamp when a p.d. of 12 V across it causes a current of 4 A?

2. Calculate the p.d. across a 10 Ω resistor carrying a current of 2 A.

3. What is the current flowing through a 4 Ω resistor when a p.d. of 8 V is applied across it?

4. What is the effective total resistance of three 10 Ω resistors connected as in Figure 84.9?

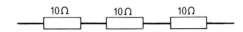

Figure 84.9

5. Calculate the effective resistance between X and Y in Figure 84.10.

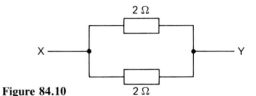

Figure 84.10

6. What is the effective resistance in Figure 84.11 between A and B?

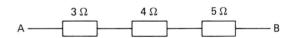

Figure 84.11

7. Figure 84.12 shows three resistors. What is their combined resistance?

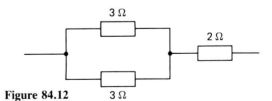

Figure 84.12

8. A 4 Ω coil and a 2 Ω coil are connected in parallel and a total current of 3 A passes through the coils.

(a) What is their combined resistance?
(b) What current flows through the 2 Ω coil?

85 ELECTRIC CELLS : E.M.F.

In an electric cell chemical energy becomes electrical energy.

Simple (voltaic) cell

This was the first kind of cell to be made. It has a voltage or electromotive force (e.m.f.—see later) of about 1·0 V but stops working after a short time due to 'polarization', i.e. the collection of hydrogen bubbles on the copper plate. The cell is depolarized by adding potassium dichromate which oxidizes the hydrogen to water. A second defect is 'local action'. This is due to impurities in the zinc and results in the zinc being used up even when current is not supplied. The simple cell is no longer used.

Dry (Leclanché) cell

This cell is used in torches and transistor radios and has a voltage (e.m.f.) of 1·5 V (page 357). The source of energy is chemical action between zinc and ammonium chloride, but as a result hydrogen is produced which collects at the carbon rod and 'polarizes' the cell. Manganese dioxide acts as the depolarizer. It can supply a steady current, e.g. 0·3 A, for some time.

Lead-acid accumulator

This accumulator supplies much larger currents than a dry cell and is called a 'secondary' cell because it can be recharged by passing a current through it in the opposite direction. Simple and dry cells are 'primary' cells. A car battery consists of six accumulators in series (Figure 85.1).

The positive plate is lead dioxide (brown) and the negative one is lead (grey), with sulphuric acid as the liquid. During discharge when current is being supplied, both plates change to lead sulphate (white) and the acid becomes more dilute. When fully charged the relative density of the acid (measured by a hydrometer) is 1·25 which falls to 1·18 at full discharge.

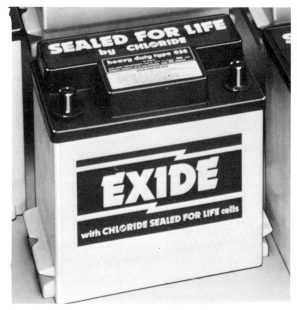

Figure 85.1

A charging circuit is shown in Figure 85.2. The supply must be direct current (d.c.) of greater voltage than that of the accumulator to be charged. The *positive* of the supply is joined to the *positive* of the battery and the current is adjusted by the rheostat to the recommended value. Before charging, the accumulators should be 'topped up' with distilled water unless they are sealed.

The capacity of an accumulator is stated in *ampere-hours* (Ah) for a 10 hour discharge time. A 30 Ah accumulator would supply 3 A for 10 hours but whilst 1 A would be supplied for more than 30 hours, 6 A would not flow for 5 hours.

Figure 85.2

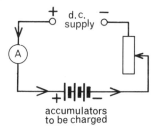

E.M.F. and terminal P.D.

A voltmeter connected across the terminals of a cell measures the *terminal p.d.* of the cell. The reading depends on the current being supplied; it is greatest when the cell is on 'open circuit', i.e. not supplying current (Figure 85.3a).

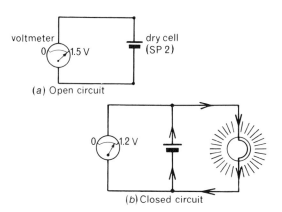

Figure 85.3

When a circuit is connected to the cell, it is on 'closed circuit' and the reading is less (Figure 85.3b). It decreases as the current supplied increases. The terminal p.d. of a cell on open circuit is thus greater than its terminal p.d. on closed circuit.

If we look upon a cell as a device which *supplies* electrical energy, we can define its terminal p.d. on open circuit as the number of joules of electrical energy it gives to each coulomb. If a voltmeter across the cell reads 1·5 V then each coulomb is supplied with 1·5 J of electrical energy and the cell is said to have an *electromotive force* (e.m.f. denoted by E) of 1·5 V.

The e.m.f. of a source of electrical energy is its terminal p.d. on open circuit.

A high resistance voltmeter must be used to measure the e.m.f. of a cell. Otherwise appreciable current flows through the voltmeter and the terminal p.d. measured is not that on open circuit.

Like p.d., e.m.f. is stated in volts and it too is often called a 'voltage': it has the same value for all cells made of the same materials.

Internal resistance

The terminal p.d. of a cell on closed circuit is also the p.d. applied to the external circuit. For example, in Figure 85.3b the voltmeter reading is the same (i.e. 1·2 V) if it is connected across the lamp terminals instead of the cell terminals. (The resistance of the connecting wires from the cell to the lamp is negligible, as is the p.d. across them.)

In an external circuit electrical energy is *changed* into other forms of energy and we regard the terminal p.d. of a cell on closed circuit as being the number of joules of electrical energy changed by each coulomb in the external circuit. If the terminal p.d. on closed circuit is 1·2 V then each coulomb changes 1·2 J of electrical energy.

Not all the electrical energy supplied by a cell to each coulomb is changed in the external circuit. The 'lost' energy per coulomb is due to the cell itself having resistance. Each coulomb has to 'waste' some energy—0·3 J in the above example—to get through the cell itself and so less is available for the external circuit. The resistance of a cell is called its *internal resistance* (r) and depends among other things on its size.

The circuit equation

Taking stock of the energy changes in a complete circuit, including the cell, we can say from the principle of conservation of energy (page 161):

energy supplied per coulomb by cell = energy changed per coulomb in external circuit + energy *wasted* per coulomb on cell resistance

Or, from the definitions of e.m.f. and p.d.:

$$\text{e.m.f.} = \text{useful p.d.} + \text{'lost' p.d.} \qquad (1)$$

Hence, the e.m.f. in a circuit equals the p.d. across the external resistance + the p.d. across the internal resistance.

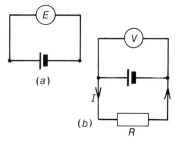

Figure 85.4

If E is the e.m.f. of a cell of internal resistance r (Figure 85.4a) and V is its terminal p.d. when sending a current I through a resistor R (Figure 85.4b) then from Equation (1):

$$E = V + v \qquad (2)$$

where v is the 'lost' p.d. It is a quantity which cannot be measured directly by a voltmeter but is only found by subtracting V from E. In Equation (2), called the *circuit equation*, we can write $V = IR$ and $v = Ir$ since the current all round the circuit is I.

Suppose $E = 1 \cdot 5 \text{ V}$, $V = 1 \cdot 2 \text{ V}$ and $I = 0 \cdot 30 \text{ A}$, then replacing in (2):

$$1 \cdot 5 = 1 \cdot 2 + v$$
$$\therefore v = 1 \cdot 5 - 1 \cdot 2 = 0 \cdot 30 \text{ V}$$

But
$$v = Ir$$
$$\therefore 0 \cdot 30 = 0 \cdot 30 \times r$$
$$\therefore r = 1 \cdot 0 \, \Omega$$

Also,
$$V = IR$$
$$\therefore 1 \cdot 2 = 0 \cdot 30 \times R$$
$$\therefore R = 1 \cdot 2 / 0 \cdot 30 = 4 \cdot 0 \, \Omega$$

The effect of internal resistance can be seen when a car starts with its lights on. Suppose the electric motor which starts the engine needs a current $I = 100 \text{ A}$ and is connected to a battery of e.m.f. $E = 12 \text{ V}$ and internal resistance $r = 0 \cdot 040 \, \Omega$. The terminal p.d. of the battery when it sends 100 A through the starter motor $= V = E - v = 12 - 4 \cdot 0 = 8 \cdot 0 \text{ V}$. If the lights are meant to be fully bright when there is a p.d. of 12 V across them, they will be dim on 8·0 V, as is observed.

Questions

1. (a) Two small lamps are connected in series with a battery of dry cells. At first the lights shine brightly. Explain why they gradually become dimmer.

(b) The battery is disconnected for half an hour. Explain why the lamps again shine brightly when the battery is reconnected.

(c) Explain how, by altering the circuit, you could increase the brilliance of the lamps, and state what effect this change would have upon the life of the battery.

2. (a) What current will flow in the circuit of Figure 85.5?

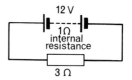

Figure 85.5

(b) What is the p.d. across the 3 Ω resistor in this circuit?

3. Figure 85.6 shows a battery of e.m.f. 6 V and internal resistance 1 Ω in series with a 5 Ω resistor. What current flows?

Figure 85.6

4. A high resistance voltmeter reads 3·0 V when connected across the terminals of a battery in open circuit and 2·6 V when the battery sends a current of 0·20 A through a lamp. What is (a) the e.m.f. of the battery, (b) the terminal p.d. of the battery when supplying 0·20 A, (c) the p.d. across the lamp, (d) the 'lost' p.d., (e) the internal resistance of the battery, (f) the resistance of the lamp?

86 ELECTRIC POWER

Power in electric circuits

Power was stated earlier (page 162) to be the *rate at which energy is changed from one form to another*. The unit of power is the *watt* (W) and equals an energy change rate of 1 joule per second, i.e. 1 W = 1 J/s. An electric lamp with a power of 100 W changes 100 J of electrical energy into heat and light each second. A larger unit is the kilowatt (kW): 1 kW = 1000 W.

It can be shown that if a current I flows from one point to another when there is a p.d. V between them, the electrical energy changed per second, i.e. the power P, is:

$$P = V \times I$$

In units,

$$\text{WATTS} = \text{VOLTS} \times \text{AMPERES}$$

Also, since $V = IR$ we can write:

$$P = IR \times I = I^2 R$$

That is, if the current is doubled in a resistor R four times as much heat is produced per second.

Electric lighting

(*a*) *Filament lamps* (Figure 86.1a). The filament is a small coil of tungsten wire (Figure 86.1b) which becomes white-hot when current flows through it. The higher the temperature of the filament the greater is the proportion of electric energy changed to light and for this reason it is made of tungsten, a metal with a high melting point (3400 °C).

Most lamps are gas-filled and contain nitrogen or argon, not air. This reduces evaporation of the tungsten which would otherwise condense on the bulb and blacken it. The coil, being compact, is cooled less by convection currents in the gas.

(*b*) *Fluorescent lamps.* A filament lamp changes only 10 per cent of the electrical energy supplied into light; the other 90 per cent becomes heat. Fluorescent lamps are five times as efficient and may last 3000 hours compared with the 1000 hour life of filament lamps. They cost more to install but running costs are less.

A simplified diagram of a fluorescent lamp is shown in Figure 86.2. When the lamp is switched on, the mercury vapour emits ultraviolet radiation (invisible) which makes the powder on the inside of the tube fluoresce (glow), i.e. light (visible) is emitted. Different powders give different colours.

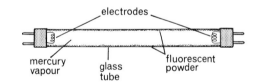

Figure 86.2

Figure 86.1

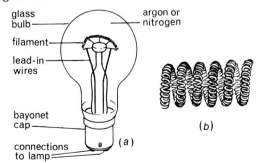

Electric heating

(*a*) *Heating elements.* In domestic appliances such as electric fires, cookers, kettles and irons, the 'elements' (Figure 86.3) are made from nichrome wire. This is an alloy of nickel and chromium which does not oxidize (and become brittle) when the current makes it red-hot.

The elements in *radiant* electric fires are at red heat (about 900 °C) and the radiation they emit is directed into the room by polished reflectors. In *convector* types the element is below red heat

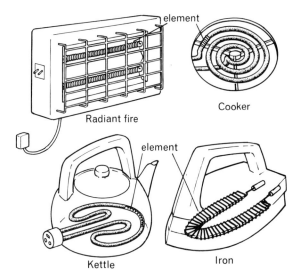

Figure 86.3

(about 450 °C) and is designed to warm air which is drawn through the heater by natural or forced convection (page 181). In *storage* heaters the elements heat fireclay bricks during the night using 'off-peak' electricity. On the following day these cool down, giving off the stored heat to warm the room.

(b) *Fuses.* A fuse is a short length of wire of material with a low melting point (often tinned copper), which melts and breaks the circuit when the current through it exceeds a certain value. Two reasons for excessive currents are 'short circuits' due to worn insulation on connecting wires and overloaded circuits. Without a fuse the wiring would become hot in these cases and could cause a fire.

Two types of fuse are shown in Figure 86.4. *Always switch off before replacing a fuse.*

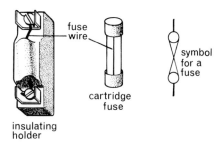

Figure 86.4

House circuits

Electricity usually comes to our homes by an underground cable containing two wires, the *live* (L) and the *neutral* (N). The neutral is earthed at the local sub-station. The supply is alternating current (a.c.) and the live wire is alternately positive and negative. Study the modern house circuit shown in Figure 86.5.

Figure 86.5

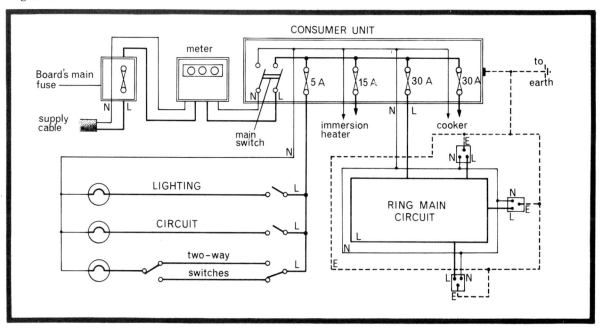

(a) *Circuits in parallel.* Every circuit is connected in parallel with the supply, i.e. across the live and neutral, and receives the full mains p.d. of 240 V.

(b) *Switches and fuses.* These are always in the live wire. If they were in the neutral, lamp and power sockets would be 'live' when switches were 'off' or fuses 'blown'. A shock (fatal) could then be obtained by, for example, touching the element of an electric fire when it was switched off.

(c) *Staircase circuit.* The lamp is controlled from two places by the two two-way switches.

(d) *Ring main circuit.* The live and neutral wires each run in two complete rings round the house, and the power sockets, each rated at 13 A, are tapped off from them. Thinner wires can be used since the current to each socket flows by two paths, i.e. in the whole ring. The ring has a 30 A fuse and if it has ten sockets all can be used so long as the total current does not exceed 30 A.

(e) *Fused plug.* Only one type of plug is used in a ring main circuit. It is wired as in Figure 86.6 and has its own cartridge fuse, 3 A (blue) for appliances with powers up to 720 W and 13 A (brown) for those between 720 W and 3 kW.

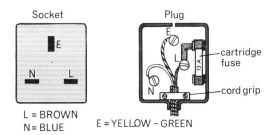

Figure 86.6

(f) *Earthing and safety.* A ring main has a third wire which goes to the top sockets on all power points (Figure 86.6) and is earthed by being connected either to a *metal* water pipe in the house or to an earth connection on the supply cable. This third wire is a safety precaution to prevent electric shock should an appliance develop a fault.

The earth pin on a three-pin plug is connected to the metal case of the appliance which is thus joined to earth by a path of almost zero resistance. If then, for example, the element of an electric fire breaks or sags or touches the case, a large current flows to earth and 'blows' the fuse. Otherwise the case becomes 'live' and anyone touching it receives a shock *which might be fatal*, especially if they were 'earthed' by, say, standing on a concrete floor or holding a water-tap.

Paying for electricity

Electricity companies charge for the *electrical energy* they supply. A joule is a very small amount of energy and a larger unit, the *kilowatt-hour* (kWh) is used.

A kilowatt-hour is the electrical energy used by a 1 kW appliance in 1 hour.

A 3 kW electric kettle working for 2 hours uses 6 kWh of electrical energy—usually called 6 'units'. Electricity meters are marked in kWh: the reading on the one in Figure 86.7 is 87939. When

Figure 86.7

the pointer is between two figures the lower one is read, except when the figures are 9 and 0. At present a 'unit' costs about 5 p.

Typical powers of some appliances are:

Lamps	60, 100 W	Fire	1, 2, 3 kW
Fridge	150 W	Kettle	2–3 kW
TV set	200 W	Immersion heater	3 kW
Iron	750 W	Cooker	8 kW

Questions

1. How much electrical energy in *joules* does a 100 W lamp change in (a) 1 s, (b) 5 s, (c) 1 minute?

2. What is the power of a lamp rated at 12 V 2 A?

3. What is the largest number of 60 W lamps which can safely be run from a 240 V supply with a 5 A fuse?

4. What is the maximum power in kilowatts of the appliance(s) that can be connected safely to a 13 A 240 V mains socket?

5. If a current of 3 A passes through a 2 Ω resistor, calculate (a) the p.d. between the ends of the resistor, (b) the power used by the resistor.

6. A car bulb is marked 12 V 24 W. If it is used on a 12 V supply, (a) what current flows through it, (b) what is its resistance?

7. A 3 kW electric heater is used for 10 hours.
(a) How many units (kWh) are used?
(b) What is the cost if electricity is 5 p per unit?

8. An electric cooker has an oven rated at 3 kW, a grill rated at 1 kW and four rings each rated at 500 W. What is the cost of operating all parts for 30 minutes if electricity costs 5 p per unit?

87 ELECTROMAGNETS

Oersted's discovery

In 1819 Oersted accidentally discovered the magnetic effect of an electric current. His experiment can be repeated by holding a wire over and parallel to a compass needle which is pointing north and south (Figure 87.1). The needle moves when the current is switched on. Reversing the current causes the needle to point in the opposite direction.

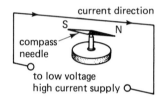

Figure 87.1

Evidently around a wire carrying a current there is a *magnetic field*. As with the field due to a permanent magnet, we represent the field due to a current by *field lines* or *lines of force*. Arrows on the lines show the direction of the field, i.e. the direction in which a N pole points.

Different field patterns are given by differently shaped conductors.

Field due to a straight wire

If a straight vertical wire passes through the centre of a piece of card held horizontally and a current is passed through the wire, iron filings sprinkled on the card set in concentric circles when the card is tapped (Figure 87.2).

Plotting compasses placed on the card set along the field lines and show the direction of the field at different points. When the current is reversed, the compasses point in the opposite direction showing that the direction of the field reverses when the current reverses.

If the current direction is known, the direction of the field can be predicted by the *right-hand screw rule*.

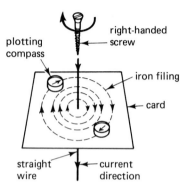

Figure 87.2

If a right-handed screw moves forward in the direction of the current (conventional), the direction of rotation of the screw gives the direction of the field.

Field due to a circular coil

The field pattern is shown in Figure 87.3. At the centre of the coil the field lines are straight and at right angles to the plane of the coil. The right-hand screw rule again gives the direction of the field at any point.

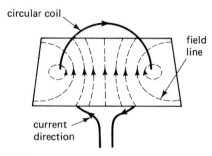

Figure 87.3

Field due to a solenoid

A solenoid is a long cylindrical coil. It produces a field similar to that of a bar magnet; in Figure 87.4 end A behaves like a N pole and end B like a S

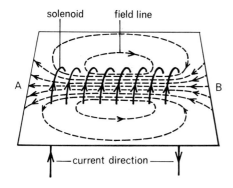

Figure 87.4

pole. The polarity is found as before by applying the right-hand screw rule to a short length of one turn of the solenoid. Alternatively the *right-hand grip rule* can be used (page 341).

The field inside a solenoid can be made very strong if it has a large number of turns and a large current flows. Previously we used it to magnetize materials (page 341): permanent magnets are now made by allowing the molten metal to solidify in such fields.

Electromagnets

An electromagnet consists of a coil of wire wound on a soft iron core. The magnetism of an electromagnet is *temporary* and can be switched on and off, unlike that of a permanent magnet. It has a core of *soft iron* which is magnetized only when current flows in the surrounding coil.

The strength of an electromagnet increases if:

(*i*)   the current in the coil increases
(*ii*) the number of turns on the coil increases
(*iii*) the poles are closer together.

In C-core (or horseshoe) electromagnets, condition (*iii*) is achieved (Figure 87.5). Note that the coil is wound in *opposite* directions on each limb of the core.

Figure 87.5

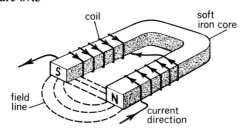

Figure 87.6

Apart from being used as a crane to lift iron objects (Figure 87.6), scrap iron, etc., an electromagnet is an essential part of many electrical devices.

Electric bell: door chimes

(*a*) *Bell* (Figure 87.7). When the circuit is completed, current flows in the coils of the electromagnet which becomes magnetized and attracts the soft iron bar (the armature). The hammer hits the gong but the circuit is now broken at the point C of the contact screw.

Figure 87.7

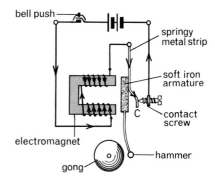

363

The electromagnet loses its magnetism and no longer attracts the armature. The springy metal strip is then able to pull the armature back, remaking contact at C and so completing the circuit again. This cycle is repeated so long as the bell push is depressed and continuous ringing occurs.

(b) *Chimes.* The circuit is shown in Figure 87.8. Work out what happens when the push switch is (*i*) pressed, (*ii*) released.

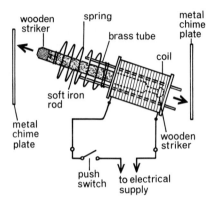

Figure 87.8

Telephone

A telephone contains a microphone at the speaking end and a receiver at the listening end.

(a) *Carbon microphone* (Figure 87.9). When someone speaks, sound waves (page 220) cause the diaphragm to move backwards and forwards. This varies the pressure on the carbon granules between the front carbon block which is attached to the diaphragm and the back one which is fixed. When the pressure increases, the granules are squeezed closer together and their electrical resistance decreases. A decrease of pressure has the opposite effect. If there is a current passing through the microphone (from a battery), this too will vary in a similar way to the sound wave variations.

Figure 87.9

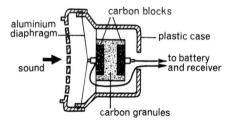

(b) *Receiver.* A microphone changes sound energy into electrical energy which travels along the telephone cables and is changed back into sound by the receiver (Figure 87.10).

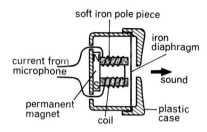

Figure 87.10

The varying current from the microphone passes through the coils of an electromagnet. This pulls the diaphragm towards it, by a distance which depends on the current. As a result, the diaphragm moves in and out and produces sound waves that are a copy of those that entered the microphone.

Questions

1. The vertical wire in Figure 87.11 is at right angles to the card. In what direction will a plotting compass at A point when (*a*) there is no current in the wire, (*b*) current flows upwards?

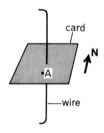

Figure 87.11

2. A solenoid is wrapped round a soft iron core (Figure 87.12). When current (conventional) flows in the direction shown, will end A be a N pole or a S pole?

Figure 87.12

3. (*a*) Why is soft iron a better material than steel to use for the core of an electromagnet?
(*b*) State *two* ways in which an electromagnet can be made more powerful.

88 ELECTRIC MOTORS AND METERS

Electric motors are used in devices such as fans, lifts, washing machines and locomotives.

The motor effect

A wire carrying a current in a magnetic field experiences a force. If the wire can move it does.

(a) *Demonstration.* In Figure 88.1 the flexible wire is loosely supported in the strong magnetic field of a C-shaped magnet (permanent or electro).

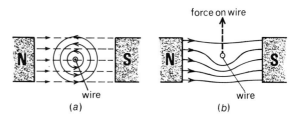

Figure 88.2

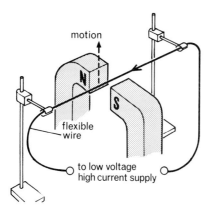

Figure 88.1

When the switch is pressed, current flows in the wire which jumps upwards as shown. If either the direction of the current or the direction of the field is reversed, the wire moves downwards.

(b) *Explanation.* Figure 88.2a is a side-view of the magnetic field lines due to the wire and the magnet. Those due to the wire are circles and we will suppose their directions are as shown. The dotted lines represent the field lines of the magnet and their direction is to the right.

The resultant field obtained by combining both fields is shown in Figure 88.2b. There are more lines below than above the wire since both fields act in the same direction below but in opposition above. If we *suppose* the lines are like stretched elastic, those below will try to straighten out and in so doing will exert an upwards force on the wire.

Fleming's left-hand rule

The direction of the force or thrust on the wire can be predicted by this rule (Figure 88.3).

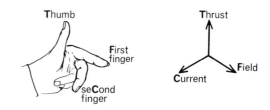

Figure 88.3

Hold the thumb and first two fingers of the left hand at right angles to each other with the First finger pointing in the direction of the Field and the seCond finger in the direction of the Current, then the Thumb points in the direction of the Thrust.

If the wire is not at right angles to the field, the force is smaller and is zero if it is parallel to the field.

Simple d.c. electric motor

A simple motor to work from direct current (d.c.) consists of a rectangular coil of wire mounted on an axle which can rotate between two poles of a C-shaped magnet (Figure 88.4). Each end of the coil is connected to half of a split ring of copper,

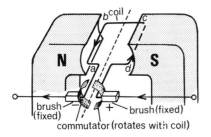

Figure 88.4

called the *commutator*, which rotates with the coil. Two carbon blocks, the *brushes*, are pressed lightly against the commutator by springs. The brushes are connected to an electrical supply.

If Fleming's left-hand rule is applied to the coil in the position shown, we find that side *ab* experiences an upward force and side *cd* a downward force. (No forces act on *ad* and *bc* since they are parallel to the field.) These two forces form a *couple* which rotates the coil in a clockwise direction until it is vertical.

The brushes are then in line with the gaps in the commutator and the current stops. However, because of its inertia, the coil overshoots the vertical and the commutator halves change contact from one brush to the other. This reverses the current through the coil and so also the directions of the forces on its sides. Side *ab* is on the right now, acted on by a downward force, whilst *cd* is on the left with an upward force. The coil keeps rotating clockwise.

Practical motors

Practical motors have:

a *A coil of many turns wound on a soft iron cylinder or core* which rotates with the coil. This makes it more powerful. The coil and core together are called the *armature*.
b *Several coils* each in a slot in the core and each having a pair of commutator segments. This gives increased power and smoother running. The motor of an electric drill is shown in Fig. 88.5.
c *An electromagnet* (usually) to produce the field in which the armature rotates.

Experiment 1   A model motor

Make one from a kit of parts as shown in Figure 88.6.

1 Wrap Sellotape round one end of the metal tube which passes through the wooden block.

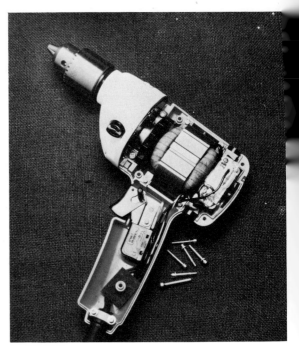

Figure 88.5

2 Cut two rings off a piece of narrow rubber tubing; slip them on to the Sellotaped end of the metal tube.
3 Remove the insulation from one end of a 1½ m length of SWG 26 PVC-covered copper wire and fix it under both rubber rings so that it is held tight against the Sellotape. This forms one end of the coil.
4 Wind ten turns of the wire in the slot in the wooden block and finish off the second end of the coil by removing the PVC and fixing this too under the rings but on the *opposite* side of the tube from the first end. The bare ends act as the *commutator*.
5 Push the axle through the metal tube of the wooden base so that the block spins freely.
6 Arrange two ½ m lengths of wire to act as *brushes* and leads to the supply, as shown. Adjust the brushes so that they are vertical and each touches one bare end of the coil when the plane of the coil is horizontal. *The motor will not work if this is not so.*
7 Slide the base into the magnet with *opposite poles facing*. Connect to a 3 V battery (or other low voltage d.c. supply) and a slight push of the coil should set it spinning at high speed.

Moving coil galvanometer

A galvanometer detects small currents of the order of milliamperes (mA). (1 mA = 1/1000 A = 0·001 A). Some are even more sensitive.

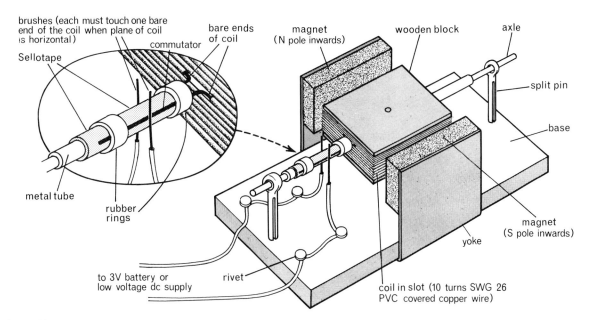

Figure 88.6

In the moving coil *pointer-type* meter, a coil is pivoted on jewelled bearings between the poles of a permanent magnet (Figure 88.7a). Current enters and leaves the coil by hair springs above and below it. When current flows, a couple acts on the coil (as in an electric motor), causing it to rotate until stopped by the springs. The greater the current the greater the deflection which is shown by a pointer attached to the coil.

Figure 88.7

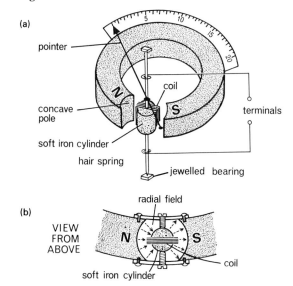

The soft iron cylinder at the centre of the coil is *fixed* and along with the concave poles of the magnet it produces a *radial* field (Figure 88.7b), i.e. the field lines are directed to the centre of the cylinder. The scale on the meter is then even or linear, i.e. all divisions are the same size.

The sensitivity of a galvanometer is increased by having:

(i) more turns on the coil
(ii) a stronger magnet
(iii) weaker hair springs or a wire suspension
(iv) as a pointer, a long beam of light reflected from a mirror on the coil.

The last two are used in *light-beam* meters which have a full scale deflection of a few microamperes (μA). (1 μA = 1/1 000 000 A = 0·000 001 A).

A moving coil galvanometer measures d.c. not alternating current (a.c.) unless it is used with a rectifier.

Ammeters and shunts

A galvanometer can be modified for use as an ammeter. Suppose a moving coil meter has resistance 5 Ω (due largely to the coil) and gives a full scale deflection (f.s.d.) when 1 mA (0·001 A) passes through it.

To convert it to an ammeter to read 0–1 A, a resistor of low value is connected in parallel with it as a by-pass. The resistor is called a *shunt* and, when the current to be measured is 1 A, it must allow only 0·001 A to go through the meter and the rest, i.e. 0·999 A, must flow through the shunt (Figure 88.8).

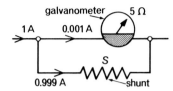

Figure 88.8

The meter and shunt are in parallel, and so:

p.d. across meter = p.d. across shunt
$$\therefore 0\cdot 001 \times 5 = 0\cdot 999 \times S$$

where S is the resistance of the shunt in Ω.

$$\therefore S = \frac{0\cdot 001 \times 5}{0\cdot 999} = \frac{5}{999} = 0\cdot 005\,\Omega$$

Shunts consist of short lengths of thick manganin wire or strip. Manganin is an alloy whose resistance does not change when it is warmed by the current.

An ammeter is placed in *series* in a circuit and must have a *low resistance* compared with the rest of the circuit, otherwise it changes the current to be measured. The combined resistance of this meter and shunt is below $0\cdot 005\,\Omega$.

Voltmeters and multipliers

A moving coil galvanometer of resistance $5\,\Omega$ and f. s. d. 1 mA can be converted to a voltmeter to read 0–1 V by connecting a resistor of high value in series with it.

The resistor is called a *multiplier* and it must be such that when a p.d. of 1 V is applied across it and the meter together, the current is 0·001 A, i.e. the meter records a f.s.d. (Figure 88.9).

If M is the resistance of the multiplier in Ω, then since it is in series with the meter:

p.d. across multiplier and meter at f.s.d.
$$= 0\cdot 001 \times (M + 5) = 1$$
$$\therefore M + 5 = 1/0\cdot 001 = 1000$$
$$\therefore M = 995\,\Omega$$

Figure 88.9

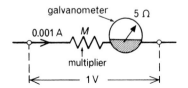

Multipliers consist of long lengths of insulated manganin wire wound on a bobbin. A moving coil galvanometer with shunts and multipliers is shown in Figure 88.10.

A voltmeter is placed in *parallel* with the part of the circuit across which the p.d. is to be measured. It should have a *high resistance* compared with the

Figure 88.10

resistance across which it is connected. Otherwise the total resistance of the whole circuit is reduced so changing the current and the p.d. required. The ideal voltmeter has an infinite resistance.

Moving iron meter

In the repulsion type (Figure 88.11) when current flows in the coil the two small soft iron rods P and

Figure 88.11

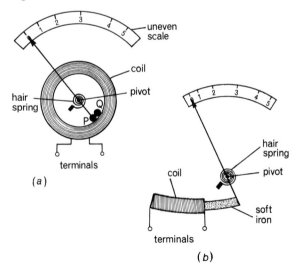

are magnetized in the same direction. They repel. Q is fixed but P is attached to a pivoted pointer and moves away from Q until stopped by the hair spring. The larger the current the greater the repulsion.

It can be used as an ammeter or a voltmeter with a suitable shunt or multiplier. It works whichever way current passes and so will measure a.c. or d.c. A disadvantage is its uneven scale; it is closer at the start.

Figure 88.12

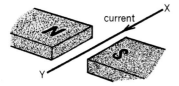

Questions

1. What is the direction of the force on XY in Figure 88.12?

2. In Figure 88.13 does the coil rotate clockwise or anticlockwise as seen from A?

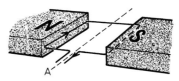

Figure 88.13

3. A galvanometer of resistance $20\,\Omega$ gives a full scale deflection with current of $5\,mA$. How should it be changed for use as (*a*) a $1\cdot 0\,A$ ammeter, (*b*) a $10\,V$ voltmeter?

89 GENERATORS AND TRANSFORMERS

GENERATORS

An electric current creates a magnetic field. The reverse effect of producing electricity from magnetism was discovered in 1831 by Faraday and is called *electromagnetic induction*. It led to the construction of generators (dynamos) for producing electrical energy in power stations.

Electromagnetic induction

Here are two ways of investigating the effect.

(*a*) *Straight wire and U-shaped magnet* (Figure 89.1). First the wire is held at rest between the poles of the magnet and the galvanometer observed. It is then moved in each of the six directions shown. Only *when it is moving* upwards

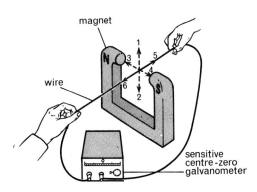

Figure 89.1

(direction 1) or downwards (direction 2) is there a deflection on the galvanometer, indicating an induced current in the wire. The deflection is in opposite directions in each case and only lasts while the wire is in motion.

(*b*) *Bar magnet and coil* (Figure 89.2). The magnet is pushed into the coil, one pole first, then held still inside it. It is next withdrawn. The galvanometer shows that current is induced in the coil in one direction as the magnet *moves in* and in the opposite direction as it is *removed*. There is no deflection when the magnet is at rest. The results are the same if the coil is moved instead of the magnet, i.e. only *relative motion* is needed.

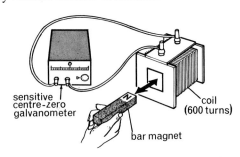

Figure 89.2

Faraday's law

To 'explain' electromagnetic induction Faraday suggested that an e.m.f. is induced in a conductor whenever it 'cuts' magnetic field lines, i.e. moves *across* them, but not when it moves along them or is at rest. If the conductor forms part of a complete circuit, an induced current is also produced.

Faraday found, and it can be shown with apparatus like that in Figure 89.1, that the induced e.m.f. increases with increases of:

(*i*) the speed of motion of the magnet or coil
(*ii*) the number of turns on the coil
(*iii*) the strength of the magnet.

These facts led him to state this law.

> *The size of the induced e.m.f. is directly proportional to the rate at which the conductor cuts magnetic field lines.*

Lenz's law

The direction of the induced current can be predicted by a law due to the Russian scientist, Lenz.

> *The direction of the induced current is such as to oppose the change causing it.*

In Figure 89.3a the magnet approaches the coil, N pole first. According to Lenz's law the induced current should flow in a direction which makes the coil behave like a magnet with its top a N pole. The downward motion of the coil will then be opposed.

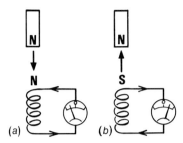

Figure 89.3

When the magnet is withdrawn, the top of the coil should become a S pole (Figure 89.3b) and attract the N pole of the magnet, so hindering its removal. The induced current is thus in the opposite direction to that when the magnet approaches.

Lenz's law is an example of the principle of conservation of energy (page 161). If the currents caused opposite poles to those that they do, electrical energy would be created from nothing. As it is, mechanical energy is provided by whoever moves the magnet, to overcome the forces that arise.

For a straight wire moving at right angles to a magnetic field a more useful form of Lenz's law is *Fleming's right-hand rule* (Figure 89.4).

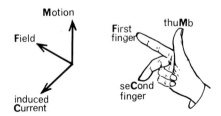

Figure 89.4

Hold the thumb and first two fingers of the right hand at right angles to each other with the First finger pointing in the direction of the Field and the thuMb in the direction of Motion of the wire, then the seCond finger points in the direction of the induced Current.

Simple a.c. generator (alternator)

The simplest alternating current (a.c.) generator consists of a rectangular coil between the poles of a C-shaped magnet (Figure 89.5a). The ends of the coil are joined to two *slip rings* on the axle and against which carbon *brushes* press.

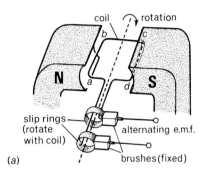

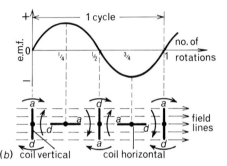

Figure 89.5

When the coil is rotated it cuts the field lines and an e.m.f. is induced in it. Figure 89.5b shows how the e.m.f. varies over one complete rotation.

As the coil moves through the vertical position with *ab* uppermost, *ab* and *cd* are moving along the lines (*bc* and *da* do so always) and no cutting occurs. The induced e.m.f. is zero.

During the first quarter rotation the e.m.f. increases to a maximum when the coil is horizontal. Sides *ab* and *dc* are then cutting the lines at the greatest rate.

In the second quarter rotation the e.m.f. decreases again and is zero when the coil is vertical with *dc* uppermost. After this, the direction of the e.m.f. reverses because, during the next half rotation, the motion of *ab* is directed upwards and *dc* downwards.

An alternating e.m.f. is generated which acts first in one direction and then the other; it would cause a.c. to flow in a circuit connected to the brushes. The *frequency* of an a.c. is the number of complete cycles it makes each second (c/s) and is measured in *hertz* (Hz), i.e. 1 c/s = 1 Hz. If the coil rotates twice per second, the a.c. has frequency 2 Hz. The mains supply in Britain is a.c. of frequency 50 Hz. In actual generators several coils are wound in evenly-spaced slots in a soft iron cylinder and electromagnets usually replace permanent magnets.

Mutual induction

When the current in a coil is switched on or off or changed, an e.m.f. and current are induced in a neighbouring coil. The effect, called *mutual induction*, is an example of electromagnetic induction not using a permanent magnet and can be shown with the arrangement of Figure 89.6. Coil A is the *primary* and coil B the *secondary*.

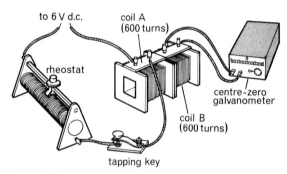

Figure 89.6

Switching on the current in the primary sets up a magnetic field and as its field lines 'grow' outwards from the primary they 'cut' the secondary. An e.m.f. is induced in the secondary until the current in the primary reaches its steady value. When the current is switched off in the primary, the magnetic field dies away and we can imagine the field lines cutting the secondary as they collapse, again inducing an e.m.f. in it. Changing the primary current by *quickly* altering the rheostat has the same effect.

The induced e.m.f. is increased by having a soft iron rod in the coils, or better still, by using coils wound on a complete iron ring. More field lines then cut the secondary due to the magnetization of the iron.

Experiment 1 Mutual induction with a.c.

An a.c. is changing all the time and if it flows in a primary coil, an alternating e.m.f. and current are induced in a secondary coil.

Connect the circuit of Figure 89.7. The 1 V high current power unit supplies a.c. to the primary and the lamp detects the secondary current.

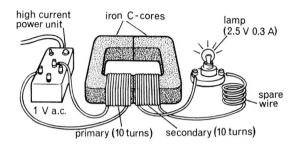

Figure 89.7

Find the effect on the brightness of the lamp of:

(*i*) pulling the C-cores apart slightly
(*ii*) increasing the secondary turns to 15
(*iii*) decreasing the secondary turns to 5.

Transformer equation

A transformer transforms (changes) an *alternating p.d.* (voltage) from one value to another of greater or smaller value. It has primary and secondary coils wound on a complete soft iron core, either one on top of the other (Figure 89.8*a*) or on separate limbs of the core (Figure 89.8*b*).

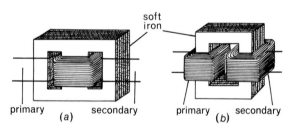

Figure 89.8

An alternating voltage applied to the primary induces an alternating voltage in the secondary whose value can be shown to be given by:

$$\frac{\text{secondary voltage } (V_s)}{\text{primary voltage } (V_p)} = \frac{\text{secondary turns } (t_s)}{\text{primary turns } (t_p)}$$

A 'step-up' transformer has more turns on the secondary than the primary and V_s is greater than V_p (Figure 89.9a). For example, if the secondary has twice as many turns as the primary, V_s is about twice V_p. In a 'step-down' transformer there are fewer turns on the secondary than the primary and V_s is less than V_p (Figure 89.9b).

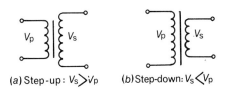

(a) Step-up: $V_s > V_p$ (b) Step-down: $V_s < V_p$

Figure 89.9

At Area Control Centres engineers direct the flow and re-route it when breakdown occurs. This makes the supply more reliable, and cuts costs by enabling smaller, less efficient stations to be shut down at off-peak periods.

(b) *Use of high alternating p.ds.* If 400 000 W of electrical power has to be sent through cables it can be done as 400 000 V at 1 A or 400 V at 1000 A (since watts = amperes × volts). But the amount of electrical energy changed to unwanted heat (due to the resistance of the cables) is proportional to the *square of the current* and so the power loss (I^2R) is less if transmission occurs at high voltage (high tension) and low current. On the other hand, high p.ds need good insulation. The efficiency with which transformers step alternating p.ds. up and down accounts for the use of a.c. rather than direct current (d.c.).

Transmission of electrical power

(a) *Grid system.* This is a network of cables, most of which are supported on pylons, which connect about 160 power stations throughout Britain to consumers. In the largest modern stations, electricity is generated at 25 000 V (25 kilovolts = 25 kV) and stepped up at once in a transformer to 275 or 400 kV to be sent over long distances on the Supergrid. Later, the p.d. is reduced by sub-station transformers for distribution to local users (Figure 89.10).

Questions

1. State what is observed when a bar magnet is (a) pushed into a coil connected to a centre-zero galvanometer so that one pole enters first, (b) held at rest inside the coil, (c) pulled out from inside the coil.

2. A simple generator is shown in Figure 89.11.
 (a) What are A and B called and what is their purpose?
 (b) What changes can be made to increase the e.m.f. generated?

Figure 89.10

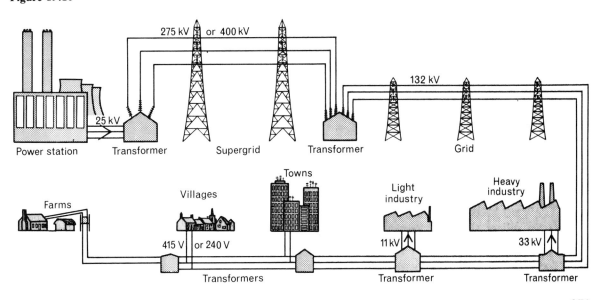

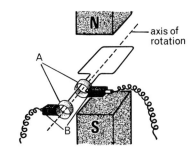

Figure 89.11

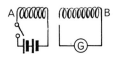

Figure 89.12

3. Two coils of wire, A and B, are placed near one another (Figure 89.12). Coil A is connected to a switch and battery. Coil B is connected to a centre-reading moving coil galvanometer.

(a) If the switch connected to coil A were closed for a few seconds and then opened, the galvanometer connected to coil B would be affected. Explain and describe, step by step, what would actually happen

(b) What changes would you expect if a bundle of soft iron wires were placed through the centre of the coils? Give a reason for your answer.

(c) What would happen if more turns of wire were wound on coil B?

4. A step-down transformer designed to work a 12 V lamp from a 240 V supply has 1000 turns on the primary. What is the number of turns on the secondary?

Atomic Physics

90 ELECTRONS

The discovery of the electron was a landmark in physics and led to great technological advances.

Thermionic emission

The evacuated bulb in Figure 90.1 contains a small coil of wire, the *filament*, and a metal plate called the *anode* because it is connected to the positive of the 400 V direct current (d.c.) power supply. The negative of the supply is joined to the filament which is also called the *cathode*. The filament is heated by current from a 6 V supply (a.c. or d.c.).

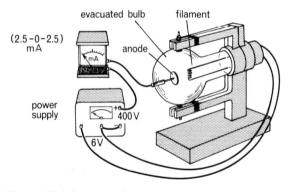

Figure 90.1

With the circuit as shown, the meter deflects, indicating current flow in the circuit containing the gap between anode and cathode. The current stops if *either* the 400 V supply is reversed to make the anode negative, *or* the filament is not heated.

This demonstration supports the view that negative charges, in the form of electrons, escape from the filament when it is hot. The process is known as *thermionic emission*. The electrons are attracted to the anode if it is positive and are able to reach it because there is a vacuum in the bulb.

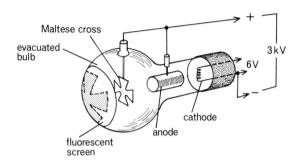

Figure 90.2

Cathode rays

Streams of electrons moving at high speed are called *cathode rays*. Their properties can be studied using a *Maltese cross tube* (Figure 90.2). Electrons emitted by the hot cathode are accelerated towards the anode but most pass through the hole in it and travel on along the tube. Those that miss the cross, cause the screen to fluoresce with a green light and cast a shadow of the cross on it. The cathode rays evidently travel in straight lines.

If the N pole of a magnet is brought up to the neck of the tube, the rays (and the fluorescent shadow) move upwards. The rays are deflected by a magnetic field and, using Fleming's left-hand rule (page 365), we see that they behave like conventional current (positive charge flow) travelling from anode to cathode, i.e. like negative charge moving from cathode to anode.

Cathode ray oscilloscope (C.R.O.)

The C.R.O. is one of the most important scientific instruments ever to be developed. Like a television set, it contains a cathode ray tube which has three main parts (Figure 90.3).

(a) Electron gun. This consists of a *heater* H, a *cathode* C, another electrode called the *grid* G and two or three *anodes* A. G is negative with respect to C and controls the number of electrons passing through its central hole from C to A; it is the

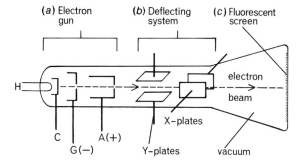

Figure 90.3

brightness control. The anodes are positive relative to C; they accelerate the electrons along the highly evacuated tube and also *focus* them into a narrow beam.

(b) *Deflecting system*. Beyond A are two pairs of deflecting plates to which p.d.s. can be applied. The *Y-plates* are horizontal but deflect the beam vertically. The *X-plates* are vertical and deflect the beam horizontally.

(c) *Fluorescent screen*. A bright spot of light is produced on the screen where the beam hits it.

The C.R.O. is often used as a 'graph-plotter' to display a waveform showing how a p.d. changes with time. The p.d. is applied to the Y-plates and a circuit is switched on in the C.R.O., called the *time base*, which generates a p.d. across the X-plates. This sweeps the spot steadily and horizontally across the screen until it 'flies back' very rapidly to the left of the screen and repeats the motion. Figure 90.4 shows traces for (*i*) time base on X only, (*ii*) a.c. on Y only, and (*iii*) a.c. on Y and time base on X.

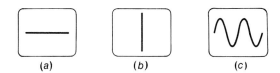

Figure 90.4

Experiment 1 Using a C.R.O.

The instructions refer to the C.R.O. in Figure 90.5.

a *Preliminary adjustments*. Before switching on, set the controls as follows.

Figure 90.5

BRIGHTNESS—to OFF; FOCUS—mid-position
TIME BASE RANGE switch—to OFF
A.C.—D.C. switch to D.C.; Y-SHIFT—mid-position
Y-GAIN—fully anticlockwise (minimum)

Switch on at the BRIGHTNESS control and after 30 s, turn it clockwise till the spot appears. Centralize it with the Y-SHIFT.

Adjust BRIGHTNESS and FOCUS to give a small, sharp spot but *do not have it too bright* when the time base is off or the screen will be damaged.

b *Measuring p.ds*. Connect a 1·5 V cell to the input (marked HIGH and LOW) and adjust the Y-GAIN so that the spot is deflected 1·5 divisions on the scale. Now apply two cells in series (3 V) and note the deflection. Repeat with three cells (4·5 V). The C.R.O. is now calibrated as a voltmeter.

Disconnect the cells and apply 1 V a.c. to the INPUT. The length of the trace is twice the peak p.d.

- What is the value of the peak p.d.?

c *Studying waveforms*. With 1 V a.c. still applied to the INPUT, set the TIME BASE RANGE to '2' and increase Y-GAIN to obtain a wave trace. See the effect of changing the TIME BASE VARIABLE control.

Disconnect the 1 V a.c. and connect a crystal microphone. Set the Y-GAIN fully clockwise and observe the waveforms when you speak or whistle.

Questions

1. A simplified diagram of a C.R.O. is shown in Figure 90.6.

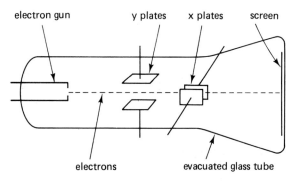

Figure 90.6

(a) Why is the electron gun necessary? Explain in two or three sentences how it works.
(b) Why must the tube have a good vacuum?
(c) Copy the diagram but show the Y-plates connected to a battery with the upper plate to the positive and the lower to the negative. What will now happen to the spot of light on the screen? On your diagram draw the new path taken by the electron beam.
(d) What difference would it make to the light spot if (i) the battery is reversed, (ii) two batteries are used in series?
(e) If the battery is disconnected and an a.c. supply is connected to the Y-plates what happens to (i) the beam, (ii) the 'picture' on the screen?

2. The time base on a C.R.O. is adjusted to work at a frequency of 50 Hz. Figure 90.7 shows the picture on the screen when there is no input to the Y-plates. Draw another three circles and show what happens when each of the following is joined to the input sockets.

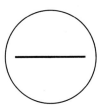

Figure 90.7

(a) A battery which makes the upper plate negative.
(b) An a.c. supply of 50 Hz.
(c) An a.c. supply of 100 Hz.

91 RADIOACTIVITY

The discovery of radioactivity in 1896 by the French scientist Becquerel was accidental. He found that uranium compounds emitted radiation which (*i*) affected a photographic plate even when wrapped in black paper, and (*ii*) ionized a gas. Soon afterwards Madame Curie discovered radium. Today radioactivity is used widely in industry, medicine and research.

We are all exposed to *background radiation* caused partly by radioactive materials in rocks, the air and our bodies, and partly by cosmic rays from outer space.

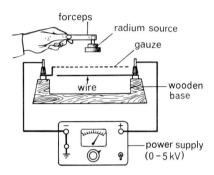

Figure 91.2

Ionizing effect of radiation

A charged electroscope discharges when a lighted match or a radium source (*held in forceps*) is brought near the cap (Figure 91.1a, b).

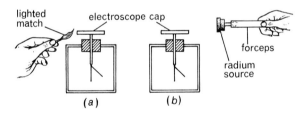

Figure 91.1

In the first case the flame knocks electrons out of surrounding air molecules; in the second case radiation causes the ionization. The resulting positive air ions are attracted to the cap if it is negatively charged; if it is positively charged the electrons are attracted. As a result, the charge on the electroscope is neutralized, i.e. it loses its charge.

Radiation detectors

The ionizing effect is used to detect radiation.

(*a*) *Spark counter* (Figure 91.2). A p.d. is applied between the wire and gauze till sparking occurs. It is then reduced until it *just* stops (usually about 4·5 kV).

When a radium source is held (*in forceps*) 1 or 2 cm above the gauze, sparks are seen and heard at irregular intervals due to the radiation ionizing the air between the gauze and the wire. The range of the radiation can be found by raising the source till sparking stops.

(*b*) *Geiger-Müller (G-M) tube* (Figure 91.3). When radiation enters the tube, either through a thin end-window made of mica, or, if it is very penetrating, through the wall, it creates argon ions and electrons. These are accelerated towards the electrodes and cause more ionization by colliding with other argon atoms.

On reaching the electrodes, the ions produce a current pulse which is amplified and fed either to a *scaler* or a *ratemeter*. A scaler counts the pulses and

Figure 91.3

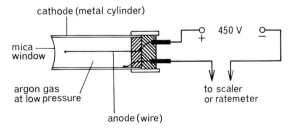

379

shows the total received in a certain time. A ratemeter has a meter marked in 'counts per second (or minute)' from which the average pulse rate can be read. It usually has a loudspeaker which gives a 'click' for each pulse.

Alpha, beta, gamma rays

Experiments to study the penetrating power, ionizing ability and behaviour in magnetic and electric fields, show that a radioactive substance emits one or more of three types of radiation—called alpha (α), beta (β) and gamma (γ) rays.

Penetrating power can be investigated as in Figure 91.4 by observing the effect on the count rate of placing in turn between the G-M tube and the lead, a sheet of (*i*) thick paper (the radium source, lead and tube must be *close together* for this part), (*ii*) aluminium 2 mm thick, and (*iii*) lead 2 cm thick. Other sources can be tried, e.g. plutonium, americium, strontium and cobalt.

streams of *high-energy electrons*, like cathode rays, emitted with a range of speeds, up to that of light.

Strontium (Sr 90) emits β only.

The magnetic deflection of β particles can be shown as in Figure 91.5. With the G-M tube at A and without the magnet, the count rate is noted. Inserting the magnet reduces the count rate but it increases again when the G-M tube is at B.

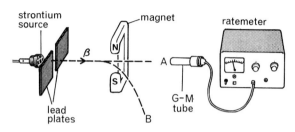

Figure 91.5

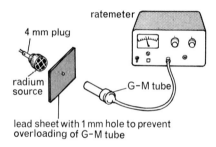

Figure 91.4

(*a*) *Alpha rays* are stopped by a thin sheet of paper and have a range in air of a few centimetres. They cause intense ionization in a gas and are deflected by electric and *strong* magnetic fields in a direction and by an amount which suggests they are helium atoms minus 2 electrons, i.e. *helium ions with a double positive charge*. From a particular substance, they are all emitted with the same speed (about one-twentieth of that of light).

Plutonium (Pu 239) and americium (Am 241) may be used as pure α sources.

(*b*) *Beta rays* are stopped by a few millimetres of aluminium and some have a range in air of several metres. Their ionizing power is much less than that of alpha particles. As well as being deflected by electric fields, they are more easily deflected by magnetic fields and measurements show they are

(*c*) *Gamma rays* are the most penetrating and are stopped only by many centimetres of lead. They ionize a gas even less than β particles and are not deflected by electric and magnetic fields. They give interference effects and are *electromagnetic radiation* travelling at the speed of light. Their wavelengths are those of very short X-rays (page 219) from which they differ only because they arise *in* atomic nuclei whereas X-rays come from energy changes in the electrons *outside* the nucleus.

Cobalt (Co 90) is a pure γ source. Radium (Ra 226) emits α, β and γ rays.

A G-M tube detects β and γ rays and energetic α particles; a spark counter detects α only, as does a charged electroscope. All three types of rays cause fluorescence.

Cloud chambers

When air containing vapour, e.g. alcohol, is cooled enough, saturation occurs. If ionizing radiation passes through the air, further cooling causes the saturated vapour to condense on the air ions created. The resulting white line of tiny liquid drops shows up as a track when illuminated.

(*a*) *Expansion type* (Figure 91.6). A rapid expansion is produced by withdrawing *quickly* the piston of a bicycle pump having the leather washer reversed so that it removes air. As a result the

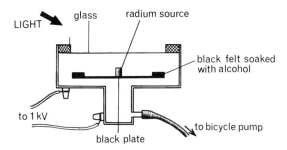

Figure 91.6

alcohol vapour in the chamber is cooled and condensation occurs on the ions produced by a weak radium source in the chamber. A high p.d. (e.g. 1 kV) between the top and bottom of the chamber provides an electric field which clears away unwanted ions. The tracks seen are therefore those of rays that leave the source as the expansion occurs.

(b) *Diffusion type* (Figure 91.7). Vapour from alcohol in the felt ring diffuses downwards, is cooled by the 'dry ice' (solid carbon dioxide at −78 °C) in the lower section and condenses near

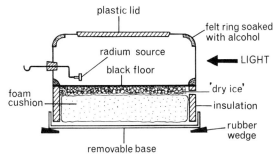

Figure 91.7

the floor on air ions formed by radiation from the source in the chamber. Tracks are produced continuously which are sharp if an electric field is created by frequently rubbing the plastic lid of the chamber with a cloth.

Alpha particles give straight, thick tracks (Figure 91.8a). Very fast β particles produce thin, straight tracks; slower ones give short, twisted, thicker tracks (Figure 91.8b). Gamma rays eject electrons from air molecules: the electrons behave like β particles and produce their own tracks spreading out from the rays.

Figure 91.8a

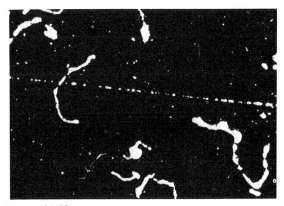

Figure 91.8b

The *bubble chamber*, in which the radiation leaves a trail of bubbles in liquid hydrogen, has now replaced the cloud chamber in research work.

Radioactive decay: half-life

Radioactive atoms change or *decay* into atoms of different elements when they emit α or β particles (page 384). These changes are spontaneous and cannot be controlled; also, it does not matter whether the material is pure or combined chemically with something else.

The *rate of decay* is unaffected by temperature but every radioactive element has its own definite decay rate, expressed by its *half-life*. This is *the time for half the atoms in a given sample to decay*. It is

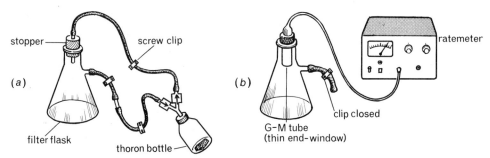

Figure 91.9

difficult to know when a substance has lost all its radioactivity, but the time for its activity to fall to half its value can be found more easily.

Half-lives vary greatly. For radium it is 1600 years so that starting with 1 g of pure radium, ½ g remains as radium after 1600 years, ¼ g after 3200 years, and so on.

The half-life of the α-emitting gas *thoron* can be found as in Figure 91.9a by squeezing the thoron bottle three or four times to transfer some thoron to the flask. The clips are then closed, the bottle removed and the stopper replaced by a G-M tube so that it seals the top (Figure 91.9b).

When the ratemeter reading has reached its maximum and started to fall, the count rate is noted every 15 s for 2 minutes and then every 60 s for the next few minutes. (The G-M tube is left in the flask for at least 1 hour until the radioactivity has decayed.)

A graph of *count rate* (from which the back-

Figure 91.10

ground count rate, found separately, has been subtracted) against *time* is plotted and the half-life (52 s) estimated from it.

USES OF RADIOACTIVITY

Radioactive substances called *radioisotopes* (page 384) are now made in nuclear reactors (page 385) and have many uses.

(*a*) *Thickness gauge.* If a radioisotope is placed on one side of a moving sheet of material and a G-M tube on the other, the count rate decreases if the thickness increases. This technique is used to control automatically the thickness of paper, plastic and metal sheets during manufacture (Figure 91.10).

(*b*) *Tracers.* The progress of a small amount of a weak radioisotope injected into a system can be 'traced' by a G-M tube or other detector. The method is used in medicine to detect brain tumours, in agriculture to study the uptake of fertilizers by plants and in industry to measure fluid flow in pipes.

(*c*) *Radiotherapy.* Gamma rays from strong cobalt radioisotopes are replacing X-rays in the treatment of cancer.

(*d*) *Archaeology.* A radioisotope of carbon, present in the air, is taken in by plants and trees and is used to date archaeological remains of wood and linen.

Dangers and safety

The danger from α particles is small unless the source enters the body. Beta and γ rays can cause radiation burns (i.e. redness and sores on the skin) and delayed effects such as cancer and eye cataracts. Fall-out from atomic explosions contains highly active elements (e.g. strontium), with long half-lives, that are absorbed by the bones.

The weak sources used at school *should always be lifted with forceps, never held near the eyes and should be kept in their boxes when not in use.* In industry, sources are handled by long tongs and transported in thick lead containers. Workers are protected by lead and concrete walls.

Questions

1. Which type of radiation from radioactive materials (*a*) has a positive charge, (*b*) is the most penetrating, (*c*) is easily deflected by a magnetic field, (*d*) consists of waves, (*e*) causes the most intense ionization, (*f*) has the shortest range in air, (*g*) has a negative charge, (*h*) is not deflected by an electric field?

2. What is the effect of radiation on (*a*) a charged electroscope, (*b*) a photographic film, (*c*) a fluorescent screen?

3. (*a*) What do you understand by 'background radiation'? State two sources of this radiation.
   (*b*) State three safety precautions to be observed when using radioactive sources.
   (*c*) Describe briefly two uses of radioactive sources.
   (*d*) How would you test to distinguish between two radioactive sources, one of which emits only α particles and the other which emits only β particles?

4. A radioactive sample has a mass of 16 g and a half-life of 10 days. What mass of the original sample remains after (*a*) 10 days, (*b*) 20 days, (*c*) 40 days?

5. If the half-life of a radioactive gas is 2 minutes, then after 8 minutes the activity will have fallen to a fraction of its initial value. This fraction is: **A** $\frac{1}{4}$, **B** $\frac{1}{6}$, **C** $\frac{1}{8}$, **D** $\frac{1}{16}$, **E** $\frac{1}{32}$.

92 NUCLEAR ENERGY

Atomic structure

Protons and neutrons are in the nucleus of the atom and are called *nucleons*. Together they account for the mass of the nucleus (and most of that of the atom); the protons account for its positive charge. In a neutral atom the number of protons equals the number of electrons surrounding the nucleus.

The table below shows the particles in some atoms. Hydrogen is simplest with 1 proton and 1 electron. Next is the inert gas helium with 2 protons, 2 neutrons and 2 electrons. The soft white metal lithium has 3 protons and 4 neutrons.

	Hydrogen	Helium	Lithium	Oxygen	Copper
Protons	1	2	3	8	29
Neutrons	0	2	4	8	34
Electrons	1	2	3	8	29

The atomic number Z of an atom is the number of protons in the nucleus. It is also the number of electrons in the atom. The electrons determine the chemical properties of an atom and from chemistry you will know that when the 104 elements (89 natural and 15 man-made) are arranged in atomic number order in the periodic table (page 51), they fall into chemical families.

The mass number A of an atom is the number of nucleons in the nucleus.

Atomic *nuclei* are represented by symbols. Hydrogen is written ^1_1H, helium ^4_2He, lithium ^7_3Li and in general atom X is written ^A_ZX where A is the mass number and Z the atomic number.

The *mass number A* of an atom is not the same as its *relative atomic mass* A_V (formerly the *atomic weight*) which is defined by:

$$A_V = \frac{\text{mass of atom}}{\frac{1}{12} \text{ mass of } ^{12}_6C \text{ atom}}$$

Atomic masses used to be compared by taking the mass of the hydrogen atom as 1. Since 1960 the carbon 12 ($^{12}_6C$) scale has been adopted.

Isotopes

Isotopes of an element are atoms which have the same number of protons but different numbers of neutrons.

That is, their atomic numbers are the same but not their mass numbers.

Isotopes have identical chemical properties since they have the same number of electrons and occupy the same place in the periodic table. (In Greek, *isos* means same and *topos* means place.)

Few elements consist of identical atoms; most are mixtures of isotopes. Chlorine has two isotopes; one has 17 protons and 18 neutrons (i.e. $Z=17, A=35$) and is written $^{35}_{17}Cl$, the other has 17 protons and 20 neutrons (i.e. $Z=17, A=37$) and is written $^{37}_{17}Cl$. They are present in ordinary chlorine in the approximate ratio of three atoms of $^{35}_{17}Cl$ to one atom of $^{37}_{17}Cl$, giving chlorine an average atomic mass of 35·5.

Hydrogen has three isotopes. ^1_1H with 1 proton, *deuterium* ^2_1D with 1 proton and 1 neutron and *tritium* ^3_1T with 1 proton and 2 neutrons. Ordinary hydrogen contains 99·99 per cent of ^1_1H atoms. Water made from deuterium is called heavy water (D_2O); it has a density of 1·108 g/cm³, it freezes at 3·8 °C and boils at 101·4 °C.

Radioactive decay

The emission of an α or β particle from a nucleus produces an atom of a different element, called the *decay* product, which may itself be unstable. After a series of changes a stable end-element is formed.

(a) *Alpha decay.* An α particle is a helium nucleus having 2 protons and 2 neutrons and when an atom decays by α emission, its mass number decreases by 4 and its atomic number by 2. For example, when radium of mass number 226 and atomic number 88 emits an α particle, it decays to radon of mass number 222 and atomic number 86. The change is written:

$$^{226}_{88}Ra \longrightarrow {}^{222}_{86}Rn + {}^{4}_{2}He$$

(b) *Beta decay*. Here a neutron changes to a proton and an electron. The proton remains in the nucleus and the electron is emitted as a β particle. The new nucleus has the same mass number, but its atomic number increases by one since it has one more proton. Radioactive carbon, called carbon 14, decays by β emission to nitrogen.

$$^{14}_{6}C \longrightarrow {}^{14}_{7}N + {}^{0}_{-1}e$$

(c) *Gamma emission*. After emitting an α or β particle some nuclei are left in an 'excited' state. Rearrangement of the protons and neutrons occurs in the nucleus and a burst of γ rays is released. Radium 226 does this.

Energy from the nucleus

When a nucleus decays, energy is released as (*i*) heat, and (*ii*) energy of the radiation. Where does this energy come from? Careful measurements have shown that when a large nucleus is broken down (fissioned) some mass 'disappears' and reappears as energy.

(a) $E = mc^2$. Einstein actually predicted that mass and energy are equivalent, with an 'exchange rate' given by the equation:

$$E = mc^2$$

energy released = mass change × (speed of light)²

i.e. if 1 g (0·001 kg) of mass is released as energy, the number of joules available is;

$$E = 0 \cdot 001 \times 3 \times 10^8 \times 3 \times 10^8 = 9 \times 10^{13} \text{ J}.$$

This is about the same amount of energy as we would obtain by burning 2000 tonnes of fuel oil!

(b) *Fission*. The mass changes when radioactive materials decay are very small indeed, so they cannot be used as sources of energy on a large scale. The isotopes of some heavy elements can, however, be broken down (fissioned) by bombardment with neutrons. The uranium isotope $^{235}_{92}U$ (or uranium 235) is one which behaves this way. Some atoms of the isotope decay quite naturally, throwing out high-speed neutrons. If one of these hits the nucleus of a neighbouring uranium 235 atom (being uncharged the neutron is not repelled by the nucleus), this may break (*fission*) into two nearly equal radioactive nuclei, often of barium and krypton, with the production of 2 or 3 more neutrons.

$$^{235}_{92}U + {}^{1}_{0}n \longrightarrow {}^{144}_{56}Ba + {}^{90}_{36}Kr + 2{}^{1}_{0}n$$

neutron   fission fragments   neutrons

The mass defect is large and appears mostly as K.E. (page 159) of the fission fragments. These fly apart at great speed, colliding with surrounding atoms and raising their average K.E., i.e. their temperature. Heat is therefore produced.

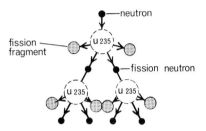

Figure 92.1

If the fission neutrons split other uranium 235 nuclei, a *chain reaction* is set up (Figure 92.1). In practice some fission neutrons are lost by escaping from the surface of the uranium before this happens. The ratio of those escaping to those causing fission decreases as the mass of uranium 235 increases. This must exceed a certain *critical* value for a chain reaction to start.

(c) *Nuclear reactor*. In a nuclear power station a nuclear reactor produces the steam for the turbines instead of a coal- or oil-burning furnace. Figure 92.2 is a simplified diagram of a reactor.

The chain reaction occurs at a steady rate which is controlled by inserting or withdrawing neutron-absorbing rods of boron among the uranium rods. The graphite core is called the *moderator* and slows down the fission neutrons: fission of uranium 235 occurs more readily with slow rather than with fast neutrons. Carbon dioxide gas is pumped through the core and carries off heat to the heat exchanger where steam is produced. The concrete shield gives protection from γ rays and neutrons.

In an *atomic bomb* an increasing uncontrolled chain reaction occurs when two pieces of uranium 235 come together and exceed the critical mass.

(d) *Fusion*. The union of light nuclei into heavier ones can also lead to a loss of mass and, as a result, the release of energy. At present, research is being done on the controlled fusion of isotopes of hydrogen (deuterium and tritium) to give helium.

385

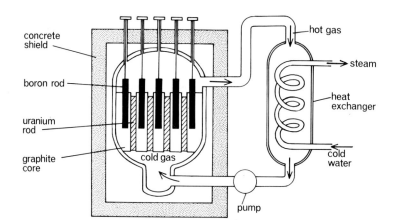

Figure 92.2

Temperatures of about 100 million °C are required. Fusion is believed to be the source of the sun's energy.

Questions

1. What are the particles you would expect to find in an atom? Give some idea of their relative masses and state what electrical charge, if any, each kind has.

2. (a) An atom of cobalt has an atomic number of 27 and a mass number of 59. Describe simply the structure of the cobalt atom.
   (b) What are isotopes?
   (c) Why are isotopes difficult to separate by chemical methods?

3. Uranium 238 and uranium 235 are 'isotopes' of uranium and have the same atomic number, 92.
   (a) What do the numbers 238 and 235 represent?
   (b) (i) What does the number 92 tell you about the nucleus of either of these two atoms?
       (ii) What else does the number 92 tell you about the atom as a whole?
   (c) In what way does the nucleus of uranium 238 differ from the nucleus of uranium 235?

4. What changes, if any, occur in the atomic number of a radioactive atom if the nucleus emits (a) an α particle, (b) a β particle, (c) a γ ray?

5. Atomic bombs and atomic reactors both provide large quantities of energy through chain reactions.
   (a) Explain what is meant by a chain reaction.
   (b) Explain how the reaction, which is violent in the case of a bomb, is slowed down in the reactor.

ANSWERS

3 Mixing and joining
1. (a) Law of Constant Composition (b) Iodine (c) 2 g (d) 20 g (e) Law of Conservation of Mass

5 Investigating air
3. (a) 4/5
   (b) Some carbon dioxide produced which prematurely extinguishes candle

31 Measurement
1. (a) 10 (b) 40 (c) 5 (d) 67 (e) 1000
2. (a) 3 (b) 5·50 (c) 8·70 (d) 0·43 (e) 0·1
3. (a) 53·3 mm (b) 95·8 mm
4. (a) 2·31 mm (b) 14·97 mm
5. 40 cm³; 5
6. 80
7. (a) 252 cm³ (b) 72 cm³
8. (a) (i) 0·5 g (ii) 1 g (iii) 5 g (b) (i) 10 g/cm³ (ii) 3 kg/m³ (c) (i) 2 cm³ (ii) 5 cm³
9. (a) 8 g/cm³ (b) 8000 kg/m³
10. 15 000 kg
11. 130 kg
12. (a) 200 cm³ (b) 200 cm³ (c) 2·5 g/cm³
13. 1·2

33 Forces and turning effects
1. (a) and (b) Turn clockwise (c) Balanced
2. (a) No (b) 25 N
3. 600 N (60 kg)
4. (a) 40 N (b) 10 N

34 Centre of gravity
2. (a) B (b) A (c) C
3. Tips to right

35 Forces and pressure
2. (a) (i) 25 Pa (ii) 0·5 Pa (iii) 100 Pa (b) 30 N
3. 200 Pa
4. (a) 200 m³ (b) 400 000 kg (c) 4 000 000 N
   (d) 200 000 Pa (e) (i) None (ii) Less
5. (a) 100 Pa (b) 200 N
7. 1 150 000 Pa ($1·15 \times 10^6$ Pa) (ignoring air pressure)
10. (c) 76 cm
13. (a) 10 N (b) 8 N (c) 6 N (d) 5 N
14. (a) 30 g (b) 30 g (c) 30 cm³

36 Forces and motion
1. (a) 5 km/h (b) 8 km/h (c) 15 km/h
2. 120 km; 3 h
3. 2 m/s²
4. (a) 2 km/h/s (b) 74 km/h (c) 17·5 s after reaching 70 km/h
5. 50 s
7. (a) 100 m (b) 20 m/s (c) Slows down
8. (a) 1·25 m/s² (b) (i) 10 m (ii) 45 m (c) 22 s
9. (a) (i) 10 m/s (ii) 20 m/s (iii) 30 m/s (iv) 40 m/s (v) 50 m/s
   (c) (i) 5 m (ii) 20 m (iii) 45 m (iv) 80 m (v) 125 m
10. (a) 10 m/s (b) 0
11. D
12. 20 N
13. (a) 5000 N (b) 15 m/s²
14. (a) 2 m/s² (b) 1000 N (c) Because of friction
15. (a) Friction between turntable and coin
    (b) Gravitational attraction of earth for capsule

37 Energy
2. 180 J
3. (a) 500 N (b) 1000 J (c) Kinetic energy of 1000 J
4. (a) 3200 J (b) 25 000 J
5. 400 W

38 Simple machines
1. (a) (iii) (b) (iv)
2. (a) 5000 J (b) 7500 J (c) 67%
3. (a) (i) 10/7 (ii) 2 (iii) 71%
   (b) (i) 2 (ii) 3 (iii) 67%
   (c) (i) 10/3 (ii) 6 (iii) 56%
4. (a) 4 (b) 5 (c) 80%
5. (a) 5 (b) 10 (c) 50%
6. $2\frac{1}{2}$

40 Thermometers
1. (a) 37 °C (b) 0 °C (c) 20 °C (d) 1600 °C

42 Gas laws
1. (a) $\frac{1}{4}$ previous volume (b) 4 times previous pressure
   (c) Nothing
2. (a) 15 cm³ (b) 6 cm³
3. (a) 4 m³ (b) 1 m³
4. (a) 450 cm³ (b) 50 cm³ (c) 100 cm³

43 Conduction, convection and radiation
3. E
4. E

45 Measuring heat
1. 15 000 J
2. 60 000 J
3. A = 2000 J/kg °C B = 200 J/kg °C C = 1000 J/kg °C
4. 88 000 J
5. (a) 20 000 J (b) 120 000 J
6. 1050 W
7. 45 kg
8. (a) 3400 J (b) 6800 J
9. (a) 2750 J (b) 8480 J
10. 680 s
11. (a) 0 °C (b) 45 g
12. (a) 9200 J (b) 25 100 J
13. 157 g

46 Light rays
1. Larger, less bright
2. (a) 4 images (b) Brighter, but blurred

47 Reflection of light
1. (a) 60 ° (b) 60 °C
2. (a) (i) 50 ° (ii) 40 °
3. (a) 15 cm (b) Reflects some of it and behaves as a poor mirror
4. 12·45
6. 5·2 cm from the mirror, 1·5 cm high

48 Refraction of light
3. D

49 Lenses
2. 17 cm beyond lens, 7·0 mm high, real
3. 4

50 Optical instruments
1. Farther away
2. (a) Larger, blurred, less bright (b) Moved closer to the slide
3. (a) 4 cm (b) 8 cm behind lens, virtual, magnification 2

51 Dispersion and colour
3. (a) Black (b) Blue

52 Waves
1. (a) 1 cm (b) 1 Hz (c) 1 cm/s
3. (a) Speed of ripple depends on depth of water
   (b) AB, since ripples travel more slowly towards it, therefore water shallower in this direction

54 Sound waves
1. 1650 m (about 1 mile)

56 Charged particles
4. (a) +ve (b) −ve (c) Copper ions blue; Permanganate ions purple
   (d) Charge on copper ions twice the size of charge on permanganate ions

57 Chemical formulas
1. (i) NaI (ii) $Ca(OH)_2$ (iii) CuCl (iv) $FeSO_4$ (v) LiOH (vi) PbI_2 (vii) $NaHCO_3$ (viii) K_2CO_3 (ix) $(NH_4)_2SO_4$ (x) $Fe(NO_3)_3$
2. (a) 6·5 g (b) 0·1 mol Zn (c) 0·2 mol I (d) ZnI_2 (e) $ZnCl_2$, $ZnBr_2$
3. $CaBr_2$
4. C_4H_{10}–molecular formula
   C_2H_5–empirical formula

58 Equations
1. (a) 2Zn(s), 2ZnO(s)
   (b) 2HCl (aq)
   (c) $2AgNO_3$(aq), 2AgCl(s)
   (d) $2Zn(NO_3)_2$(s), 2ZnO(s), $4NO_2$(g)
   (e) 3ZnO(s), 2Al(s), 3Zn(s)

59 Calculations using equations
1. 4·8 litres H_2 (g)
2. (a) 92 kg Na(s) (b) 48 000 litres Cl_2 (g)
3. (a) 28 (b) CH_2 (c) Covalent bonds (e) 24 litres CO_2 (g)

62 Chemicals from limestone
3. Impure because expected volume of CO_2(g) would be 24 litres

69 Rates of chemical reactions
2. (a) and (c) speed up reaction
3. (b) (i) Powdered CuO (iii) Powdered PbO_2

80 Magnets and magnetic fields
1. C
2. (a) Steel (b) N
3. Y
4. Because the induced poles at their points are of opposite polarities
5. (a) A (b) A
6. A, C, E, F—north; B, D—south

82 Electric current
2. (b) Brighter than normal (c) Normal brightness (d) Brighter than normal
3. A_1 same; A_2, A_3 and A_4 all greater than 0·2 A because there are three batteries and only two lamps
5. A_2 and A_3 both read 0·25 A. If lamp X is removed A_1 and A_2 will read 0·25 A and A_3 zero
6. A connected wrong way round
7. All read 0·25 A
8. (a) 5 C (b) 50 C (c) 1500 C
9. (a) 5 A (b) 0·5 A (c) 2 A

83 Potential difference
1. (b) Brighter than normal (c) Normal brightness
2. (a) V_2 reads 6 V
3. (a) 12 J (b) 60 J
4. (a) 6 V (b) (i) 2 J (ii) 6 J

84 Resistance
1. 3 Ω
2. 20 V
3. 2 A
4. 30 Ω
5. 1 Ω
6. 12 Ω
7. 3½ Ω
8. (a) 4/3 Ω (b) 2 A

85 Electric cells: e.m.f.
2. (a) 3 A (b) 9 V
3. 1 A
4. (a) 3·0 V (b) 2·6 V (c) 2·6 V (d) 0·4 V (e) 2 Ω (f) 13 Ω

86 Electric power
1. (a) 100 J (b) 500 J (c) 6000 J
2. 24 W
3. 20
4. 3·12 kW
5. (a) 6 V (b) 18 W
6. (a) 2 A (b) 6 Ω
7. (a) 30 kWh (b) 150p
8. 15p

87 Electromagnets
1. (a) North (b) East
2. South

88 Electric motors and meters
1. Vertically upwards
2. Anticlockwise
3. (a) 0·1 Ω shunt (b) 1980 Ω multiplier

89 Generators and transformers
4. 50

91 Radioactivity
4. (a) 8 g (b) 4 g (c) 1 g
5. D

INDEX

Absolute
  scale of temperature 178
  zero 177
absorption (of digested food) 297–8
acceleration
  definition 150
  due to gravity 154
accommodation (vision) 328–9
acids 52, 61–3
active transport 101
adaptation
  for fruit dispersal 115–16
  of leaf for photosynthesis 86–7
aerobic respiration 80
air
  active part of 18–20
  composition of 20
  fractional distillation of 21
  pollution of 267–8, 281–2
  spaces in leaves 86–7, 92–3, 100, 104
  spaces in soil 31–2
alimentary canal 295
alkali metals 48
alkaline earth metals 50
alkalis 52
allotropy 39
alloys 44–5
alpha particles 380
alternating current 371
altimeter 148
alveoli 306, 308
amino acids 90, 292, 297–9
ammeters
  moving coil 366
  moving iron 368
ammonia 263, 271–3
amnion 318–19
ampere 348
amplitude 214
ampulla 330
amylase 80
animal dispersal (of seeds) 116
anions 228
anode 58
anther 111–13
anticline (in rocks) 284
anus 295, 298
aorta 302–3, 310
appendix 295, 297
aqueous humour 327
arteries 303
arteriole 303–4, 310, 314
atom 10, 224
atomic
  bomb 385
  mass, *see* relative atomic mass 384
  number 384
  structure 384
atoms 14, 381, 384

atrium (heart) 302–3

Bacteria 97–8
  in soil 31
balance
  lever 126
  posture 330–1, 336
  top pan 126
ball and socket joint 323
barometer
  aneroid 148
  mercury 147
battery 351
bell, electric 363
Benedict's solution 294, 299, 300
beta particles 380
biceps muscle 324, 334, 336
bicuspid valve 302–3
bile 297, 299
bimetallic strip 175
birth 319–20
Biuret test 294
bladder 310–11, 315–16, 320
blind spot 327
block and tackle 164
blood
  circulation 303
  composition 301–2
  functions 305
  sugar 299
body weight 299
boiling point
  and impurities 185
  and pressure 185
bonds 246
  between atoms 249–50
  between ions 248–9
bone marrow 301
bones 293
boreholes 284
Bourdon gauge 147
Bowman's capsule 310–12
Boyle's law 177–8
brain 327, 332, 334–6
breast bone 321
breathing 306–9
bromine 50
bronchi 306
Brownian motion 15
buds 94
burning 18, 97

Caecum 295, 297
calcium (in diet) 293
calipers, vernier 125
cambium 91–2
camera
  lens 210
  pin-hole 193
candle, burning of 281

capillaries (blood) 297–8, 304, 306, 308, 310–11, 313–14, 319
capillarity 134
capillary attraction (soil) 31–3
carbohydrates 271, 292–3
carbon 39
carbonate 66–7, 71, 256–9
carbon cycle 97–8, 281
carbon dioxide 8, 20, 256–9, 310
  in carbon cycle 97–8
  in photosynthesis 86–7, 89
  from respiration 80–2
  transport in blood 304–5, 308
cartilage 323
catalysts 20, 79, 288–9
  in industry 267, 272
  poisoning of 267
cathode 58
  ray oscilloscope 376
  rays 376
cations 228
caustic curve 198
cell body 332–6
cell division 76–7, 94–5, 318
cell membrane 75–6
cells
  biology 75–8
  dry 278, 356
  electrical 277–9
  lead–acid 356
  Leclanché 356
  simple 356
cell sap 76, 102, 104
Celsius scale 172
cell wall 76–7, 102–4
cellulose 76–7, 90, 102–4
  in the diet 292–3, 297–8
censer mechanism 115
centrum 322
cerebral hemispheres 336
cervix 315, 318–19
centre of gravity 139
centripetal force 157
chalk 260
charge, electric 346
Charles' law 177–8
chemical energy 276–9
chemical reaction 7, 11
chlorine 49, 264–5
chlorophyll 86, 88, 90
chloroplasts 86–7, 92–3
chromatography 5
ciliary muscle 327–9
circuit
  diagrams 348
  equations 357
circular motion 157
circulation of the blood 303
clavicle 321–2
clay 31–2

389

cloud chamber
  diffusion 381
  expansion 380
coal 283
cochlea 330–1
cohesion 134
coleoptile 118, 120
coleorhiza 118, 120
collar bone 321–2
collecting tubule 311–12
colour
  complementary 213
  filters 212
  mixing 213
  primary 213
  secondary 213
  vision 327
compounds 10, 11–13
concave
  lens 207
  mirror 198
conductors of
  electricity 347, 353
  heat 180
cones (eye) 327
conservation of energy 161, 371
conservation of Mass, Law of 8
constant composition, Law of 13
contact process 267
control (experimental) 80–2, 87
convection 181
convex
  lens 207
  mirror 198
co-ordination 324, 336
copulation 316
cornea 327–8
cortex
  brain 336
  kidney 310–11
  root 95
  stem 91–2
cotyledons 114, 117–19
coulomb 349
covalent bonds 249–50
critical angle 204
crop rotation 99
cross pollination 113
crude oil 6, 281, 283–5
crumb structure (soil) 99
crystallization 3
crystals, growth of 14, 30
Curie, Marie 379
cytoplasm 75–7

Deamination 299
decay 97–8
declination, magnetic 344
dehydrating agent 63
denatured 80
dendrite 332–3
denitrifying bacteria 98
density
  bottle 128
  definition 127
  measurements 127
  relative 128
deoxygenated blood 302–3, 305
dermis 313, 326
destarching 88
detergents 54
detoxication 299
diaphragm 295, 307–8
diatomic molecules 224
diet 292–3, 299

dietary fibre 293
diffusion 14–15, 100–1, 106
  in gases 100
  in leaf 93
  in lungs 308
  in solutions 101
digested food
  absorption 297
  storage 298
  transport in blood 305
digestion 295–6
  experiments 299–300
dip, angle of 345
dispersal (fruits and seed) 114–16
displacement reactions (metals) 42–3
dissolving 2, 101
distance–time graphs 153
distillation 3
  fractional 4, 6
dorsal root ganglion 335–6
drainage (soil) 32–3
duodenum 295, 297
dynamo 371

Ear 330
  bones 330
  drum 330
earthing and safety 361
earth's magnetic field 344
echoes 220
eclipses 193
effector 334–5
efficiency of machines 163
egestion 296
egg membrane 315
Einstein 385
ejaculation 316–17
elasticity 133
elbow joint 323–4
electric current, effects of 348
electrical energy 351, 356, 359
electricity from reactions 277–9
electrodes 58
electrolysis 58
electrolytes 58
electromagnetic
  induction 370
  spectrum 217
  waves 217
electromagnets 362
electromotive force (e.m.f.) 357
electrons 225, 376
  arrangements in atoms 226–7, 246–7
  mass and charge of 225
  outer 246–7
elements
  classification of 38–9, 47–51
  definition of 10
embryo (human) 318–20
  plant 114, 117–18
endosperm 118, 120
endothermic reactions 276
energy
  changes 160
  conservation 161
  forms of 159
  for photosynthesis 86
  from food 292–3
  from respiration 79, 80–1, 83
  kinetic 159
  levels in atoms 226–7, 246–7
  nuclear 384
  potential 159
  unit of 161

engines
  internal combustion 167
  jet 168
  petrol 167
enzymes 79–80, 86
  characteristics 80
  classes 79
  digestive 295–7, 299, 300
  extra-cellular 80
  intra-cellular 80
epidermis
  human 313, 326
  plants 91–3
epididymis 316
epiglottis 295–6, 306
epithelium 297–8, 306, 308
equations
  calculations based on 244–5
  from gas volumes 240–1
  using atomic symbols 238
  using ions 241–3
  using words 7
equilibrium 137
evaporation 3, 106–7, 185
excretion 74, 310–12
excretory organs 310
exercise 303, 336
exhaling 307–8
exothermic reactions 11, 276
expansion of
  gases 176
  solids 174
  water 176
experimental design 87
explosive fruits 116
eye 327–9

Faeces 298, 310
falling bodies 153
Faraday's law 370
fats
  in the diet 292–3
  digestion 297–8
  storage 298–9
  test for 249
  use in the body 298
fatty acids 297–8
faults (rocks) 284
femur 321, 323–4
ferromagnetic substances 340
fertilization
  human 315, 317–18
  plants 114
fertilizers 99, 273–4
fibre (dietary) 293, 298
fibrous root 94
fibula 321, 323
field, magnetic 343
filament (stamen) 111–12
filtration 3
  kidney 311–12
fission 385
fixation of nitrogen 98
fixed points 172
flame test 25
flavour 326
Fleming's
  left-hand rule 365
  right-hand rule 371
floating and sinking 129
flower structure 111–13
focal length of
  lenses 207
  mirrors 198
follicle 316–17

ood 292–4
  chain 96
  chemicals from 270–4
  classes 292
  digestion 295–300
  production of 273–4
  pyramids 96
  tests 294
  web 96
force
  and pressure 142
  moment of 136
  on a conductor 365
  unit of 156
formulas 232
  and the periodic table 232–3
  empirical 237
  molecular 237
  of electrolytes 233–4
fossil fuels 283
fossils 29
fovea 327
fractional distillation, see distillation
free-fall 156
french bean 117
  germination 119, 122
frequency 214, 221, 371
friction 133
fruit 111–12
  formation 114
  and seed dispersal 114–16
fuels 280–2, 283–5
  comparison of 280
  and air pollution 281–2
fungi 96
fuses 361
fusion 385

Galileo 153
gall bladder 295, 297, 299
galvanometer, moving coil 366
gametes 315–16
gamma rays 380
ganglion 334–5
gas
  laws 177
  pressure 179
gaseous exchange (lungs) 308
gastric juices 296
gears 165
Geiger-Muller tube 379
generator, a.c. 371
germination 117–22
  conditions for 120–1
  experiments 118, 121–2
glomerulus 310–12
glucose 79, 297–9
  absorption 298
  in respiration 80
  storage 298
  test for 294
  use in body 298
glycerol 297–8
glycogen 292–3, 299
gold-leaf electroscope 347
granite 29
gravity 154
grey matter 335–7
grid system 373
grip rule, right hand 341
group (in periodic table) 51
growth 74, 76, 102–3, 292
  of roots 121
  of shoots 122
guard cell 87, 92–3

gullet 295–6

Haber process 272–3
haemoglobin 299, 301
hair 313–14
half-life 381
halogens 48
heart 302–3
heat
  and temperature 173
  equation 188
  receptor 326
  from respiration 80–1, 83
  transport in blood 305
heating elements 359
heavy soil 31–2
hepatic portal vein 298, 303
Hertz 214, 221, 371
hinge joints 323–4
hip joint 323
hooked fruits 116
Hookes' law 133
hormones 299
  transport in blood 305
house circuit 360
humerus 321, 323–4
humus 31–2
hydrated crystals 9
hydraulic machines 143
hydrocarbons 281
hydrochloric acid 41–2, 63, 264
hydrogen carbonates 259
hydrogen chloride 262–4

Images
  in curved mirrors 199
  in lenses 207
  in plane mirrors 197
  multiple 204
  real 197
  virtual 197
impulses (nervous) 324
inclination, magnetic 345
inclined plane 165
indicators 52
inertia 155
infrared radiation 217
ingestion 295
inner ear 330
insect pollination 113
insoluble substances 2
insulators
  electrical 347
  heat 180
insulin 305
intercellular spaces 93, 106
intercostal muscles 307–8
internal
  combustion engine 167
  resistance 357
intervertebral discs 322–3
iodine 48
  in diet 293
  test for starch 299, 300
ionic bonds 248–9
ionizing radiation 379
ions, 228–31, 379
  charges on 234
  discharge of 229–30
  movement of 228–9
iris (eye) 327
iron
  in diet 293
  storage in body 299
isotopes 384

Jack, screw 165
jet engine 168
joints (skeleton) 323–4
joule 161

Kelvin scale 178
kidneys 303, 305, 310–12, 315–16
  function 311–12
  structure 310–13
  tubule 311
kilogram 127
kilowatt-hour 361
kinetic
  energy 159
  theory 15–16, 179
knee cap 321
knee jerk 334

Labour (childbirth) 319
lacteal 298
lamps, electric 359
large intestine 295, 298
larynx 306
latent heat 184, 189, 190, 314
lateral bud 91, 94
lateral inversion 197
lateral root 91, 94–5, 119–20
leaching 98
leaf
  adaptation to photosynthesis 86–7
  blade 92
  stalk 92
  structure 87, 92–3
  testing for starch 88
  water movement 103–4, 106
leguminous plants 98
length 124
Lenz's law 370
levers 136
ligaments 323
light
  and chemical reactions 277
  pipes 205
  rays 192
  reflection of 195
  refraction of 202
  waves 217
light soil 31–2
limbs (skeleton) 323–4
lime 32
limestone 256–60
  caves 259
lines of force (field lines)
  magnetic 343
litmus 52
liver 295, 298–9
loam 32
local gin 4
locomotion 323–4
longitudinal waves 220
long sight 329
loudness of a note 221
lungs 295, 303, 306–8, 310
  ventilation 307–8
lupin 112–14, 116
lymphatic system 298
  vessels 304–5

Machines 163
Magdeburg hemispheres 146
magnetic
  materials 340
  saturation 343
magnetism
  earth's 344

391

magnetism—*cont.*
  induced 342
  test for 341
  theory of 342
magnets
  demagnetizing of 342
  electro- 362
  making 341
  permanent 340
  properties 340
  storing 343
magnification, linear 208
magnifying glass 211
mains electricity supply 360, 373
maize
  fruit structure 118
  germination 120
Maltese cross tube 376
mammals (in seed dispersal) 116
manometer, U-tube 147
mass
  definition 126, 155
  number 384
  spectrometer 234
matter, states of 14, 16
mechanical advantage 163
medulla
  brain 335–6
  kidney 310–11
Meissner's corpuscle 326
melting point
  and impurities 185
  and pressure 184
mesophyll 87, 92–3, 106
metals 38–9
  corrosion of 45–6, 268–9
  extraction of 252–5
  reactivity series for 41–3, 57, 71
  uses of 44–6
meters
  moving coil 366
  moving iron 368
methane 281–2
micrometer 125
microphone 364
micropyle 114, 117
midbrain 336
middle ear 330
middle lamella (leaf) 77
midrib 92
milk 294
mineral 23
  description of 23–5
  in diets 293
  salts in soil 31
mirrors
  concave 198
  convex 198
  parabolic 200
  plane 196
mixtures 2, 112
molar gas volume 240
mole
  of atoms 235
  of molecules 236
molecular mass, *see* relative molecular mass
molecule 100, 224
  in gases 16
  in liquids 16
  in solids 16
moments 136
monatomic molecules 224
motion
  circular 157
  laws of 154

motor effect 365
motor fibres 334–6
motor impulse 332, 335
motor neurone 332–4
motors, electric 365
movement 74, 323
multiplier 368
multipolar neurone 332–3, 336
muscle 302, 304, 307–8, 313, 323–4, 327
  attachment 322–4
  cells 77
  fibres 333
musical notes 221
mutual induction 372

Nasal cavity 296, 326
natural gas 281, 283–5
nectar 113
nectary 111, 113
nerve 332
  cells 77, 332–3
  endings 326
  fibre 332–3, 335
  impulse 324, 326–7, 330, 332–3
neural arch 322
neurone 332–3
neutralization 53, 66, 273
neutral point 344
neutron 226, 384
newton 132, 156
Newton's
  disc 213
  laws of motion 154
nitrates 71
  in plants 90
  in soil 97–9
nitric acid, 61–3, 272
nitrifying bacteria 98
nitrogen 20
  cycle 98, 274
  fixation 98
  in food substances 271, 293
nitrogen fixing bacteria 98
noble gases 20
  electronic structures of 248
noise 221
non-metals 38–9
non-reducing sugar 294
nuclear
  atom 384
  energy 385
  reactor 385
nucleons 384
nucleus 384
  of cell 75, 76–7, 114, 315, 317

Oersted's discovery 362
Ohm 353
Ohm's law 354
oil, *see* crude oil
optic nerve 327–8
ores 25
  extraction of metals from 252–5
  investigation of 25–6
organ 77
organism 77
oscilloscope 376
osmo-regulation (kidney) 312
osmosis 101–4, 106
ossicles 330
outer ear 330
oval window 330
ovary
  human 315–17
  plant 111–15

oviduct 315–18
ovulation 316
ovule 111–14
ovum 315–18
oxidation
  as addition of oxygen 35
  as removal of hydrogen 265
oxides 20–1, 55–7
oxidizing agent 35, 61–2
oxygen 8, 19–21, 41, 55, 280–2
  in germination 120
  from photosynthesis 86–7, 89
  for respiration 80–1
  transport in blood 301, 304–6, 308
oxygenated blood 302–3, 305
oxyhaemoglobin 301

Pacinian corpuscle 326
pain 326
palisade cells 87, 92–3
pancreas 295, 297
parachute fruits and seed 115
parallel circuits 349
pascal 142
pelvic girdle 320–4
pelvis
  kidney 310–11
  skeleton 322–4
pendulum 129
penis 316
pepsin 80, 296, 299, 300
peptides 296–7
periscope 196, 204
peristalsis 296
permeability (soil) 33
petals 111–12
petrol 282
petrol engine, four stroke 167
petroleum *see* crude oil
pH
  scale 53
  effect on enzymes 80
phloem 91–2, 95, 108
phosphates (in soil) 99
phosphorus (in diet) 293
photosynthesis 86–90, 93, 96–7, 100
pigments 213
pin-hole camera 193
pitch (sound) 221, 330
pith 91–2
placenta 318–20
plant structure 91–5
plasma 301–2
platelets 301
pleural membranes 307–8
plumule 117–20
pollen 111–14
pollination 113
pore (skin) 313–14
posture 331
potassium (in soil) 99
potential
  difference 351
  energy 159
potometer 107–8
power
  electrical 359
  mechanical 162
precipitation reactions 69–70, 242–3
predator 96
pregnancy 318–19
pressure
  atmospheric 146
  definition 142
  gauges 147

law (gases) 178
liquid 142
sense of 326
transmission of 143
prism
   refraction by 203
   totally reflecting 204
projector 210
prostate gland 316
proteins
   in diet 292–3
   digestion of 296–7
   test for 271, 294
proton 225, 384
puberty 316
pulleys 163
pulmonary artery and vein 302–3, 306, 308
pump
   bicycle 145
   force 145
   lift 144
   vacuum 145
pyloric sphincter 295

Quality of musical note 221
quarries 257

Radiation, heat
   absorbers of 182
   emitters of 182
radiation, nuclear
   background 379
   dangers 383
   detectors 379
radicle 117–21
radioactivity
   decay 381, 384
   half-life 381
   uses of 383
radio waves 218
radius (skeleton) 321, 323–4
rate of reaction 287–9
   factors affecting 288–9
ray diagrams 199, 208
reactivity series 41–3
   and classification of oxide 57
   and decomposition of compounds 71
   and extraction of metals 253
receptacle (flower) 112–13
receptor 326, 334–6
rectum 295, 298
red blood cells 301, 308
reducing agent 35
reduction
   as addition of hydrogen 265
   as extraction of metal from ore 252
   as removal of oxygen 35
reflection of
   sound waves 220
   water waves 215
reflection of light
   at curved mirrors 198
   at plane mirrors 195
   laws of 195
   total internal 203
reflex 334–6
refraction of light
   and apparent depth 203
   by a prism 203
   in eye 327–8
   laws of 203
refraction of water waves 215
refractive index 203
refrigerator 186
regelation 185

relative atomic mass 235
   list of values (inside back cover)
relative molecular mass 236
renal artery and vein 303, 310–11
renal tubule 310–12
reproduction 74
   human 315–20
   plant 111–14
resistance
   definition 353
   in parallel 354
   in series 354
   internal 357
   measurement 353
respiration 74, 80–3, 86, 90, 97, 100, 292
retina 327–9
reversible reactions 9
ribs 307–8, 321–2
ripple tank 214
rockets 168
rocks 27–30, 283–5
   igneous 29
   impermeable 283
   sedimentary 27
root
   cap 94–5
   function 94–5
   growth 121
   hairs 31, 95, 103, 119–20
   nodules 98
   structure 94–5
   system 91, 94
   uptake of water 103
rotation of crops 99
roughage 293, 298
rusting, see corrosion of metals

Sacral vertebrae 323
saliva 296, 299
salivary amylase 296
salivary glands 295–6
salts 65
   decomposition of 71–2
   in diet 293
   in excretion 310–12
   movement in plants 106–7
   from natural waters 70, 261–2
   preparation of 53, 64–7
   in soil 31
   solubility of 68–9
sand particles 31–2
sandstone 28–9, 283
saprophytes 96
saturated solutions 68
scapula 322, 324
sclera 327
screw jack 165
scurvy 294
seed 112, 114, 117–22
   dispersal 114–16
   formation 114
   structure 117
seismic survey 284
selective permeability 101
self-fertilization 114
self-pollination 113
semen 316–17
semi-circular canals 330–1
semilunar valve 302–3
seminal vesicle 316
sense organs 332
senses 326–30
sensitivity 74
sensory
   cells 326–7

fibre 334–6
impulse 332, 335
nerve endings 326, 330
neurone 332–3
receptor 333–4, 336
series circuits 349
sexual reproduction
   human 315–18
   plants 111–14
shadows 193
shivering 314
shoot (plant) 91
   growth 122
short sight 329
shoulder 323–4
shunt 367
sight 327–9
siphon 146
skeleton (human) 321–4
   functions 322
skin 313–14
   senses 326
skull 321–2
small intestine 295, 297–8
smell (sense) 326
Snell's law 203
soap 53
sodium chloride (rock salt) 3, 5, 65, 261–3
soil 31–3
solenoid 362
soluble substances 2
solute 3
solvent 3
sound
   loudness 221
   pitch 221, 330
   quality 221
   reflection 220
   speed of 221
   waves 220
sour taste 326
spark counter 379
specialization (in cells) 77
specific capacity, heat 188
specificity (of enzymes) 80
spectrum
   electromagnetic 217
   pure 212
speed 150
sperm 315–18
   duct 316
   production 316
spinal column 307–8
   cord 322, 332–4, 335–7
   nerve 332, 334–6
   reflex 334–6
spongy mesophyll 87, 92–3
springs 133
stability 141
staircase circuit 361
stamens 111–13
starch 109
   in diet 292
   digestion 296–7, 299
   in a leaf 88–90
   in photosynthesis 88–9
   stored in seeds 117–18, 120
   synthesis 79
   test for 88, 294
states of matter 14
static electricity 346
steam turbine 169
stem 91–2
sternum 307–8, 321
stigma 111–14

stimuli 326
stomach 295
  digestion in 296
stomata 87, 92–3, 100, 109
storage (food in body) 292, 298
stretch receptor 334–5
style 111–14
sub-atomic particles 225
sucrose 90, 109
  in diet 292
  test for 294
sugar
  in diet 292
  test for 294
sulphur 266
sulphur dioxide 266–8
sulphuric acid 62–3, 267
sulphur trioxide 267
sunlight in photosynthesis 86–9, 96
surface tension 134
suspensory ligament 327–9
swallowing 295–6, 337
sweat 314
  duct 313–14
  gland 313–14
synapse 333–6
synovial fluid 323
synthesis 11, 34, 272–3
syringe 144

Tap root 94
tape charts 151
taste buds 326–7
taste sense 326
teeth 293
telephone 364
temperature
  absolute 178
  effect on enzymes 80
  effect on germination 120
  effect on transpiration 107
  regulation 313–14
  scale 178
tendons 323
terminal
  p.d. 357
  velocity 157
testa 114, 117, 119
testis 315–16
thermal decomposition 9
thermionic emission 376
thermometers 172
Thermos flask 183
thermostats 175
thorax 306
tibia 321, 323–4
tickertape timer 151
timbre 221
tissue 75, 77
  fluid 304–5
tongue 295–6, 326

toppling 140
total internal reflection 203
touch 326
trachea 306
transformer 372
transmission of
  electrical power 373
  pressure 143
transpiration 106–9
  conditions affecting 107–8
  experiments 107–9
  rate 107–9
  stream 106–7
transverse waves 221
triceps muscle 324
tricuspid valve 302–3
turbines 169
turgor 102, 104, 106

Ulna 321, 323–4
ultraviolet radiation 218
umbilical cord 318–20
umbilical vein and artery 319
universal indicator 53
uranium 385
urea 299, 310–11
  transport in blood 305
ureter 310–11, 315–16
urethra 310–11, 315–16, 320
uric acid 310–11
urine 311
uterus 315, 318–20

Vacuole 76–7, 102–4
vacuum
  flask 183
  pump 145
vagina 315–16, 318, 320
vascular bundle 91–2, 95, 109
vaso-constriction 315
vaso-dilation 314
vein 304–5
velocity
  definition 150
  ratio 163
  terminal 157
vena cava 302–3, 310
ventilation (of lungs) 307–8
ventral root 334, 336
ventricle 302–3
venule 304
vernier scales 125
vertebra 321–3
vertebral column 321–4
vessel (in plants) 92, 104, 109
vibration of
  air columns 221
  strings 221
villi 297–8
vision 327–9
vitamins 293–4

vitreous humour 327
voluntary actions 335
volatile substances 3
volcanoes 30
volt 352
voltage 351
voltaic pile 277
voltmeter 351
volume 126
vulva 315, 320

Water
  absorption 298
  as a product of burning 34
  in diet 293
  excretion 312
  expansion 176
  in germination 120
  movement in a leaf 103–4, 106
  in photosynthesis 86–7
  from respiration 80–1
  salts from 70, 261–2
  in soil 31–3
  supply system 142
  synthesis of 34
  tests for 2
  transpiration of 106, 108
  uptake by roots 103
water of crystallization 9
water sac 318–19
watt 162, 359
wave equation 214
wavelength 214
waves
  describing 214
  electromagnetic 217
  light 217
  longitudinal 220
  sound 220
  transverse 214
  water 214
weight 132
wheel and axle 164
white blood cells 301
white matter 335, 337
wilting 102
wind
  dispersal 115
  pollination 113
windpipe 295–6, 306
winged fruits 115
wiring, of house, 360
womb 315
wood, thermal decomposition of 7
word equations, *see* equations
work 161

X-rays 219
xylem 91–2, 95, 109

Yield of reaction 244